Energiepolitik und Klimaschutz
Energy Policy and Climate Protection

Reihe herausgegeben von
L. Mez, Berlin, Deutschland
A. Brunnengräber, Berlin, Deutschland

Diese Buchreihe beschäftigt sich mit den globalen Verteilungskämpfen um knappe Energieressourcen, mit dem Klimawandel und seinen Auswirkungen sowie mit den globalen, nationalen, regionalen und lokalen Herausforderungen der umkämpften Energiewende. Die Beiträge der Reihe zielen auf eine nachhaltige Energie- und Klimapolitik sowie die wirtschaftlichen Interessen, Machtverhältnisse und Pfadabhängigkeiten, die sich dabei als hohe Hindernisse erweisen. Weitere Themen sind die internationale und europäische Liberalisierung der Energiemärkte, die Klimapolitik der Vereinten Nationen (UN), Anpassungsmaßnahmen an den Klimawandel in den Entwicklungs-, Schwellen- und Industrieländern, Strategien zur Dekarbonisierung sowie der Ausstieg aus der Kernenergie und der Umgang mit den nuklearen Hinterlassenschaften.
Die Reihe bietet ein Forum für empirisch angeleitete, quantitative und international vergleichende Arbeiten, für Untersuchungen von grenzüberschreitenden Transformations-, Mehrebenen- und Governance-Prozessen oder von nationalen „best practice"-Beispielen. Ebenso ist sie offen für theoriegeleitete, qualitative Untersuchungen, die sich mit den grundlegenden Fragen des gesellschaftlichen Wandels in der Energiepolitik, bei der Energiewende und beim Klimaschutz beschäftigen.

This book series focuses on global distribution struggles over scarce energy resources, climate change and its impacts, and the global, national, regional and local challenges associated with contested energy transitions. The contributions to the series explore the opportunities to create sustainable energy and climate policies against the backdrop of the obstacles created by strong economic interests, power relations and path dependencies. The series addresses such matters as the international and European liberalization of energy sectors; sustainability and international climate change policy; climate change adaptation measures in the developing, emerging and industrialized countries; strategies toward decarbonization; the problems of nuclear energy and the nuclear legacy.
The series includes theory-led, empirically guided, quantitative and qualitative international comparative work, investigations of cross-border transformations, governance and multi-level processes, and national "best practice"-examples. The goal of the series is to better understand societal-ecological transformations for low carbon energy systems, energy transitions and climate protection.

Reihe herausgegeben von
PD Dr. Lutz Mez
Freie Universität Berlin

PD Dr. Achim Brunnengräber
Freie Universität Berlin

Weitere Bände in der Reihe http://www.springer.com/series/12516

Karoline Steinbacher

Exporting the Energiewende

German Renewable Energy Leadership and Policy Transfer

Springer VS

Karoline Steinbacher
Berlin, Germany

Dissertation Freie Universität Berlin, 2016

D 188

OnlinePlus material to this book is available on
http://www.springer.com/978-3-658-22496-7

Energiepolitik und Klimaschutz. Energy Policy and Climate Protection
ISBN 978-3-658-22495-0 ISBN 978-3-658-22496-7 (eBook)
https://doi.org/10.1007/978-3-658-22496-7

Library of Congress Control Number: 2018946543

Springer VS

Printed on acid-free paper

This Springer VS imprint is published by the registered company Springer Fachmedien Wiesbaden GmbH part of Springer Nature
The registered company address is: Abraham-Lincoln-Str. 46, 65189 Wiesbaden, Germany

Acknowledgments

Completing the doctoral research this book is based on has been a formidable journey. It has become possible thanks to travel companions who invested their time and support with overwhelming generosity. I owe a particular debt of gratitude to those who generously shared their insights in numerous interviews in Germany, Morocco, South Africa, and California.

I am grateful to my doctoral supervisors, Prof. Miranda Schreurs and Prof. Markus Lederer, for their guidance and advice on this research project. In Berlin, fellow researchers at FFU and from the Potsdam Institute for Climate Impact Research have played an important role in helping me to look at my research from different angles and strengthen my arguments.

I had the great privilege of finalizing this piece of research as a Giorgio Ruffolo Research Fellow with the Harvard Kennedy School's Belfer Center for Science and International Affairs and am grateful for the invaluable support of my faculty mentor Prof. Henry Lee and colleagues at Harvard.

From Sacramento to Rabat, and from Pretoria to Berlin, energy systems are being fundamentally transformed. Getting a glimpse into these transitions as they unfold and being able to observe how ideas travel around the globe, spark debate, and provoke thought has been a true privilege.

Berlin, 2018

Overview of Contents

Table of Contents

Appendix: To access the book's appendix, please follow the URL given in the imprint or visit www.springer.com and search for the author's name.

List of Tables

List of Figures

List of Abbreviations

ADEME	French Environment and Energy Management Agency [Agence de l'Environnement et de la Maîtrise de l'Energie]
ADEREE	Moroccan National Agency for the Development of Renewable Energies and Energy Efficiency [Agence Nationale pour le Développement des Energies Renouvelables et de l'Efficacité Energétique]
AFD	French Development Agency [Agence Française de Développement]
AFDB	African Development Bank
AHKs	German Chambers of Commerce [Auslandshandelskammern]
AMISOLE	Moroccan Wind and Solar Industries Association [Association Marocaine des Industries Solaires et Eoliennes]
B90/Greens	Alliance 90 – The Greens [Bündnis 90 – Die Grünen, German Green Party]
BMU	German Federal Ministry for the Environment, Nature Conservation, and Nuclear Safety [Bundesministerium für Umwelt, Naturschutz und Reaktorsicherheit | until 2013]
BMUB	German Federal Ministry for the Environment, Nature Conservation, Building, and Nuclear Safety [Bundesministerium für Umwelt, Naturschutz, Bau und Reaktorsicherheit | since 2013]
BMWi	German Federal Ministry for Economic Affairs and Energy [Bundesministerium für Wirtschaft und Energie]
BMZ	German Federal Ministry for Economic Cooperation and Development [Bundesministerium für Wirtschaftliche Zusammenarbeit und Entwicklung]
CAISO	California Independent System Operator
CALSEIA	California Solar Energy Industries Association
CEC	California Energy Commission
CDER	Center for the Development of Renewable Energy [Centre de Développement des Energies Renouvelables]
CDU	Christian Democratic Union of Germany [Christlich-Demokratische Union Deutschlands]
CO_2	Carbon dioxide
COP	Conference of the Parties [of the UNFCCC]
CPUC	California Public Utilities Commission
CSP	Concentrated Solar Power

CSU	Christian Social Union in Bavaria [Christlich-Soziale Union in Bayern]
CTF	Clean Technology Fund
DKTI	German Climate Technology Initiative [Deutsche Klimatechnologie-Initiative, also: Deutsche Klima- und Technologieinitiative]
DoE	South African Department of Energy
EEG	Renewable Energy Sources Act [Erneuerbare-Energien-Gesetz]
EU	European Union
FDP	German Free Democratic Party [Freie Demokratische Partei]
FiT	Feed-in Tariffs
GDP	Gross Domestic Product
GEF	Global Environment Facility
GHG	Greenhouse Gas Emissions
GIZ	German Agency for International Cooperation [Deutsche Gesellschaft für Internationale Zusammenarbeit GmbH |since 2011]
GTZ	German Agency for International Cooperation [Deutsche Gesellschaft für Technische Zusammenarbeit |until 2011]
GW(h)	Gigawatt (hours)
HI	Historical Institutionalism
IEA	International Energy Agency
IEPR	Integrated Energy Policy Report
IFIC	International Feed-in Cooperation
IKI	International Climate Initiative [Internationale Klimaschutzinitiative]
IKLU	Initiative für Klima und Umweltschutz [Initiative for Climate and Environmental Protection]
[I]NDCs	[Intended] Nationally Determined Contributions
IRENA	International Renewable Energy Agency
IRESEN	Institute for Solar and New Energies Research [Institut de Recherche en Energie Solaire et Energies Nouvelles]
IRP	Integrated Resource Plan
IPPs	Independent Power Producers
kWh	Kilowatt hour
KfW	German Development Bank [Kreditanstalt für Wiederaufbau (KfW) Entwicklungsbank, abbreviated "KfW" in this thesis, part of KfW Group]
LABC	Los Angeles Business Council
LADWP	Los Angeles Department of Water and Power
MASEN	Moroccan Agency for Solar Energy
MEMEE	Moroccan Ministry for Energy, Mining, Water and the Environment [Ministère de l'Energie, des Mines, de l'Eau et de l'Environnement]
MENA	Middle East North Africa

MW(h)	Megawatt (hours)
NERSA	National Energy Regulator of South Africa
NT	National Treasury [South African Department of Finance]
ODA	Official Development Assistance
OECD	Organization for Economic Cooperation and Development
ONEE	Moroccan National Energy and Water Office [Office National de l'Electricité et de l'Eau potable]
PERG	Moroccan Rural Electrification Program [Programme d'Electrification Rurale Global]
PG&E	Pacific Gas and Electric Company
PPA	Power Purchase Agreement
ProBEC	Programme for Biomass Energy Conservation
PURPA	Public Utility Regulatory Policy Act
PV	Photovoltaics
RCREEE	Regional Center for Renewable Energy and Energy Efficiency
REFIT	South African Renewable Energy Feed-in Tariff
REN21	Renewables 21 Network
REIPPPP	Renewable Energy Independent Power Producer Procurement Programme
RPS	Renewables Portfolio Standard
SAGEN	South African-German Energy Programme
SAPVIA	South African Photovoltaic Industry Association
SASTELA	South African Solar Thermal and Electricity Association
SAWEA	South African Wind Energy Association
SCE	Southern California Edison
SDG&E	San Diego Gas & Electric
SEIA	Solar Energy Industry Association
SMUD	Sacramento Municipal Utility District
SPD	Social Democratic Party of Germany [Sozialdemokratische Partei Deutschlands]
StrEG	Electricity Feed-in Law [Stromeinspeisungsgesetz]
TERNA	Technical Expertise for Renewable Energy Application project
UNFCCC	United Nations Framework Convention for Climate Change
USAID	United States Agency for International Development

Abstract

With its Energiewende, Germany has embarked on an ambitious long-term transition toward an energy system largely based on renewable energy. The Energiewende has sparked interest and triggered lively reactions, from enthusiasm to skepticism, well beyond the country's borders. An aim of finding followers has been part of the Energiewende since its beginnings and is a recurrent theme in political discourse in Germany. In addition to being a unilateral leader by setting an example, the country has also taken a range of steps for active leadership by communicating its vision, building transfer channels, and supporting the uptake of renewables abroad. The effectiveness of these efforts in facilitating the transfer of renewable energy policies remains to be assessed, which is the empirical research gap this book addresses.

The role of pioneering countries as active policy promoters has so far been neglected in the policy transfer literature that this book speaks to. To contribute to the further development of the field, three diverse recipient cases – Morocco, South Africa, and California – were selected to explore transfer and the role of Germany's leadership therein. The three cases cooperate closely with Germany but have not implemented its most iconic policy instrument to promote renewables, feed-in tariffs. This makes them cases of apparent non-transfer, which are underrepresented in the literature but promise to yield rich insights, both in theoretical and in empirical terms, into what shapes transfer processes and outcomes. The heterogeneity of the three cases in terms of economic development, regulatory traditions, and the setup of the renewable energy policy network, permits a focus on common features of German leadership and policy transfer patterns. The cases also cover diverse transfer channels, from informal encounters to institutionalized energy partnerships, providing insights into the effectiveness of different forms of leadership.

Findings presented in this book, which is based on PhD research completed at Freie Universität Berlin and the Harvard Kennedy School in 2016, rely on interviews and background conversations with close to 200 decision-makers, government officials, utility and industry stakeholders as well as transfer agents in Morocco, South Africa, California, and Germany. In addition to coded interview material, qualitative content analysis and ranking exercises are used to triangulate findings and trace the role of transfer in renewable energy policy debates and decisions.

Across the cases, the Energiewende – as a reference point – and Germany – as an active sender – play important roles as catalysts for renewable energy policy, refuting the idea of non-transfer in the absence of convergence on policy instruments. The use of transferred knowledge appears as a political exercise, with actors viewing and using transferred evidence differently, depending on their role in the policy network, objectives, normative positions, and prospective evaluations of gains and losses to be expected from transfer. Depending on the relative power of proponents and opponents to transfer, ideas are adapted and modified to fit the recipient context, emphasizing the need for a differentiated view on the consideration, use, and effect of transferred knowledge.

Regardless of convergence, the value of policy transfer and German leadership is recognized by stakeholders in recipient states. A better understanding of what makes leadership through transfer effective is of particular relevance in the context of the polycentric climate governance architecture enacted by the Paris Agreement. This book illustrates important policy lessons for the effectiveness of these efforts, such as a sound understanding and the adaptation of outreach to recipient policy objectives, clarity about Germany's agenda for leadership, and reinforced and transparent communication on the Energiewende's achievements and challenges.

Zusammenfassung

Die deutsche Energiewende mit ihrem dreifachen Ziel der Dekarbonisierung des Energiesektors, der Reduzierung des Primärenergieverbrauchs und des Ausstiegs aus der Nuklearenergie, ruft weltweit kontroverse Reaktionen hervor. Der Anspruch, international Nachahmer zu finden, ist seit jeher eng mit der Umsetzung der Energiewende in Deutschland verbunden und spiegelt sich in der Kooperation mit Partnerstaaten wider. Die tatsächliche Rolle der Energiewende als Vorbild, d.h. Deutschlands "unilateraler Leadership", und der verschiedenen Ausformungen "aktiver Leadership" – von Kommunikation zu Anreizen und Kapazitätenbildung in Empfängerstaaten – für den Transfer von Politiken ist bislang kaum erforscht. Das vorliegende Buch, welches auf einer an der Freien Universität Berlin und der Harvard Kennedy School im Sommer 2016 fertiggestellten Dissertation beruht, leistet einen Beitrag zur Schließung dieser empirischen Forschungslücke.

Im Rahmen dreier Fallstudien zu Marokko, Südafrika und Kalifornien wird erörtert, wie die Energiewende als Modell in Betracht gezogen und genutzt wird, und sich in Politikentscheidungen widerspiegelt. Über eine Betrachtung der passiven Vorbildwirkung der Energiewende hinaus, setzt die Arbeit einen Schwerpunkt auf Deutschlands Interaktion mit Empfängerstaaten. Die Arbeit schlägt somit eine Brücke zwischen der wachsenden Politiktransfer-Literatur, in der Pionierstaaten als Sender eine bislang stark untergeordnete Rolle spielen, und der vorrangig im Bereich internationaler Verhandlungen verankerten Leadership-Literatur.

Die drei untersuchten Fälle zeichnen sich trotz ihrer engen Zusammenarbeit mit Deutschland durch die Ablehnung des zentralen Politikinstruments der deutschen Erneuerbaren-Politik, des Einspeisetarifs, aus. Während sich die Politiktransfer- und Politikdiffusionsliteratur lange auf Konvergenz zwischen Sendern und Empfängern konzentrierte, versprechen gerade Fälle von negativem bzw. selektivem Transfer einen entscheidenden Beitrag zur Weiterentwicklung des Forschungsgebiets. Um derartige komplexe Transferprozesse zu erfassen, basiert die Arbeit auf qualitativen Methoden, insbesondere knapp 200 in Deutschland, Marokko, Südafrika und Kalifornien durchgeführten Interviews und Hintergrundgesprächen. Diese wurden codiert, mit den Ergebnissen qualitativer Dokumentenanalyse und Rankings verbunden, und für die Nachzeichnung von Politikprozessen im Rahmen der drei Fallstudien genutzt.

Trotz der großen Heterogenität der untersuchten Fälle im Hinblick auf Faktoren wie wirtschaftliche Entwicklung, Kapazitäten, und regulatorische Traditionen, treten gemeinsame Muster von Leadership und Transfer in den Vordergrund. Die Ergebnisse unterstreichen die politische Dimension von Transfer im Politikfeld nachhaltige Energie: transferiertes und von transfer agents "übersetztes" Wissen mit Bezug zur Energiewende wird selektiv rezipiert, und von Akteuren strategisch dazu genutzt, eigene Positionen zu untermauern, Politikwandel anzustoßen, oder im Gegenteil zu verhindern. Kräfteverhältnisse innerhalb von Politiknetzwerken verändern sich durch Transfer, sowohl durch die Verfügbarkeit neuer Informationen, als auch durch die Präsenz und Aktivitäten deutscher transfer agents. Der Energiewende als Quelle positiver und negativer Erfahrungen und Deutschland als aktivem Sender kommt in den drei Fällen eine wichtige Rolle als Katalysatoren energiepolitischer Entscheidungen zu. Auch wenn der Transfer von Politikinstrumenten aufgrund wahrgenommener Konflikte mit regulatorischen Traditionen und unterschiedlicher Politikziele selektiv erfolgte, wird der Einfluss von Transfer auf die Ausrichtung von Politiken sowie auf regulatorischer, inkrementeller Ebene deutlich.

Im Hinblick auf die entscheidende Rolle nationaler Politiken für die Umsetzung des 2015 in Paris vereinbarten Klima-Abkommens, ist ein besseres Verständnis der Mechanismen von Politiktransfer und der Rolle aktiver Sender-Staaten besonders bedeutsam. Für effektive Leadership erscheinen im Licht der Forschungsergebnisse Transparenz im Hinblick auf verfolgte Ziele, die Anpassung von Strategien an Prioritäten im Empfängerstaat und umfassende Kommunikation zu Erfolgen und Herausforderungen der Energiewende relevant.

1 Introduction

1.1 The Energiewende and Its International Dimension

1.1.1 Research questions and aim

The Energiewende, Germany's ongoing energy system transformation, has received attention well beyond the country's borders, triggering reactions from enthusiastic praise to outright disbelief. It has been lauded as "a world-saving achievement" (Friedman 2015) and as a "model for how we'll get power in the future" (Kunzig 2015), while others see in it "an expensive gamble" (Karnitschnig 2014) and even "an energy market disaster" (Clark 2014). As a far-reaching transition away from nuclear and fossil fuels and toward renewable energies and efficiency, the Energiewende model stands out, especially among industrial countries (IRENA 2014, Morgan & Weischer 2013, Stigson et al. 2013). Its aim is threefold – decarbonizing the power sector, cutting primary energy consumption, and phasing out nuclear power. These objectives translate into ambitious long-term targets, such as 80-95% CO_2 reduction and a goal of at least 80% renewables in the electricity mix by 2050 (Bundesregierung 2010a).

Germany's move toward a more sustainable energy system has had an international dimension since its beginnings, well before entering a brighter global spotlight following the Fukushima nuclear accident in 2011 (e.g., Bundesregierung 1982, Bundesregierung 1988, Bundesregierung 1996, Wieczorek-Zeul 2009 (2001)). German decision-makers regularly acknowledge the international ambition of Germany's energy transition and their desire to gain followers to the Energiewende. Chancellor Angela Merkel affirmed in a 2014 governmental address that "there is no other comparable country in the world that tackles such a radical change of its energy supply", and therefore, "the world is looking at it with a mix of incomprehension and curiosity [...]. If we succeed, [the Energiewende] will become – I am convinced of this – another German export hit"[1] (Merkel 2014).

1 "Es gibt kein weiteres vergleichbares Land auf dieser Welt, das eine solch radikale Veränderung seiner Energieversorgung anpackt. [...] Die Welt schaut mit einer Mischung aus Un-

K. Steinbacher, *Exporting the Energiewende*, Energiepolitik und Klimaschutz. Energy Policy and Climate Protection, https://doi.org/10.1007/978-3-658-22496-7_1

Germany's aim of inspiring policy action for sustainable energy abroad is the main empirical motivation for this research. The effectiveness of "Energiewende leadership" in promoting policy transfer to particular recipient cases abroad has not been systematically addressed so far. The present book aims to contribute to closing this gap by looking at renewable energy policy transfer to three relevant recipient states, namely to Morocco, South Africa, and California.[2] Indeed, much is at stake for Germany in having other countries join its objective to spur a global energy transition, from creating technology export opportunities to the reinforcement of soft power, and, above all, contributing to global climate protection efforts. With 2.36% of global greenhouse gas (GHG) emissions generated within German borders (Germanwatch n.d.), the impact of the Energiewende strongly depends on whether it can also facilitate action elsewhere around the globe. As Vice-Chancellor and Minister for Economic Affairs and Energy Sigmar Gabriel therefore emphasized, "With the Energiewende, we wanted to generate followers. Of course we did not want to do climate protection [only] in Germany, but persuade others to join us"[3] (Gabriel 2014: 2699). By massively deploying renewable energy within Germany, the country has contributed to driving down the cost of technologies, especially in the area of solar photovoltaics (PV), and thereby facilitated their uptake (IRENA 2015c: 28, Morris 2016, Pahle et al. 2012, Weiss 2014: 3). Lower technology prices alone, however valuable, are nevertheless unlikely to suffice for a global energy transition to happen at the necessary scale and speed to rein in climate change (IEA-RETD 2016). The effect of Germany's Energiewende leadership on the transfer of *policies* for renewable energy is therefore the main question this book addresses:

What determines the effectiveness of German Energiewende leadership in promoting renewable energy policy transfer to Morocco, South Africa, and California?

---

verständnis und Neugier darauf [...]. Wenn sie [die Energiewende] uns gelingt, dann wird sie – davon bin ich überzeugt – zu einem weiteren deutschen Exportschlager."

2 See section 1.3.1 and Chapter 4 for case selection rationale.

3 "Wir wollten mit der Energiewende Nachahmer erzeugen. Wir wollten doch nicht Klimaschutz in Deutschland machen, sondern wir wollten andere dafür gewinnen, dass sie mitmachen."

This book explores two related aspects of Germany's international Energiewende leadership. The first aspect concerns the role of *unilateral leadership,* or leadership by example. In other words, does the Energiewende example trigger lesson drawing among stakeholders abroad or even spur policy action? Based on its domestic energy policy model and its aim to find followers, Germany also engages in various forms of *active leadership*, including the creation of bilateral partnerships, international fora, and the provision of technical and financial support to promote the diffusion of renewable energies abroad. The questions to be explored here concern the effectiveness of active leadership in promoting policy transfer related to the Energiewende model: how are German programs, transfer channels, and transfer agents affecting the spread of renewable energy policies to recipients?

On the basis of three in-depth case studies with qualitative interviews, background conversations, qualitative document analysis, and ranking exercises, this book traces the *consideration*, *use*, and *effect* of ideas related to the Energiewende in policy making for renewable energy. It thereby provides insights into the neglected role of pioneering countries as senders, investigates the conditions for effective leadership through transfer, and explores how transferred knowledge affects policy decisions in heterogeneous contexts. By adopting an inclusive view of the transfer process – from the Energiewende as a source to Germany as an active sender to policy making in the recipient context – this book brings together heretofore separated strands of research in a novel empirical field.

1.1.2 The need for a global energy transition

At the December 2015 UNFCCC Conference of the Parties in Paris (COP21), the representatives of 195 countries ended years of stalemate in international negotiations by reaching a landmark climate agreement (UNFCCC 2015). Built around a system of Intended Nationally Determined Contributions (INDCs), the Paris Agreement institutionalizes a longstanding reliance on national efforts in global climate governance. The effectiveness of this "polycentric" governance architecture, based on policy action in diverse locations rather than on a top-down approach, will decisively depend on whether linkages and spillover effects between different national (and subnational) policy models are leveraged (Bodansky et al. 2015, Cole 2015, Jordan & Huitema 2014, Ostrom 2009, Ostrom 2012). In such a context, lessons from the success and failures of policies implemented elsewhere can increase the knowledge base for decision-makers, reduce uncertainty, send signals for action, and avoid the constant reinvention of the "policy wheel". The relevance of policy transfer and diffusion, i.e., the spread of policies across borders, for the global adoption of climate and environmental policies has indeed been established (Busch & Jörgens 2012, Fankhauser et al. 2015, Holzinger et al. 2010, Jänicke 2013: 8). Whereas leading countries have been found to play decisive roles in multilateral negotiations (Saul & Seidel 2011) and in providing innovative policy models (Jänicke 2005), strikingly little is known about how leaders can promote the international transfer of their own policy models outside institutionalized negotiation frameworks (Biedenkopf 2012: 111).

The relevance of understanding how the spread of policies for sustainable energy can be supported is evident by looking at global energy and climate challenges. According to the International Energy Agency's (IEA) 2015 *World Energy Outlook*, 1.3 billion people worldwide still lack secure and reliable access to the most basic energy services (IEA 2015b: 23). About 2.6 billion people use traditional biomass for cooking, exposing themselves to serious health risks and causing severe harm to the environment (BMZ 2014: 15). The strain on energy resources is likely to grow, with the IEA (2015b: 53) predicting a 70% increase in electricity demand worldwide by 2040. At the same time, the battle against climate change is becoming more urgent every day. In early May 2013, the threshold of 400ppm of CO_2 in the atmosphere was crossed for the first time in history, prompting the United Nations Framework Convention on Climate Change's (UNFCCC) executive secretary to urge the world to "wake up" and work toward "a policy response which truly rises to the challenge" in the face of "clear and present danger"(UNFCCC 2013).

Accounting for 42% of global CO_2 emissions, the energy sector is a primary target for climate mitigation efforts (IEA 2015a: 5). Among low-carbon technology options for electricity generation, renewable energy technologies have experienced the biggest growth by far. The IEA, often criticized for its conservative outlook on renewables (Metayer et al. 2015), projects that more than half of additional electricity capacity until 2040 will consistently come from renewables and that renewables will overtake coal as the largest source of electricity by the early 2030s (IEA 2015b: 53). In 2015, renewable energy investments in developing countries exceeded investments in industrialized countries for the first time (Frankfurt School-UNEP Centre & BNEF 2016: 14).

By 2015, 164 countries had renewable energy targets of some sort and at least 145 of them also had measures in place to promote renewables specifically, hinting at a major global spread of policies in this field (REN21 Secretariat 2015: 18). In an assessment of INDCs submitted for COP21 in Paris, Stephan et al. (2016: 3) find that "142 INDCs mention renewable energy, 108 name the increase in renewable energy as one of their mitigation actions, of which 75 include quantified goals", confirming expectations that renewables will continue to play a central role in global climate protection efforts and the implementation of the Paris Agreement.

Renewable energy technologies have experienced dramatic price declines in recent years. Photovoltaics and wind have reached "grid parity", that is to say, price competitiveness with new fossil or nuclear plants, in solar- and wind-rich areas, and further price cuts are expected (IRENA 2015b: 27). Nevertheless, important path dependencies and non-economic barriers in energy sectors around the globe still render the adoption of dedicated policy and regulatory frameworks for the uptake of renewable energy sources necessary (IEA-RETD 2016). Energy policy decisions taken today create path dependencies for decades, given the long amortization periods of grid infrastructure and power plants (Mattauch et al. 2015, Unruh 2000). A better understanding of how pioneering policies in the field of sustainable energy can impact policy choices elsewhere is therefore of relevance beyond a purely academic interest. Against this background, this book dives into international policy transfer processes linked to a particularly prominent national effort to transform an energy system, the German Energiewende.

1.1.3 Germany, a sustainable energy pioneer

Chancellor Angela Merkel's decision to revise her former position and accelerate nuclear phase-out in Germany following the 2011 Fukushima disaster brought the term "Energiewende" – literally translated as "energy turnaround" – to the attention of a broader global audience. But Germany's move toward renewables and away from nuclear energy has a much longer history (Hake et al. 2015: 535, Lauber & Jacobsson 2016, Lauber & Mez 2004, Schreurs 2013). Proponents of renewable energy and nuclear phase-out in Germany have used the term "Energiewende" since the 1980s (Krause et al. 1980). Following several unsuccessful attempts to introduce a first renewable energy act, the Electricity Feed-In Law (*Stromeinspeisungsgesetz,* StrEG) was adopted in 1990. It laid the ground for an acceleration of the deployment of renewable energy in Germany and for future policy developments, in particular Germany's iconic Renewable Energy Sources Act (*Erneuerbare-Energien-Gesetz*, EEG) adopted in 2000 (Hierl 2011, Hirschl 2008, Lauber & Jacobsson 2016, Lauber & Mez 2004).

Although it is by far not the only example of a country engaging in an energy transition, Germany is widely acclaimed as a pioneer and leader with regard to sustainable energy (e.g., IRENA 2014: 141, IRENA 2015c, Jänicke 2011, Stigson et al. 2013). The country's early support for renewables despite high technology prices; policy experimentation at the local level; ambitious long-term targets; a high share of decentralized, citizen-owned renewable energy installations; and the phasing-out of nuclear power are cornerstones of the Energiewende that make Germany's model stand out (Agentur für Erneuerbare Energien 2014b, IRENA 2015c, Lauber & Jacobsson 2016, Leidreiter 2012, Weidner & Mez 2008).

The Energiewende's long-term targets, set in the 2010 Energy Concept and the 2011 Energiewende legislative package, include an 80 to 95% reduction of CO_2 emissions compared to 1990 levels, a 50% reduction of primary energy consumption compared to 2008, 80% renewables in the electricity mix and a 60% share of renewables in primary energy consumption – all by 2050 – in addition to complete nuclear phase-out by 2022 (Bundesregierung n.d., Bundesregierung 2010a). It is important to underline that the 2010 Energy Concept (Bundesregierung 2010a) already contains these long-term targets and only the timeline was adjusted following the Fukushima disaster, in a "reversal of the reversal" of nuclear phase-out (Schreurs 2013: 94). The 2011 Energiewende decisions are therefore to be seen as a continuation of earlier developments, in particular the initial decision to phase out nuclear power taken in 2000 and rapid growth of renewables in Germany throughout the 2000s (Lauber & Jacobsson 2016).

The promotion of renewable energy in Germany has long been synonymous with one iconic policy instrument, the feed-in tariff.[4] The 2000 EEG (Bechberger

4 Feed-in tariffs grant producers of electricity from renewable sources a fixed compensation for each kilowatt-hour (kWh) of electricity over a certain period. Feed-in tariffs can be designed in a variety of ways and include different design features (Jacobs 2014); the German feed-in tariff is addressed in more detail in Chapter 5.

2000, Bundestag 2000) incarnates this policy approach and has led to dramatic increases in renewable energy capacity in Germany since the early 2000s (BMWi 2016g). Despite its long history, the Energiewende is still a "transformation in the making". Its policy landscape is not an easily intelligible master plan, but a mosaic of political goals, laws, acts and ordinances (BMWi 2015d). These have evolved to reflect the development of renewables from a niche to the single largest source of electricity in Germany (Agora Energiewende 2015a, Hoppmann et al. 2014). Recently, the promotion of renewable energy in Germany has been shifting toward a policy mix, with larger renewable energy plants having to compete for support in auctions (Appunn 2016, BMWi 2015a, BMWi 2016f).

Despite positive developments in other areas, the Energiewende's most significant progress is today mainly visible in the area of renewables in the electricity mix, in which they had a share of 32.5% in 2015 (Agora Energiewende 2016a, Graichen et al. 2015). Important challenges remain, however, regarding the achievement of Germany's climate goals and have stimulated a debate on next steps to be taken (Löschel et al. 2015). In particular, the question of whether and how to schedule a phasing-out of coal capacities, which still accounted for 42% of electricity generation in 2015, is occupying a growing space in discussions (Agora Energiewende 2016b, Hecking et al. 2016).

The term "Energiewende" is used in this book in a broad sense and covers the entire German energy transition process toward, in particular, renewable energies. What the German energy policy model means for stakeholders in recipient contexts, what elements are promoted by Germany, and which ones are considered and used were fundamental questions addressed in empirical research. Interview results confirm the focus on renewables in the power sector, the most visible and advanced pillar of the Energiewende and an area of tremendous importance for global climate protection. Other policy segments, in particular, energy efficiency and general climate policy, are taken into account where they emerged from interviews in recipient states.

1.1.4 An aim for international "Energiewende leadership"

Gaining followers on the way to a more sustainable energy system has been a main rationale for the Energiewende since its beginnings and German decision-makers often claim leadership in this field (Altmaier 2012b, Gabriel 2014: 2699,

Merkel 2014). As Hermann Scheer, a main sponsor of the EEG, put it, the law was intended to "be an example for everyone, globally"[5] (Bundestag 1999: 7273). For Environment Minister Barbara Hendricks (SPD, 2013-present), the "world is looking at Germany, when it comes to decoupling wealth and growth from the consumption of fossil resources. We are the model many countries use to devise their climate protection policy"[6] (BMUB 2014b). For the minister, in the "transformation process the whole world will have to tackle, Germany has a leadership role. The German word 'Energiewende' is more and more used in the English language. We want to support other states, who are not as advanced to achieve this by their own efforts"[7] (Bundestag 2014e: 7385). Rainer Baake, Secretary of State at the Ministry for Economic Affairs and Energy (BMWi), stressed that "we need to continue by writing this success story globally. Only if the global Energiewende succeeds, will we be able to mitigate climate change"[8] (BMWi 2015f). The desire to facilitate similar developments abroad through the Energiewende was also mentioned in more than one-third of 101 parliamentary debates analyzed for this book (see section 5.4.1, also Steinbacher & Pahle 2015: 9). Parliamentarians affirmed that Germany had the "responsibility to make the Energiewende a success in Germany, so others follow"[9] (MP Hubertus Heil, SPD, Bundestag 2014c) and frequently used this aim for international leadership to point to the success of the Energiewende or underpin their demands for policy reform.

From its beginnings, the issue of renewables in Germany has been linked to the motivation of bringing down the cost of technologies, especially for developing countries, by deploying them at scale in Germany (Bundesregierung 1988,

5 "[…] daß wir mit diesem Entwurf ein Gesetz, novelliert und angepaßt an neue Herausforderungen, vorlegen, das weltweit ohne Beispiel ist. Deshalb kann es weltweit beispielhafte Zeichen für alle setzen."

6 "[…] nach wie vor schaut die Welt auf Deutschland, wenn es darum geht, Wohlstand und Wachstum vom Verbrauch fossiler Ressourcen zu entkoppeln. Wir sind das Vorbild, an dem viele Staaten ihre Klimaschutzpolitik ausrichten."

7 "In dem Transformationsprozess, der der ganzen Welt bevorsteht, nimmt Deutschland eine Vorreiterrolle ein. Das deutsche Wort „Energiewende" findet immer mehr Einzug in die englische Sprache. Wir wollen andere Staaten unterstützen, die noch nicht so weit sind, dies aus eigener Kraft zu schaffen."

8 "Diese Erfolgsgeschichte müssen wir nun weltweit fortschreiben […] Nur wenn die globale Energiewende gelingt, werden wir den Klimawandel eindämmen können."

9 "Wir haben eine Verantwortung dafür, dass die Energiewende in Deutschland gelingt, damit andere uns folgen."

Röhrkasten 2015: 165–166, Wieczorek-Zeul 2009 (2001)). This vision is reflected in a longstanding focus on sustainable energy in Germany's development cooperation activities, where it occupies the largest share of the budget of the Ministry for Economic Cooperation and Development (BMZ) (BMZ 2014: 9). Germany's interest in gaining followers for a global Energiewende is also expressed in the numerous international and bilateral initiatives for renewable energy it initiated and supports (see Ch.5 and additional online material for an overview). Finally, there is a strong economic component in the Energiewende's international dimension. The early creation of a domestic market for renewable energy technologies in Germany has put its companies in the position of becoming leading exporters of renewable energy and efficiency technologies (Eckermann 2016, Gehrke & Schasse 2015, Jänicke 2011, Jänicke 2013). Encouraging the creation of regulatory frameworks in other countries, which increase demand for clean technologies, is therefore of interest for German industry in the framework of a "lead market" approach (Beise & Rennings 2005, Jänicke & Jacob 2004).

In the light of clearly expressed ambitions for leadership, calls for Germany to strengthen its "Energiewende diplomacy" and its outreach to potential followers have multiplied (Fischer & Geden 2011, Li 2016, Messner & Morgan 2013, Messner et al. 2014, Morgan & Weischer 2013, Quitzow et al. 2016, Steinbacher & Pahle 2015, Tänzler & Wolters 2014, Westphal 2012). A thorough assessment of the link between Germany's efforts and actual policy transfer to recipient cases is, however, still missing. This book contributes to closing this research gap by providing an in-depth account of how German Energiewende leadership affects policy transfer to and policy making in recipient states in the area of renewable energy.

1.1.5 Active Energiewende leadership and policy promotion

This book addresses the link between Energiewende leadership and policy transfer along several mechanisms. First, the implementation of the Energiewende in Germany (i.e., "unilateral leadership") may trigger interest among stakeholders abroad, which can lead to policy transfer through mechanisms such as learning (e.g., Dolowitz & Marsh 1996, Dunlop 2009), emulation (e.g., Meseguer 2005: 73), and lesson drawing (Rose 1991, 2005). But instead of relying on the power of the example alone, Germany's desire to find followers translates into a range of "active leadership" measures. These can result in the creation of new evidence and information, the entry of new members into recipient policy networks, strengthening institutions, building capacity, or providing incentives that render the perspective of adopting particular policies more feasible in the eyes of followers.

The availability of information about the Energiewende appears as a first, basic condition for transfer and diffusion. Communication by Germany on its energy policy model is thus important to increase its "visibility" and "observability" (Heiden & Strebel 2012). German governmental departments have reinforced efforts with a view to making information about the Energiewende available to stakeholders abroad (BMWi 2016a, BMWi 2016b). For example, the Federal Foreign Office (*Auswärtiges Amt*) created a unit dedicated to the international aspects of Germany's energy concept in 2013, promotes the communication of the Energiewende abroad, and regularly invites delegations to Germany (Auswärtiges Amt 2014; see online material related to this book[10] for a list of activities). Nevertheless, official information about the Energiewende in English, let alone in other languages, remains scarce and scattered compared to the scope of the project. Other organizations, such as think tanks and political foundations, e.g., the think tank Agora Energiewende (Agora Energiewende n.d.), the Heinrich Böll Foundation, which is close to the German Green party (Heinrich Böll Stiftung n.d.-b), or the Clean Energy Wire service (CLEW n.d.), have therefore intensified efforts to communicate about the Energiewende to a global audience.

Germany has initiated or contributed to the creation of a range of institutions and initiatives that can serve as "international or transnational channels through which policies and instruments are communicated" (Busch et al. 2005: 150). Bilateral energy partnerships have been established with countries including Morocco, Tunisia, Nigeria, Norway, South Africa, Russia, Brazil, and India (BMWi 2016c, Müller 2015). In 2013, the Renewables Club or "Club of Energiewende Countries" (*Club der Energiewendestaaten)* was founded as a spearhead of leaders and aspiring leaders in sustainable energy (BMU 2013). International and regional initiatives for renewable energy such as the REN21 network, the International Feed-in Cooperation (IFIC), the French-German Renewables

10 To access the book's appendix, please follow the URL given in the imprint or visit www.springer.com and search for the author's name.

Office, the Transatlantic Climate Bridge or the Regional Center for Renewable Energy and Energy Efficiency (RCREEE) in the Middle-East North Africa (MENA) region, have been initiated and are supported by Germany. The most prominent consequence of these efforts is the International Renewable Energy Agency (IRENA), whose foundation is a result of German leadership (Land 2009, Röhrkasten 2015: 126, Röhrkasten & Westphal 2013). The roots of many of these renewable energy initiatives go back to the "*renewables 2004*" conference held in Bonn, which demonstrated Germany's willingness to exercise leadership in the field of sustainable energy early on (BMZ 2010, Hirschl 2008). As Tänzler and Wolters (2014: 149) note, the multitude of fora and alliances initiated by Germany in the field of sustainable energy, taken together, constitute a "normative power".

Development cooperation occupies a major role in Germany's global sustainable energy activities, with projects aiming to make sustainable energy technologies and policies better known, more implementable, and affordable for partner countries (BMZ 2014, Rzepka 2013). The amount spent on sustainable energy projects with developing countries is set to increase to at least €3.6 billion anually by 2030 (BMZ 2014: 11). Projects are mainly implemented by the German Agency for International Cooperation (*Gesellschaft für Internationale Zusammenarbeit*, GIZ) and KfW Development Bank (*Kreditanstalt für Wiederaufbau – Entwicklungsbank,* KfW*)*. With almost €6 billion spent between 2004 and 2011, Germany is one of the most important donors in the field of sustainable energy globally (BMZ 2014: 12), and global climate finance (Kowalzig 2015). In addition to supporting the financing of renewable energy projects through loans and grants administered by KfW (KfW 2014), non-monetary support can have a decisive impact on the likelihood of renewable energy policies being considered and adopted, as this book shows. This non-financial support includes capacity building measures, renewable energy potential studies, training sessions, pilot projects, policy advice and other technical assistance (GIZ 2013, GIZ 2014a, Beucker et al. 2014). To the same extent that the Energiewende cannot be regarded as a clear-cut policy model, German leadership efforts also encompass a variety of actors, levels, methods, and directions that are explored in more detail in Chapter 5 and in the case studies on Morocco, South Africa, and California, presented in Chapters 6-8.

1.2 Sending and Receiving: Integrating Leadership and Policy Transfer

1.2.1 State of the literature and research gaps

Whereas the international dimension of Germany's energy transition is readily acknowledged by decision-makers and experts, an assessment of the Energiewende's actual *consideration* by agents, its *use* in, and its *effect* on policy debates and decisions in recipient states and of Germany's efforts of promoting the model abroad has not been undertaken systematically. The present book addresses this gap.

The spread of feed-in tariffs has been used as a proxy to infer the Energiewende's global effect (Jacobs & Mez 2012). The feed-in tariff is a policy instrument that is at the base of Germany's success in deploying renewables (Lauber & Jacobsson 2016). As of today, 108 countries and states worldwide (REN21 Secretariat 2015: 153) have also adopted feed-in tariffs to promote renewable energy. The EEG is widely considered, as Vice Chancellor Gabriel put it, "a German export hit that does not appear in trade balances" (Gabriel 2008). A causal link between the German feed-in tariff example and the instrument's diffusion is frequently drawn across the renewable energy community in Germany – from NGOs, think tanks, and renewable energy advocates (Agentur für Erneuerbare Energien 2016, BUND 2014, Fell 2014, Greenpeace 2007: 4, Klima-Allianz Deutschland 2013, Klimaretter 2007, Leidreiter 2012, Rosenkranz & WWF Germany 2015: 7) to politics (Altmaier 2012c, BMU 2004, BMU 2007: 20–21, BMWi 2014, Bündnis 90/Die Grünen Bayern 2015, Gabriel 2008, Gabriel 2014, Krischer 2010, Land 2009), to industry (EUROSOLAR 2012, BSW 2008), to research (Diekmann et al. 2012, DIW 2010, Hirschl 2008: 562, Jacobs & Mez 2012: 264), to development cooperation (Kückmann 2014, Wieczorek-Zeul 2009 (2001)) and specialized media (Petersen 2010, Solarify 2014). Research establishing a link between the EEG and the global diffusion of feed-in tariffs is nevertheless surprisingly rare given the prominence of the claim (for exceptions, see Busch 2003, Jacobs 2012, Jacobs 2014, Jacobs & Mez 2012). Other studies survey the perception of the Energiewende among experts abroad and its reception in international media (Agentur für Erneuerbare Energien 2014a, Hirsch 2015, Konrad Adenauer Stiftung 2013, Konrad Adenauer Stiftung 2014).

In addition to the empirical research gap regarding the effectiveness of Energiewende leadership, this book addresses a series of under-researched areas in the analysis and theory of international policy transfer. First, questions remain open regarding *how* and *why* "program evidence from foreign countries enters policy debate and informs action" (Bennett 1991: 33, see also Gilardi 2013b, Majone 1991: 104). In particular, the political dimension of transfer and the strategic use of transferred knowledge in policy networks by different actors and groups are underexplored (Daviter 2015, Dolowitz & Marsh 1996: 355) and emerged as important factors throughout empirical research. Secondly, the role of pioneering countries in contributing to the spread of their own policies has hardly been addressed in policy transfer studies (Biedenkopf 2012, Gilardi 2013a). As outlined below, countries as senders are considered implicitly in this literature, as passive providers of policy models, or as leaders in negotiation settings. Policies as transfer objects are often studied as "highly plastic and timeless entities" in disconnection from their source (Saint-Martin 2010: 5) and the possible contribution of sending countries to policy transfer through "leadership by diffusion" or leadership through transfer "lacks empirical investigation almost entirely" (Biedenkopf 2012: 111).

A third research gap addressed in this book concerns case selection in the transfer literature, which is so far strongly biased toward OECD countries, especially European countries and federal states in the United States (Evans 2004b, Nedley 2004, Stone 1999). The choice of a heterogeneous set of cases for this research aims to explore the effectiveness of leadership through transfer beyond these groups of states. Another shortcoming of the transfer literature is its focus on "positive" cases, where policy transfer not only visibly took place, but also led to convergence between the sender's model and followers. Although the absence of cases of negative and selective transfer in the literature is widely deplored, more complex transfer outcomes, in which foreign models are used selectively, are rarely accounted for despite their likely prevalence and important potential contribution to theory-building (Dobbin et al. 2007: 463, Genovese et al. 2015, Klingler-Vidra & Schleifer 2014, Klingler-Vidra 2014, Marsh & Sharman 2009: 270, Radaelli 2005, Solingen 2012). Policy transfer as an outcome is, however, by no means limited to the copy-pasting of the leader's model, but it can also signify the transfer of a principle, such as climate-friendly energy systems, with followers adapting transferred knowledge to shape their own models

(Busch & Jörgens 2005a: 4, Mossberger & Wolman 2003: 431), as findings in this book show. Cases of selective and negative transfer can also provide highly relevant policy lessons for senders wishing to promote their policies abroad.

1.2.2 When policies travel: Studying transfer

The study of the international spread of ideas and policies has led to the emergence of a rich literature on policy convergence, policy diffusion, and policy transfer. The overlaps and differences between these strands of research haven given rise to a lively terminology debate (Marsh & Sharman 2009). Its case-study approach, a focus on micro-level processes, and its interest in causal mechanisms in specific instances anchor this book in the policy *transfer* literature.

The book follows Dolowitz and Marsh's widely used definition of transfer as a "process by which knowledge about policies, administrative arrangements, institutions and ideas in one political setting (past or present) is used in the development of policies, administrative arrangements, institutions and ideas in another political setting" (Dolowitz & Marsh 2000: 5). Literature on policy diffusion and transfer is extensive and has rapidly grown in recent years. A cursory search for publications on policy transfer in the ProQuest social sciences database shows 17 entries for the decade from 1980-1989, 108 entries for 1990-1999, a steep jump to 1,157 publications for the 2000-2009 period, and 1,430 from 2010-2016, hinting at a continuing dynamic in this field (ProQuest n.d.). The diffusion and transfer literature covers diverse policy areas, including economic policies and democracy (Simmons et al. 2007, Simmons & Elkins 2004), regulatory agencies (Jordana et al. 2011), pension and social systems (Brooks 2004, Weyland 2006), health care (Holden 2009), military innovations (Horowitz 2007), enterprise zones (Mossberger 2000), smoking restrictions (Asare & Studlar 2009, Shipan & Volden 2008), and state lotteries (Baybeck et al. 2011).

Due to the cross-border nature of many of the issues the subject tackles, environmental policy, and to a lesser extent energy policy, occupy an important space in the literature. Like much of the policy transfer literature, studies in the area of renewable energy policy transfer have largely focused on policies traveling within the United States or the European Union (Busch 2003, Chandler 2009, Jacobs 2014, Matisoff & Edwards 2014, Smithwood 2011, Stoutenborough & Beverlin 2008, Vasseur 2014). Large-scale diffusion studies have traced the spread of environmental policy instruments, standards, and institutions across a large number of countries and have thereby prepared the ground for more case-based studies (Arts et al. 2008, Holzinger et al. 2007, Busch & Jörgens 2010, Sommerer 2011). While this groundwork is crucial, diffusion studies rarely allow for the establishment of causal relationships or a more detailed understanding of the role that senders, the characteristics of a specific policy model, and local filters play in the process. In particular, diffusion studies are likely to overlook cases in which knowledge was transferred, but where this is not (fully) reflected in policy outcomes because transferred knowledge is used selectively (Rose 2005: 81), where it is adapted to the local context (Klingler-Vidra & Schleifer 2014: 269), is mixed with other policy models (Evans 2009b: 246), or where negative lessons are drawn and used in policy making (Marsh & Sharman 2009: 270). Rather than considering these outcomes as "non-transfer" (Dolowitz & Marsh 2000: 9), findings in this book show that transfer can be rich and relevant, independently of whether convergence between a sender and recipients can be observed on the basis of policy instruments. Case studies like the ones presented in this book can therefore complement large-n diffusion studies in decisive respects, in particular when complex causal processes are to be assessed (Howlett & Rayner 2008, Marsh & Sharman 2009, Yin 1981).

1.2.3 Bringing senders into the picture: Leadership through transfer

As a concept in the environmental and climate field, leadership has mainly been used to describe the role countries, organizations, or individuals can play in international negotiations. By sending signals, providing cognitive resources, and steering the debate, leaders are expected to facilitate more ambitious negotiation outcomes (Karlsson et al. 2012, Saul & Seidel 2011, Schreurs & Tiberghien 2007, Schwerhoff 2013, Wurzel & Connelly 2011, Underdal 1992, Underdal 1994).

While insights into leadership in environmental negotiations are highly relevant, this book is interested in a somewhat different question. It addresses leadership from the perspective of how countries can promote the bilateral transfer of their own policy models to other constituencies. Leadership through transfer or "by diffusion" is a "distinct and important mode of external governance through which jurisdictions can contribute to raising the bar of global environmental policy, separate from international agreements and coercion" (Biedenkopf 2012:

111). Pioneering countries do appear in the transfer and diffusion literature, but mostly as passive providers of policy models, and through determinants such as the senders' prominence (Finnemore & Sikkink 1998: 906, Majone 1991: 104) or their similarity with the recipient in terms of culture or language (Bennett 1991: 51, Elkins & Simmons 2005). But unlike non-governmental organizations (NGOs) (Betsill & Corell 2001, Stone 2004), epistemic communities (Haas 1992: 3), or international organizations (Brooks 2004), the role of *pioneering countries* as promoters of policy transfer has received comparatively little attention, with few exceptions (e.g., Barabasch et al. 2009, Holden 2009, Lana & Evans 2004).

The role of countries as promoters of knowledge and policy transfer is more prominently discussed in the democracy promotion literature (Huber 2015, Scott & Steele 2011) and in the context of "knowledge-based aid" and development cooperation (e.g., Ellerman 2000, King 2005, King & McGrath 2004, Lepenies 2014, Molenaers et al. 2015, Stone & Maxwell 2005). Knowledge sharing and knowledge transfer in the framework of development cooperation have also long been recognized as mechanisms of central importance by multilateral institutions, including the World Bank in its "Knowledge bank" process (World Bank 1998) and the G20 (Task Team on South-South Cooperation 2011). As Gilardi (2013a: 5) notes, a better understanding of the mechanisms of policy transfer and diffusion inevitably requires knowing who senders are and what makes them effective: "Of course, we first need to know which units are influential and why [...] We are nowhere near being able to make such specific recommendations, but this is certainly one of the potential practical payoffs of this literature". As pointed out by Biedenkopf (2012: 110), "The pioneering jurisdiction can influence diffusion by promoting its pioneering policy", albeit only "marginally" – a limitation that is challenged in this book.

The climate and environmental leadership literature is used to devise a framework for leadership through transfer in three variants (e.g., Eckersley 2011, Eckersley 2016, Malnes 1995, Underdal 1992, Underdal 1994, Wurzel & Connelly 2011b, see also Steinbacher & Pahle 2015). The first variant is unilateral leadership, which means setting an example (i.e., the Energiewende) and aiming for others to take inspiration from it (Underdal 1994). Active leadership, the second variant, signifies translating the willingness to promote a particular policy direction by active outreach to potential followers through communication, incentives, building institutions and capacities, or the strengthening of transfer channels (see section 2.4.3). Finally, effective leadership through transfer refers to an outcome, where policy action in the recipient state is spurred, and the achievement of an overarching goal is thereby facilitated (see section 2.4.7). Importantly, this does not mean – as the outcome of case studies in this book shows – that the leader's model or even objectives have to be mimicked for leadership to be effective. The effect of transfer on policy decisions is thus an important, but only one category of transfer considered in this book.

1.2.4 Hypotheses: Three categories of transfer

This book addresses the link between Germany's Energiewende leadership and policy transfer to three relevant recipient states. The analytical framework takes into account expectations from the literature on policy transfer, diffusion, leadership and policy networks (see Chapters 2 and 3). Rather than strictly testing these expectations, the book adopts an iterative, abductive approach in which theory is continuously "played against" the data (Strauss & Corbin 1994: 273). Transfer is seen as both a dependent variable, and as an explanatory variable or as an "ingredient" to policy making. Three guiding hypotheses serve to structure the case study findings presented in Chapters 6, 7, and 8, and the overarching findings discussed in Chapter 9.

Hypothesis I, "Consideration of the Energiewende", concerns determinants influencing the consideration of the German policy model by stakeholders in the recipient context. Since stakeholders cannot be expected – as a purely rational learning approach would suggest – to take all existing energy policy models and alternatives into account, cognitive shortcuts play a role in the selection of models to take into account (Meseguer 2005, Weyland 2006). Consideration is expected to be linked to the prominence of the sender (Finnemore & Sikkink 1998: 906, Jänicke 2005: 139, Majone 1991: 104), the transferred policy model's reputation of success (Elkins & Simmons 2005: 44), the availability of information and "observability" of the model (Heiden & Strebel 2012), and the capacity that members of the recipient network can dedicate to search for and process information (Kern et al. 2001: 8, Weidner et al. 2002). Unilateral and active leadership can promote consideration in decisive ways, through communication, through the creation and reinforcement of transfer channels and capacities, and by increasing the policy model's visibility.

Although consideration is a crucial first step, leaders are likely to be more interested in how their model is used in the policy process in the recipient context. Hypothesis II, "Utilization of transferred knowledge", spells out several factors that likely affect how transferred knowledge is strategically, politically, or functionally utilized by stakeholders in policy debates (Daviter 2015, Dolowitz & Marsh 1996: 355, Dunlop 2009, Radaelli 1995). It is important to emphasize that transferred knowledge not only covers the Energiewende as it is implemented in Germany, but "translations" of the model and its general principles to the recipient context as well. It is expected that members of the policy network are more likely to use transferred knowledge if it can serve to underpin positions they hold, if it is perceived as being in line with objectives supported, if it is able to close epistemic gaps, and if the transfer of particular elements is seen as feasible in the recipient context. Stakeholders are expected to have differing views on the transferred model and utilize it differently, depending on their prospective evaluation of the consequences of transfer. Active leadership can impact utilization, by bringing new transfer agents to the policy network (Howlett & Ramesh 2002: 34) and by empowering particular members of the network through targeted information, the production of adapted evidence, or the support of capacities that make a transferred model appear more implementable.

Hypothesis III, "Effect on policy outcomes", addresses how transferred knowledge is reflected in policy outcomes in the recipient context. Transfer outcomes are only one part of the transfer process and are in fact more likely than not to differ from the initial sender's model. This book therefore addresses the reflection of transferred, Energiewende-related knowledge in policy decisions by looking at transfer as an ingredient to policy making. The impact of transfer on policy decisions depends on the structure of policy making processes and policy networks, and on the relative strength of their members, who bring different positions regarding transferred knowledge into the decision-making process (Adam & Kriesi 2007: 137, Howlett et al. 2009: 82). Ultimately, timing is of relevance as well since windows of opportunity for transfer may open at times of high pressure for policy change (Bennett 1991, Jenkins-Smith & Sabatier 1993: 42, Kingdon 2011: 166). Leadership can aim at impacting these factors by attaching incentives to policy ideas, mitigating losses for those opposed to transfer, and contributing to the recognition of windows of opportunity. Importantly, active leadership can also change the structure of policy networks by bringing new agents to the process.

1.3 Research Design and Methods

1.3.1 Case selection: Transfer to Morocco, South Africa, and California

The selection of Morocco, South Africa, and California as cases followed three steps, which were guided by research gaps in the literature and an explorative research aim. A first decision was to narrow the "universe of cases" by excluding European countries. This is because the diffusion of German-style policy instruments is comparatively better explored with regard to other countries in Europe than outside European borders. Secondly, harmonization as a powerful transfer mechanism would have made it challenging to at least attempt to isolate the role of the German model and would have limited generalizability with cases outside the European Union (EU). Thirdly, as in many areas of political science research, there is a strong lack of studies on cases outside the OECD in transfer research (Nedley 2004, Stone 2001: 19). In this sense, this book is a response to Marsh and Sharman's invitation to transfer researchers: "the previous reluctance of scholars to travel too far from their home base to examine the experiences of policy-makers on the receiving end of Western models should represent a tremendous opportunity for new detailed process tracing studies and structured inter-regional and intra-regional small-N comparisons" (Marsh & Sharman 2009: 281). Understanding renewable energy policy making through the lens of transfer can provide insights on trends in the regulation of renewable energy, including in developing and emerging countries, where most of new electricity capacity will be added in the future.

The second step in narrowing down the universe of cases was to consider states that are particularly targeted by German policy promotion efforts. For developing countries, defining this group is more straightforward because of the existence of programs such as the International Climate Initiative (IKI) and the renewable energy projects it supports. Energy partnerships or membership in the German-initiated Renewables Club were further aspects taken into account in defining countries with particularly strong ties to Germany. Regarding industrialized countries, other institutionalized formats such as the Transatlantic Climate Bridge and frequent bilateral visits hint at close transfer channels, backed up by interviews with officials from German ministries. Among particularly targeted potential followers, several cases stand out as not having adopted a German-style feed-in tariff model. At the time of case selection, Morocco and South Africa were the only cases with virtually all forms of cooperation in the energy sector with Germany, but without feed-in tariffs at the time of case selection. A similar observation can be made for California, where feed-in tariffs were only introduced in small programs at the municipal level and in a highly transformed version with a low program cap at the state level. This is thought provoking and relevant theoretically given the lack of studies of apparent non-transfer. Although the Energiewende means far more than feed-in tariffs, and countries can of course not be expected to automatically mimic states they closely cooperate with, the shared feature of not adopting this iconic policy instrument despite strong transfer channels promised relevant insights into complex transfer processes and the actual content of German Energiewende leadership.

Finally, a third concern in case selection was to achieve diversity among cases regarding commonly accepted determinants for policy transfer. Proximity with the sender, similar levels of economic development, regulatory path dependencies in renewable energy, the openness of the political system, and capacities in the policy field count among these expected determinants for transfer. The three selected cases provide a high degree of diversity on these dimensions. Choosing Morocco, South Africa and California as cases also allows covering a broad range of possible transfer channels with Germany, from highly institutionalized ones, such as energy partnerships, to looser and more informal forms of cooperation such as bilateral visits. As a result, "noise" from a multitude of potential explanatory variables can be reduced to try to identify more generalizable features of Energiewende policy transfer across systems that are "most different" (Seawright & Gerring 2008: 306).

1.3.2 Tracing transfer: Research design and methods

The research design for this book was developed with an aim of providing a deep, micro-level understanding of how knowledge reaches recipients, how it is used and instrumentalized, and how senders can influence transfer processes and their outcomes. In terms of the general methodological framework, a case-based approach with three case studies and a cross-case comparison was chosen. The goal here was to combine in-depth insights and a first step toward identifying common patterns of leadership through transfer. Within the case studies, process tracing is used as a framework. This means gathering comprehensive data in order to trace, step-by-step, how debates and transfer evolved and how decisions were made in the renewable energy field.

The most important data source is the 181 interviews and 16 background conversations carried out in Germany, Morocco, South Africa, and California between the fall of 2013 and the spring of 2015. These interviews provided crucial insights into the role of foreign evidence in policy making that is not observable from public accounts. Interview findings are triangulated with publicly accessible documents, including minutes of meetings, protocols, press releases and public speeches, depending on their availability in the different cases. Five main types of interviewees were targeted: policy makers and regulators, energy industry representatives (both incumbents and renewable energy industry), NGOs and renewable energy advocates, German "transfer agents" (technical and financial cooperation, diplomacy, political foundations, business), and third parties, for an external view (e.g., other bilateral donors, World Bank). Interview questionnaires were semi-structured, meaning that they covered common themes but were closely tailored to the expertise and position of the interviewee and were flexible so new lines of information could be integrated. Interviewees remain anonymous in the presentation of findings. In addition to it being required by many interview partners, extending anonymity to all interview partners proved helpful in gathering insights into the strategic use of information and the role of foreign evidence that could not be stated publicly. The interview material was transcribed and manually coded using dedicated software (MaxQDA), with codes being developed bottom-up from the data. The coding process allowed for the identification of patterns and recurrent themes across the three cases and to structure the wealth of data collected. Given the choice of non-standardized questionnaires, interview material is used in a narrative way in the case studies. In Morocco and South Africa, where secondary literature and public statements on energy policy objectives are rare, drivers for renewable energy policies were explored using a card-ranking method that Joas et al. (2014, 2016) applied to the case of the Energiewende in Germany. Interview partners were asked to rank small paper cards on policy objectives to be achieved through their country's renewable energy strategy, with a view to assess the possibility of policy transfer despite potentially differing policy objectives (see also Steinbacher 2015). Independently of the question of transfer, the rankings provide new insights into the drivers for beginning energy transitions in Morocco and South Africa and can prove interesting also outside the question of policy transfer.

1.4 Outline and Findings

1.4.1 Effective Energiewende leadership through transfer?

Despite the heterogeneity of the three selected cases, several strong common patterns of transfer and the effectiveness of German leadership in promoting it can be identified.

First, findings show the importance of considering the source of policy models when assessing their transfer, an element that is not always taken into account in policy diffusion and transfer research. The Energiewende as a source of lessons and Germany as an active sender played an outstanding role in Morocco, South Africa, and California. This is due to the prominence of Germany as a sending country and to the perceived disruptiveness and effectiveness of the Energiewende in bringing online renewable energy capacity, making it an example that at least had to be considered, regardless of whether particular recipients found it to be transferable. The Energiewende as an energy policy model played out through different channels, depending on the case in question. In Morocco and South Africa, it provided legitimacy, credibility, and a rich pool of evidence for German transfer agents to draw from, which they "translated" to the recipient context. The important role of these transfer agents as full members of the local policy network shows that transfer not only brings new ideas, but can also modify structures of recipient networks, and thereby affect policy debates. In California, consideration of the Energiewende was more direct, but was also facilitated by interaction with German experts. Across the cases, the dimensions of unilateral and active German Energiewende leadership are inseparably interwoven, pointing to the relevance of considering the actions and perception of pioneering countries for the prospects of policy transfer.

With regard to the question of how knowledge was used, transfer appears as an inherently political process across the cases. Transfer research, and to an even greater extent, diffusion research, have a tendency of looking at transferred policies as uniform, value-neutral items. In Morocco and South Africa, but particularly in California, the same pieces of evidence and policy ideas were perceived in wildly different ways. These perceptions depended on stakeholder's prospective evaluation of losses and gains from transfer, beliefs about what constitutes appropriate policy responses in the recipient context, and how transfer would affect personal and institutional positions. These findings need to be seen in the context of the specificities of energy as a policy field and of the Energiewende as a system-changing policy model. The differential and political nature of transfer in the cases studied not only shows that recipients are all but monoliths, but also explains why policy transfer can occur even though adopted policy instruments do not point to convergence.

In Morocco and South Africa, the information provided by and the actions of German transfer agents were instrumental in facilitating the introduction of first renewable energy policy programs. Morocco's ambitious renewable energy targets and the important role solar plays therein were facilitated by the efforts of German transfer agents joining forces with Moroccan pro-renewables agents within institutions. In South Africa, a unilateral decision was taken by the regulator to announce feed-in tariffs inspired by the German model. Although these were never implemented, they spurred renewable energy policy action after years of stalemate. The empowerment of domestic agents through targeted information and evidence needed to support arguments as well as through capacity building, training sessions, and study tours was a decisive enabling factor for transfer. Empowerment as a transfer mechanism also played a role in California, where renewables advocates pointed to the German example to underpin their arguments, but transfer regarding instruments encountered obstacles of regulatory traditions and incumbent interests.

Eventually, the degree to which transferred knowledge is *reflected* in policy outcomes depended on factors that are at play in "normal" policy making as well. The position powerful stakeholders adopted regarding a transferred model, the degree to which institutional adjustments would have been required to implement it, and whether the Energiewende model was seen as conflicting with important regulatory traditions and the positions of powerful stakeholders were main factors observed across all cases. A significant finding is that these factors play out differently depending on the "policy layer" concerned. While in all three cases the Energiewende and/or active German leadership played a role in shaping or validating overall policy orientations of shifting toward more renewables and also led to the transfer of concrete, incremental adjustments to the regulatory framework, resistance to transfer was concentrated on the intermediary level of policy instruments. This is an important finding since it means that a large share of transfer processes is likely to go unnoticed if the focus is only put on convergence concerning instruments.

1.4.2 Policy lessons

Beyond its academic contribution, this book aims to generate policy lessons for leadership through transfer, against the background of it being a potential alternative mode of governance (Biedenkopf 2012, Busch et al. 2005: 164, Busch & Jörgens 2010, Kern et al. 1999). Since international governance in the domain of renewable energy is polycentric (Cole 2015, Ostrom 2009, Ostrom 2012), that is to say, not coordinated from above, the role of pioneer countries in engaging potential followers, disseminating information and sharing experience is likely to be key. Although the dissemination of best practice in the renewable energy field is part of IRENA's mission, the organization "plays virtually no role in the daily business of (energy) politics" (Röhrkasten & Westphal 2012: 1). For this reason, "insights into the ways in which governments could make better use of [leadership by diffusion as] a mode of external governance" appear crucial (Biedenkopf 2012: 111). Exploring renewable energy policy making through the lens of transfer provides relevant practical insights into the dynamics of energy transitions in different contexts. It also allows for a better understanding of how countries "window-shop" for different policy alternatives, how transferred knowledge is used politically, and how alternatives are selected (Daviter 2015, Evans 2009b: 246, Genovese et al. 2015, Radaelli 2005).

Based on the insights gained through the case studies, several points appear relevant for effective leadership strategies. The first one is clarity about sender and follower objectives. This means, first of all, that senders should aim to understand policy drivers and the objectives of different groups in the recipient context in order to adjust their outreach. Even though transfer can occur if policy objectives between the sender and recipients differ, awareness about local goals can enhance the effectiveness of leadership in promoting it. Secondly, interviews revealed a lack of clarity on what objectives Germany pursues through its activities in the field of renewable energy, in particular in Morocco and South Africa. Whereas most interview partners concluded that Germany's agenda was likely linked to global climate protection efforts, a legitimate leadership strategy requires transparency.

Although there is no such thing as a unified and clearly defined "Energiewende foreign policy" (Tänzler & Wolters 2014) or even a defined "export strategy" for the model, Energiewende leadership emerges by taking together the numerous transfer channels and interactions between Germany and recipients in each case. Particular features of this approach contributed to its effectiveness. German transfer agents in Morocco and South Africa tended to stay in the partner country for longer periods of time and engaged more intensely with the local policy network than officials from other countries, in particular by being placed within local institutions. They thereby became members of policy networks and were able to join forces with local stakeholders who defended similar positions. Even more importantly, the effectiveness and perceived legitimacy of Germany's outreach, in all cases, was linked to Germany's domestic policy example. The political will to move first and contribute to a global effort provided credibility to transfer agents on the ground, facilitating the effectiveness of transfer.

1.4.3 Outline

This book is structured as follows: Chapter 2 presents the theoretical framework and strands of literature of relevance for the research topic, in particular literature on policy transfer and diffusion, leadership, and policy networks. Chapter 3 lays out the analytical framework derived from theory and presents the guiding hypotheses. Chapter 4 introduces the methodological framework, case selection, and methods used for data collection and analysis. Germany's Energiewende as a basis for unilateral leadership and its active leadership measures are presented in Chapter 5. The chapter also includes information gathered through interviews in Germany and presents findings from parliamentary debates regarding Energiewende leadership. The case studies of transfer to Morocco, South Africa, and California are presented in Chapters 6, 7, and 8. Analysis from the cases is compared and combined in Chapter 9, where hypotheses are discussed and further refined. A critical discussion of methods and policy lessons for effective leadership through transfer completes this chapter. Finally, conclusions are drawn and areas for further research are identified in Chapter 10.

2 Theoretical Perspectives

2.1 Chapter Overview and Research Gaps

The research gaps this book addresses concern both the empirical case studied – the German Energiewende and its transfer to Morocco, South Africa, and California – and theoretical gaps related to understanding the international spread of policies. While the reality of policies traveling across borders, including in the environmental and renewable energy field, is well established (Busch & Jörgens 2010, Sommerer 2011), how and why this happens in particular instances is less clear (Bennett 1991, Gilardi 2013b, James & Lodge 2003, Majone 1991: 104). In particular, the ways in which the source of a transferred model and active sending countries interested in finding followers[11] affect transfer, have received very little attention in the literature so far. The international dimension of the German Energiewende has mainly been approached from the angle of the global diffusion of feed-in tariffs, an iconic policy instrument that served as the basis of Germany's rapid increase in renewable energy (Jacobs & Mez 2012). How exactly Germany's policy model and its aim of facilitating sustainable energy uptake in other countries actually affect policy transfer and policy making processes abroad remains to be addressed. In particular, the ex-ante focus on one policy instrument and on adoption as an outcome excludes complex and selective transfer processes. These can generate rich insights for aspiring "leaders by transfer" and for policy analysis in general (Howlett et al. 2009: 142). This book therefore adopts a more open approach that is not limited to assessing the adoption or non-adoption of a single instrument. This requires a theoretical framework that can

11 The terms "follower" and "recipient" are used interchangeably in this thesis to designate the country/state toward which transfer takes place. Germany is referred to as the "sender" or "leader" to underline its status as a source of policies whose international spread is studied, and as an entity actively aiming to attract others to take inspiration from its policy model. These terms do not imply that recipients copy-paste the sender's model, in particular because the three cases selected in this thesis are cases of selective transfer (see section 4.2.2 on case selection). The mono-directional focus on transfer from Germany to three recipients, Morocco, South Africa, and California, was chosen based on Germany's stated aim of "finding followers" and to narrow the focus of research to generate deeper causal insights, but reverse transfer is a highly promising area for further research.

K. Steinbacher, *Exporting the Energiewende*, Energiepolitik und Klimaschutz. Energy Policy and Climate Protection, https://doi.org/10.1007/978-3-658-22496-7_2

account for what happens to transferred knowledge in the policy making process in recipient states, including the political use of transferred knowledge, an aspect that has remained largely underexplored (Daviter 2015: 494).

To close the important research gap with regard to the role of pioneering countries as active senders in policy transfer, a dedicated analytical framework for "leadership through transfer" is employed. It draws in particular from the literature on climate leadership but also takes related strands of literature such as soft power and lead markets into account to assess how active senders can impact bilateral transfer processes. To trace transfer from the source to its potential reflection in policy decisions, policy transfer is considered both as an explanatory variable for policy outcomes and as a dependent variable. While the transfer and diffusion literature provides rich insights into determinants that affect the likelihood and the mechanisms at play within transfer processes, assessing the role of transfer in policy debates requires integration with other strands of literature. The choice in this book is therefore to consider the role of transfer in energy policy making through the lens of policy networks.

This chapter puts an emphasis on the many overlaps between different strands of literature as to what determines the spread, use, and impact of foreign policy ideas. Particular attention is paid to the interdependencies between each of the elements of the transfer process, such as how senders affect the reception of ideas in the recipient state, how transferred ideas affect positions in policy debates, and how stakeholders harness transferred ideas to further preferred policy outcomes. These expectations form a theoretical framework that is refined and presented in Chapter 3, to answer this book's research question, "What determines the effectiveness of German Energiewende leadership in promoting renewable energy policy transfer to Morocco, South Africa, and California?"

2.2 Policy Diffusion and Transfer: Concepts

2.2.1 Interdependent policy making: The emergence of a field

The consideration of interdependencies between policies in different countries predates the academic study of convergence, diffusion, and transfer. Francis Galton's critique of a study on "laws of marriage and descent" (Tylor 1889) for not taking into account the possibility of imitation across borders is widely considered the earliest example of the fundamental question in which diffusion and transfer studies are interested. That is: is the adoption of policies, ideas, norms and institutions in different constituencies influenced by their adoption in other constituencies? While this book focuses on international policy transfer on a bilateral level,[12] the literature is rich in studies of transfer at the subnational or local level (e.g., Berry & Berry 1990, Hakelberg 2014, Ladi 2004, Strebel 2011), as well as between governments and inter-governmental or non-governmental bodies (e.g., Stone 2000, Stone 2004), and even of inter-temporal transfer within the same organization (Rose 1991: 78).

Literature on policy diffusion and transfer is vast and growing. Its roots lie in the observation of policy convergence, a "tendency of policies to grow more alike" (Drezner 2001: 53), which has been linked to the broader phenomenon of globalization (Drezner 2001, Evans 2004a: 2, Stone 2004). The economic and competitive pressure that states encounter in a globalized world may lead them to adopt similar policy responses while possibly taking into account policies adopted abroad. Policy transfer literature is hence part of a much larger strand of research concerned with internationalization and its mutual effects on domestic policies. The question of how domestic and international levels interact is treated in seminal works including Gourevitch's "second image reversed" on the influence of domestic politics on international relations (Gourevitch 1978, see also Risse-Kappen 1994), Keck and Sikkink's "boomerang pattern" (Keck & Sikkink 1999), Finnemore and Sikkink's "norm bandwagons" (Finnemore & Sikkink 1998), Putnam's "two-level games" on the interaction of domestic and international arenas in negotiations (Putnam 1988), and works on the influence of international regimes on domestic policies (Keohane & Victor 2010, Krasner 1982). The multiplication of communication channels, increased mobility, and membership in international or multilateral organizations and transnational knowledge, advocacy, and policy networks further facilitate the diffusion of policy knowledge (Coleman & Perl 1999, Howlett et al. 2009: 77, Sikkink 1995, Stone 2004). It is no longer only foreign policy, but also the *domestic* policy decisions of a country, which can influence domestic policy choices in other countries

12 The case of transfer between Germany and California spans different levels of government. Given California's very large room for maneuver in defining energy policies (Elliott 2013, Litz 2008), studying transfer in this policy field is not fundamentally different from studying bilateral policy transfer. For an overview of cross-level transfer possibilities, see Evans and Davies (1999: 368).

(Elkins & Simmons 2005). When diffusion and convergence take place on a large scale, interdependent policy making may lead to the emergence of "de facto" regimes where a type of policy prevails without international top-down coordination (Busch et al. 2005: 164).

From an empirical point of view, literature on policy diffusion and transfer covers virtually all policy areas, including healthcare (Holden 2009), liberal economic policies (Simmons & Elkins 2004), regulatory agencies (Jordana et al. 2011), social sector reform (Weyland 2006), pension systems (Brooks 2007; Orenstein 2008), social programs (Lana & Evans 2004), vocational training (Barabasch et al. 2009), state lotteries (Berry & Berry 1990), labor policies (Daguerre 2004), health sector reform (Tambulasi 2013), smoking bans (Asare & Studlar 2009) and many others. Environmental policies hold a particularly important place in this research field (see section 2.2.3).

2.2.2 Convergence, diffusion, and transfer: Disentangling concepts

While there is relative agreement among scholars on the meaning of policy convergence, the processes potentially – *but not necessarily* – leading to this outcome of policies growing more similar have been subject to a lively terminology debate. This debate occupies a considerable part of the literature on policy diffusion and policy transfer and has shaped different research programs.

This book addresses policy transfer rather than diffusion, following the conceptual and methodological differences discussed by Marsh and Sharman (2009). While policy diffusion aims at unveiling larger trends of policies spreading over time and space (Evans & Davies 1999: 367), policy transfer scholarship is more concerned with the why and how of *particular instances* within these movements of ideas and policies. Policy transfer, as it is understood in this book, is defined by Dolowitz and Marsh (2000: 5) as a "process by which knowledge about policies, administrative arrangements, institutions and ideas in one political setting (past or present) is used in development of policies, administrative arrangements, institutions and ideas in another political setting". The definition points to a focus on *tracing* and *understanding* processes and on exploring the use of information about foreign policy models in domestic policy processes. The question of transfer *outcomes* is therefore of interest, but not the exclusive focus of transfer studies.

Policy diffusion adopts a higher-level perspective of identifying and observing the spread of ideas, models, norms, policies (or even crisis, regimes, democracy) between political entities. Diffusion is generally measured through policy output such as laws adopted, standards set, or constitutional principles enacted. This literature tends to focus on identifying general patterns of diffusion like waves, clusters (Elkins & Simmons 2005), and the "S-curve" (Berry & Berry 1999: 227), often through the use of quantitative methods, event-history analyses or "dyadic approaches" (Gilardi & Foglister 2008). The determining factors taken into account in these types of studies are externally observable characteristics of the diffusion item or the recipient system, such as the general type of policy instrument, the level of economic development, or geographic proximity between sender and recipient.

By establishing, in multiple empirical applications, that policies do indeed cross borders, the policy diffusion literature has carried out crucial groundwork for policy transfer studies. The methods predominantly used in diffusion studies however limit the ability to explore how policies travel between political entities in particular instances (Wolman 1992). As pointed out by Gilardi (2013b: 470), "while the literature has convincingly demonstrated that policies diffuse, why that occurs remains much less clear". In particular, the causal mechanisms of situations where policies are considered but not adopted or are mixed with other inspiration can hardly be understood without qualitative case studies. Transfer research is targeted at filling these gaps. From a methodological point of view, transfer studies often use "thick description" (Evans 2009b: 254) to trace "how and why program evidence from foreign countries enters policy debate and informs action policy debate and informs action" (Bennett 1991: 33).

Drawing the conceptual boundaries of policy transfer is important not only with regard to policy diffusion and convergence (Holzinger & Knill 2005a), but also with respect to "normal forms of policy making", from which transfer differs by "[focusing] on the remarkable movement of ideas between systems of governance" (Evans 2010b: 26).

2.2.3 *Empirical applications: Transfer of environmental and energy policies*

Due to the cross-border nature of the problems it addresses, environmental policy has been one of the centers of attention of researchers in policy diffusion and transfer. Geographically speaking, many studies focus on the countries of the

European Union, federal states in the United States, or the OECD (Evans 2004c: 12). Large-n assessments of policy convergence have provided strong evidence for the reality of interdependent policy making in the environmental field (Busch & Jörgens 2010, Holzinger et al. 2010, Knill et al. 2012, Liefferink et al. 2009, Sommerer 2011). The relevance of diffusion as a driver for the widespread adoption of climate policies in the absence of a strong international climate regime has also been established (Fankhauser et al. 2015). Studies in the field of energy are less common, and the applicability of findings from the larger environmental policy diffusion literature to energy policy needs to be critically assessed.

In the context of the European Union, Ürge-Vorsatz et al. (2004) consider the spread of renewables support policies to Central and Eastern European countries as "a case of incomplete policy transfer". Jacobs (2012) finds convergence of feed-in tariffs in Europe through horizontal policy diffusion in Germany, Spain, and France. Busch (2003) identifies different diffusion mechanisms in renewable energy policy in Europe, depending on the instrument, with horizontal diffusion being relevant for the diffusion of for feed-in tariffs and a stronger role for agencies in the "institutionalized" diffusion of quota systems[13]. Federal settings naturally lend themselves to (energy) policy diffusion studies. Strebel (2011) focuses on energy policy transfer among Swiss cantons and Chandler (2009) analyzes the reasons behind the spread of renewable portfolio standards among states in the United States. Matisoff and Edwards (2014) find political culture to be a main driver to the diffusion of renewable energy and energy efficiency policies in the United States, with strong evidence for diffusion among similar groups of states regardless of the policy instrument. The authors also show that the same states are likely to consistently act as innovators in the energy and climate field and to be used as reference points, pointing to the relevance of addressing the role of leadership in transfer (Matisoff & Edwards 2014: 811). Chandler (2009) arrives at a similar conclusion for the diffusion of sustainable

13 Quota systems for renewable energy are schemes in which governments set an amount or percentage of energy to be procured through renewable sources. In general, electricity suppliers are mandated to source a percentage or amount of the electricity they sell to end-consumers from renewable sources. The California Renewable Portfolio Standards (RPS) is a case in point for such a quota system. Quota obligations can be fulfilled in some systems through the purchase of renewable energy certificates (RECs) from sources outside the country in question. For more details on regulatory promotion schemes for renewables, including quota systems, see Haas et al. (2011: 2187).

energy portfolio standards[14] in the United States, underlining the importance of ideological proximity for diffusion. Stoutenborough and Beverlin (2008) identify patterns of regional diffusion for state-level net-energy metering policies[15] for renewable energy in the United States and underline the importance of regional offices of the federal Environmental Protection Agency as "conduits" for diffusion among states. Vasseur (2014) studies the adoption of renewable energy policies among US states with a particular focus on the discrepancies between policy labels and content: while symbolic imitation is a consequence of regional diffusion, other determinants internal to the recipient shape the content of transferred policies. Based on these findings, Vasseur (2014: 1650) also stresses the importance of understanding policy adoption as a "multifaceted process" and, implicitly, of taking into account different policy layers in complex transfer outcomes beyond adoption or non-adoption of foreign models (see section 2.2.7). Smithwood (2011) identifies a "race to the top" dynamic for RPS adoption in New England, with both positive and negative lesson drawing affecting policy design in addition to internal determinants. It is noteworthy that most of the above-mentioned studies employ large-n event-history analysis to identify and explain the diffusion of renewable energy policies. They also generally use single internal variables (governing party etc.) to explore the role of national filters for transfer – an approach that is certainly highly useful for identifying patterns, but is likely to ignore many dimensions of "national determinants" (see section 2.3.6) and specificities of particular policy networks (2.5.2).

2.2.4 Technology diffusion and transfer

Although the policy diffusion literature borrows terminology from the technology diffusion literature and reaches similar conclusions – for example regarding the typical S-shape of diffusion curves – the two strands of research are largely

14 Renewable Portfolio Standards (RPS) are a form of quota system (see footnote above). The term is most commonly used in the United States to refer to a quantity-based promotion scheme for renewable energy. See also Heeter et al. (2014).

15 Net-energy metering refers to policies that allow consumers of electricity to also produce (part of) their electricity on-site, for example through solar rooftops, and thereby reduce their electricity bills. The electricity generated is deducted from the bill, either on a 1:1 basis (1 kWh generated reduces the electricity bill by the price of what 1 kWh consumed costs; the electricity meter then "turns backwards") or less. See Stoutenborough and Beverlin (2008).

independent. This is also because policy diffusion and transfer research are often carried out in policy fields where little material transfer of technologies is likely to happen, such as democratization or pension systems. Energy as a policy field is fundamentally different since the "diffusion of innovative low-carbon technologies and of innovative supporting policies are interlinked" (Jänicke 2013: 5). In particular, the hope of an accelerated diffusion of sustainable energy technologies in developing countries, allowing them to "leap-frog" into a cleaner energy system, is an aim of policy promotion efforts in this field (Haselip et al. 2011).

Links between technology and policy diffusion research exist on an empirical and a theoretical level and are relevant for this book. Understanding the conditions and the impact of the transfer of renewable energy policies provides relevant information for the conditions in which technology transfer can happen (Gallagher 2014). In the field of renewable energy, transfer-inspired policies may develop a pull-effect on an emerging market for clean technologies or can allow for the ex-post development of a regulatory framework for a technology that is already starting to diffuse in a receiver country (Jänicke 2012a: 3). Technology transfer can thus be preceded by, accompanied by, or, less likely, followed by policy transfer (Jänicke 2013: 5), and either type of transfer is in principle imaginable without the other. In the empirical field under scrutiny in this book, the diffusion of technologies is a desired consequence of policy transfer, including from Germany's perspective as an active sender. The links between the promotion of policy transfer and the creation of transfer advantages for export markets are explored in the lead market literature (section 2.4.6).

Interesting parallels can be drawn between expectations from the technology diffusion literature on the one hand and policy diffusion and transfer on the other hand. Factors such as visibility, the disruptiveness or fit with an existing system, the need to adapt existing infrastructure, and whether the innovation is identified as a convincing response to an existing problem are present in a similar form in both strands of scholarship. Rogers (2003) finds the "observability" or visibility of an innovation to be an important driver for diffusion, in line with expectations regarding the importance of availability and visibility of policy transfer items (Heiden & Strebel 2012: 348). Hekkert et al. (2007: 425) underline the political economy of technology diffusion in a way that is also useful for understanding policy transfer: "in order to develop well, a new technology has to become part of an incumbent regime, or it even has to overthrow it. Parties with vested interests will often oppose to this force of 'creative destruction'." These coalitions can become involved in putting a technology on the agenda and lobby for an advantageous framework for the new technology (Hekkert et al. 2007: 425).

The uptake of a new policy idea within a policy network can indeed be compared to the introduction of a technological innovation to an existing market. In the case of cross-border diffusion, the "proof of concept" takes place in another constituency, but new policy and new technology ideas need to compete in the receiving country's environment, generate demand, be seen as contributing to the solution of a problem, and rally coalitions that are powerful enough to overcome incumbent ideas or technologies. Given the important parallels in expectations between the technology and the international policy diffusion literatures, surprisingly few studies have taken an integrated look at both phenomena (for exceptions, see Holden 2009, Jänicke & Jacob 2007b, Jänicke 2012b, Jänicke 2013, Lindebjerg 2014). The main exception is the lead market literature, where a link between the transferability of policies and the technologies they promote has been established (Beise & Rennings 2005, Jacob et al. 2005: 79–80).

2.2.5 *Prerequisites for policy transfer*

Independent of what drives countries to consider foreign models, certain prerequisites need to be present for transfer to be possible at all: an existing policy innovation (the transfer item) and communication channels allowing for information flows between sender and recipients, as well as "transfer agents" using these channels.

A first question to address is what elements can be objects of policy transfer. Weyland (2006: 18) differentiates between a "principle" that "charts an overall direction but not a specific course of action", and a "model", which "is one specific option from the menu offered by a policy principle". Different aspects inherent to transfer items are likely to affect transfer, as discussed in section 2.3.4. The existence of a pioneering model cannot lead to diffusion and transfer if agents in the recipient country are not aware of it. While this may sound self-evident, awareness about a foreign policy model can be considered a first form of transfer. Since agents in the recipient country are unlikely to take into account all existing options in their search for information, paying attention to a particular one already means that transfer channels have been used (Bennett 1991: 33).

An iconic or innovative policy model may trigger an active search for information by agents in the recipient country, but awareness of a foreign model can also be promoted by the sending country, through various channels: "Policy ideas diffuse through professional organizations, broader networks of specialists [...], the efforts of policy entrepreneurs, the media, and chance contacts [...] Conferences and publications, for example, plant ideas that lie dormant until policy makers recognize them as potential solutions to a particular problem and then begin to engage in prospective evaluation" (Mossberger & Wolman 2003: 428). Since "diffusion presupposes communication" (Busch et al. 2005: 150), channels through which information can flow – or "opportunity structures for transfer" (Evans & Davies 1999: 371) – are hence needed. Such channels for transfer can be created and used by communities of experts, including epistemic communities (Dunlop 2009, Haas 1992), professional associations and international organizations or transnational advocacy networks (Keck & Sikkink 1999).

The users of transfer channels are called "transfer agents", a term that has been employed for international organizations, consultancies, or NGOs that transcend state boundaries to promote policy ideas and tools (Stone 2004), but synonymously also for individuals in those organizations, as well as within state bureaucracies, industry and academia (Dolowitz & Marsh 2000). Active sending countries, through the creation of "spheres of leadership" (Eckersley 2011: 9), can reinforce such opportunity structures for transfer and thereby increase the availability of their model, including by placing transfer agents in potential follower countries.

2.2.6 Structure and agency in policy transfer

Policy diffusion and transfer literature encompasses both large-n studies focusing on structure and qualitative studies interested mostly in agency. These studies' relative focus on structure (diffusion) or agency (transfer) is indeed one of the main differentiating characteristics between the two branches (Marsh & Sharman 2009: 270). The lack of policy transfer studies attempting to adopt a dialectical view on the relationship between structure and agency has been criticized (Marsh & Sharman 2009: 275). Generalizing this criticism appears unjustified insofar that agency-centered transfer studies recognize the importance of channels and structure for transfer (Evans & Davies 1999: 371, Ladi 2011).

This book strives to acknowledge both the role of structural determinants and the agency of transfer actors in an integrated manner, as called for by Evans and Davies (1999: 362). For the authors, "policy transfer must be understood within the context of the relationship between structure and agency" and its study must therefore take into account "changes in economic, technological, ideological or institutional structures" as both "facilitating the space for policy transfer and affecting the nature of the transfer process itself" (Evans and Davies 1999: 373).[16]

Pushing analysis beyond this point, the potential impact of policy transfer on the very structures that affect present and future transfer outcomes must be stressed. This appears particularly important where active senders attach incentives to the policies they promote (see section 2.3.3), where external transfer agents become members of domestic policy networks, or where policies themselves carry externalities or lead to changes in the institutional structure of the recipient network. In a policy field like energy, shaped by a multitude of economic, technical, political, and societal structures, this potential impact of transfer merits particular attention.

As far as agency is concerned, transfer studies are encompassing enough to allow for different micro-foundations of action to be taken into account, including approaches that emphasize the role of norms, ideas and beliefs about what is "good" policy according to a "logic of appropriateness" (March & Olsen 1989), rational and functional arguments that see transfer as a source of information agents use to devise better policies (Mossberger & Wolman 2003, Rose 2005), and public choice concepts of action (Gawel et al. 2016, Niskanen 1971, Olson 1971, Stigler 1971, Tambulasi 2013: 82) that emphasize the role of particular economic, political, and bureaucratic interests in policy making.

Within the structure-agency debate (e.g., Hay 2006, Hay & Wincott 1998, Mayntz & Scharpf 1995, Thelen & Steinmo 1992) that has also reached the transfer literature, this book adopts a dialectical view, in which structure is likely to shape agents' responses to transferred information and even their demand for

16 This view on the transfer process strongly recalls a historical institutionalist perspective, as discussed in section 2.5.1.

information about a foreign model, but where agency and transferred ideas can also transform structure over time (Marsh & Sharman 2009: 275). An integrated view is all the more needed for the study of transfer processes that span a long period of time (Dussauge-Laguna 2012) and may lead to a co-evolution of structure and agency through transfer. More actor-centered approaches such as learning and lesson drawing (see 2.3.2) and more structural drivers such as regulatory competition or conditionality (2.3.3) have been found to hold explanatory power in transfer research and hence cannot be ruled out by focusing a priori on either aspect. The following expectation can be drawn with a view to further hypothesis development in Chapter 3:

E1: Transfer mechanisms allow for the integration of different theories of action (rational, functional, normative) and of structure- and agency-related determinants for transfer. Transfer, especially over longer periods of time, may affect structure as well as agency.

2.2.7 Transfer outcomes beyond adoption and non-adoption

A common metric used to assess transfer outcomes is the degree of resemblance of policies adopted in the recipient state compared to the policy source, i.e., convergence (Heichel et al. 2005, Holzinger et al. 2008). When transfer is understood as an outcome rather than as a process, it often refers to binary results of adoption or non-adoption of a model (Common 2013). Recent literature, however, takes a closer look at different shades of transfer outcomes, between copy-pasting a sender's model and rejecting it entirely (Dussauge-Laguna 2013, Evans 2009b, Genovese et al. 2015, Klingler-Vidra & Schleifer 2014, Marsh & Sharman 2010, Radaelli 2005). The existence of individual and structural filters, the adaptation of transferred policies to domestic conditions and partially "negative lesson-drawing" (Rose 2005) can lead to selective transfer outcomes. Rose (2005:81-84) proposes to categorize the outcome of lesson drawing in seven types. These range from "photocopying" (minimal change to initial model) to "copying" (duplicating most of the model); "adaptation" (altering details while leaving major elements intact), "hybrid" (combining different policies from several jurisdictions), "synthesis" (creating a new model based on different sources of inspiration, possibly including domestic ones), "disciplined inspiration", and "selective imitation" (Rose 2005: 81). The latter two types of outcome merit particular attention, and, for Rose, also involve a normative difference in quality and desirability. While "disciplined inspiration" means "responding to the stimulus" of a foreign model "by creating a novel programme [that is] not inconsistent" with the foreign example, "selective imitation" means cherry-picking "attractive, but not necessarily essential" elements of a program while "leaving out awkward but essential bits", for which there is a high political price in the recipient state (Rose 2005:81). For Klingler-Vidra and Schleifer (2014), different degrees of convergence resulting from policy diffusion can be traced back to a combination of factors related to the characteristics of the transfer item, its source, the similarity between sender and recipient, and the diffusion mechanism. According to the authors, diffusion is likely to be more selective and convergence is less likely if sender and recipient are different, the policy model is unspecific, learning is the main transfer mechanism, and multiple sources exist for the diffusion item (Klingler-Vidra & Schleifer 2014: 272).

However the exact form of selective transfer is labeled, stakeholders in these cases modify and adapt foreign policy models to local objectives and structures, combine them with local or other foreign models, draw negative lessons from them to fine-tune their policies, or confirm existing policies or non-action. As Marsh and Sharman (2010: 41) point out, adaptation and selective transfer are to be considered the rule rather than the exception in transfer studies. Information about foreign models is adapted to local objectives and specific contexts, except when pure emulation, "symbolic imitation" (Meseguer 2005: 73), or mimicry (Marsh & Sharman: 34) prevail. These situations are defined by Shipan & Volden (2008: 842) as "copying the actions of another in order to look like that other" and occur when another country is deemed so prominent that its policies are to be copied without consideration of the local context. This situation is unlikely to occur in practice, especially when one insists, like Shipan & Volden (2008), that emulation is strictly driven by the desire to imitate a country, independently of what policy is concerned. Probably more frequent are instances of "anchoring", where decision-makers stick to their initial positive impression about a foreign policy model and fail to adapt it (Weyland 2004a: 255).

This means that transfer studies must take into account negative lesson drawing and selective transfer as relevant variants of transfer rather than looking at them as null-hypotheses (Simmons et al. 2007: 7). A relevant criterion to assess whether transfer took place is therefore whether "information about the policy in another country is actually used [...] The criterion does not require the 'borrowing' country adopt the policy in whole; the policy may be adopted with modifications or even rejected" (Mossberger & Wolman 2003: 431; see also Busch & Jörgens 2005a: 4). While most transfer and, to an even greater extent, diffusion studies focus on policy outputs as a metric to assess transfer outcomes, awareness of a model is in fact a first transfer outcome to be considered, since "evidence of awareness [...] must precede any argument that adoption was influenced by that evidence" Bennett (1991: 33). Exploring whether stakeholders were aware of a policy is crucial to exclude the "null hypothesis" of independent policy-making. Even if transfer is not reflected in policy output, a foreign model can be considered and even used intensively, which constitutes a transfer outcome in the understanding shared in this book.

The use of information about foreign models can be an important component of policy making, independently of how this use is reflected in transfer outcomes. Rather than speaking of "non-transfer" when "parts of an original idea or programme [are] discarded or filtered out by the subject/agent" (Evans & Davies 1999: 382), this type of "conscious taking-into-account" of foreign experience is an integral part of transfer. Overcoming the selection bias toward cases of convergence in the literature is necessary (Heiden & Strebel 2012: 347, Klingler-Vidra & Schleifer 2014, Marsh & Sharman 2009), given that "Understanding what does not diffuse should be as central as what does, entailing the ability to recognize – as some would put it – why an event or stimulus 'stayed in Vegas' (as in 'what happens there, stays there')" (Solingen 2012: 633).

A second way to look at transfer outcomes is by qualitatively evaluating the contribution of transferred ideas to "successful" or "failed" policies (Dolowitz & Marsh 2000). The expectation here is that transfer could improve the quality of policies in terms of outcomes and effectiveness because it enlarges the knowledge base of policy makers and brings previously unknown potential solutions and effects to their attention. On the other hand, when transfer amounts to blind copy-pasting or when policy models are forced upon a recipient country, a lack of fit with local objectives and structures can obviously lead to a negative impact on the quality of policies or a lack of implementation. Evaluating whether transfer improves or worsens policy outcomes is methodologically challenging given the lack of counterfactuals in a real-world setting. Also, establishing against which indicators the success or failure of policies is to be measured inevitably implies judgments of what constitutes "good" or "bad" policies and requires a deep understanding of objectives pursued by different groups in recipient countries. Despite methodological challenges, the link between transfer and the success or failure of policies is important to keep in mind from an ethical standpoint, in particular in the case of active policy promotion by senders. Promoting transfer, even more so by conditionality, puts a responsibility on senders to critically reflect on the appropriateness of a policy for the context in the receiving country and on the long-term effects of transfer (see section 2.4.7 on ethics of "sending"). With regard to the policy field studied in this book, a full evaluation of the effectiveness of transfer would need to include an assessment of the effectiveness of transfer-inspired policies in reaching final objectives, e.g, environmental effectiveness, but this is outside the scope of this book.

When active senders are brought into the equation of policy transfer, assessing transfer outcomes also provides information about the effectiveness of senders' strategies. As outlined below, the leadership literature debates whether effectively "gaining followers", i.e., countries that adopt similar policies, is an actual condition for qualifying someone as a leader. As evidenced in this book, the policies promoted by a sender need not be identical to those implemented domestically by the sender, introducing an additional degree of complexity to the study of transfer outcomes. "Exporting" a policy model one to one cannot be assumed to be the objective of senders in all cases and a much more fine-grained analysis of transfer processes is therefore needed. Rather than considering whether the sender's model was adopted in a recipient case, transfer outcomes therefore need to be compared against a sender's objectives to assess the effectiveness of leadership through transfer (section 2.4.7).

Rather than conditioning transfer upon adoption (Evans 2009b: 245), different variants of transfer exist – from awareness/consideration to utilization and possible reflection in policy outcomes – and are reflected in the hypotheses presented in Chapter 3.

E2: The selective or negative use of knowledge about foreign models leads to transfer outcomes beyond imitation. Since this outcome is likely to be frequent given individual and structural filters, the consideration and use of transferred knowledge appear as essential additional categories of transfer, in addition to the classic metric of policy outputs.

2.3 Determinants for Policy Transfer

2.3.1 Drivers for policy transfer

Determinants for policy transfer are the factors that shape transfer processes and their outcomes. These determinants can be grouped into three main categories: underlying drivers pushing countries to consider foreign experience in the first place; determinants that influence how the foreign example is used; and factors that impact the link between transfer and policy outcomes. The sender can influence many of these determinants for bilateral transfer. Discussing individual determinants for transfer separately is useful from a conceptual point of view, but the likelihood of complex transfer processes involving several causal mechanisms with explanatory power at a time needs to be taken into account (Marsh & Sharman 2009: 273). Figure 1 lists the determinants for transfer that shape how transferred knowledge enters recipient systems, how it is used and how it affects policy outcomes. These determinants will be discussed in the next sections.

A first distinction of drivers for policy transfer is generally drawn between voluntary and coercive transfer (Busch & Jörgens 2005b, Dolowitz & Marsh 2000, Evans 2009a, Heinze 2011, Holzinger & Knill 2005b, Tews 2002). Dolowitz and Marsh (2000) present different drivers for transfer on a "policy transfer continuum" ranging from coercion to processes that are solely based on the initiative and the desire to learn of agents in the receiver country. At this end of the spectrum, agents turn to foreign policy models in search of inspiration and lessons, in an effort to update their knowledge, increase the information base for decisions, and potentially reduce transaction cost and risk by implementing policy models tested elsewhere. At the other end of the spectrum, cases of fully coercive policy adoption can be imagined to occur in the aftermath of wars or have been identified in the form of "hegemonic imposition" (Wejnert 2014: 43). The discretion of governments can, however, also be limited in situations of conditionality and harmonization. Conditionality covers situations where material or immaterial threats or strong, existential incentives leave countries with few alternatives to adopting specific policies (Dobbin et al. 2007: 455, Ranis et al. 2006, Svensson 2003).

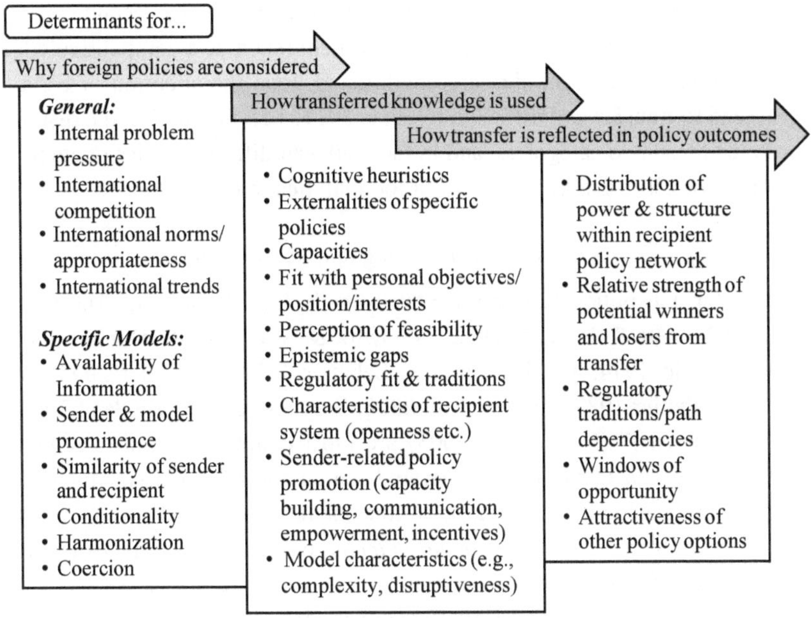

Figure 1: Overview of determinants for transfer

International harmonization is sometimes regarded as a non-voluntary form of transfer, since in these situations, policies have to be adopted in order to comply with obligations stemming from the adherence to international treaties or agreements reached within supranational entities. This diffusion mechanism has been studied extensively for the European Union (Börzel & Risse 2009, Börzel & Risse 2012, Schimmelfennig & Sedelmeier 2005, Windhoff-Héritier et al. 2001) as well as its neighborhood (Knill & Tosun 2009, Lavenex & Schimmelfennig 2009). For the cases studied in this book, strictly coercive and harmonization-driven transfers are of little relevance.[17]

Mechanisms of transfer can usefully be categorized into whether they work by "changing ideas" or by "changing incentives" (Dobbin et al. 2007: 450). In the first category, transfer affects either norms or knowledge and thereby poten-

17 Given Morocco's "advanced status" (statut avancé) of cooperation with the EU, harmonization could play a role in energy policy in Morocco (Carafa 2011). This mechanism did not emerge as being of relevance in empirical findings, possibly due to the focus on renewable energy policies rather than, e.g., liberalization.

tially leads to policy change (Finnemore & Sikkink 1998). In the second category, the externalities certain policies bring with them (such as a competitive edge over other countries or loans that are conditioned upon the adoption of certain policies) lead to transfer and diffusion (Heinze 2011: 12). In practice, the distinction between these two categories, and in fact between different logics of agency, is unlikely to be clear-cut, but it is helpful for analytical purposes.

2.3.2 Transfer driven by demand: Learning and lesson drawing

Policy transfer and policy learning are intrinsically linked concepts. The terms "learning" (Bennett & Howlett 1992, Dolowitz & Marsh 1996, Dunlop 2009, Hall 1993, Heclo 1974, Jenkins-Smith & Sabatier 1993, Meseguer 2005, Radaelli 2009, Sabatier 1987, Volden et al. 2008, Weyland 2004c, Weyland 2006), and "lesson-drawing" (Rose 1991, 2005) are most commonly used to refer to transfer driven by an intrinsically motivated search for information. Much like policy transfer in general, learning as a concept has been used to cover both processes and outcomes. Learning *as an outcome* occurs where policies change in response to additional/new information or experience, as in Hall's work on three degrees of social learning (Hall 1993) or in Heclo's work on learning as a factor for change in social policies (Heclo 1974: 306). Understood as a process rather than as an outcome, learning refers to the "updating [of] beliefs about key components of policy (such as problem definition, results achieved at home or abroad, goals, but also actors' strategies and paradigms)" (Radaelli 2009: 1146–1147). Learning can occur at the individual level or at the organizational level (Adam 2015, Bennett & Howlett 1992, Carroll & Common 2013, Jachtenfuchs 1996), but collective learning does not "imply reifying society into a discrete organic mind" but rather acknowledging that "alone and in interaction […] individuals acquire and produce changed patterns of collective action" when learning takes place (Heclo 1974: 306).

The main differentiation among learning processes is between learning that is rational, in its approach and its conclusions, and learning that is boundedly rational (Meseguer 2005). Rational learning is closely linked to evidence-based policy making (Legrand 2012) and, in its transfer-related variant, occurs when a lack of knowledge, know-how, and capacities to solve specific problems prompts agents to turn to foreign examples (Stone 2004). Studying learning as a mechanism of policy transfer thus also means contributing to "the debate about rationality in politics" (Weyland 2006: 5, also: Evans 2004a: 3, Stone 1999: 53) and understanding the role of transferred knowledge for policy makers "collectively wondering what to do" (Heclo 1974: 305–306). Understood functionally, learning is likely to be driven by "microlevel dissatisfaction with existing policy systems [...]", within "organizations [who] often do not have the expertise to tackle all the problems they confront and increasingly look outside their organizations for the answers to their problems" (Evans 2009b: 259). Radaelli (2009: 1150) calls this type of learning "instrumental learning" as it is targeted at improving instruments needed to solve an identified problem. The concept of lesson drawing developed by Rose (1991, 2005) is a more practice-oriented view on learning with the purpose of improving policy. Starting from the question of how learning can be useful for policy-makers, lesson drawing is defined as "utilizing available experience elsewhere to devise a programme that is new to the agency adopting it and attractive because of the evidence that it has been effective elsewhere" (Rose 1991: 7). In this definition, lesson drawing effectively leads to policy action in order to improve a situation, following a functional/rational logic of action.

In theory, learning is rational only when all available options, from the past and the present, within and outside the organization and the country, including the option of not acting, are thoroughly taken into account (Meseguer 2005: 2, Weyland 2005). Given the realities of resource and time scarcity in the policy making process, learning is rarely likely to be purely rational. As underlined by Meseguer (2005), it may in fact be considered rational for actors *not* to take into account all imaginable and existing options, but rather perform selective searches for information. Bounded rational learning, influenced by cognitive heuristics, might therefore be closer to the actual practice of policy making, which has long been seen as a form of "muddling through" (Lindblom 1959) rather than a purely rational endeavor. By using cognitive shortcuts (Meseguer 2005: 2, Weyland 2005: 263, Weyland 2006), actors then take into account only a limited number of foreign models in their search for information. In its most extreme form, the process is reduced to the mere emulation of models that are particularly prominent (Radaelli 2009: 1150), a transfer mechanism that is, however, different from learning (Biedenkopf 2011).

Among cognitive shortcuts influencing learning, representativeness and availability heuristics (Weyland 2006) are particularly relevant for the study of policy transfer. These heuristics concern perceptions of a model's and a sender's prominence and the degree to which information about them is readily available. Capacities in the policymaking process are additional boundaries for learning, transfer, and the subsequent implementation of environmental policies (Jänicke 1997, Kern et al. 2001: 8, Tews & Jänicke 2005: 40, Weidner et al. 2002).

Since learning is a central mechanism of policy transfer, the question of what can be learned is of essence. Naturally, important overlaps between objects of learning and objects of transfer exist. Depending on the objects agents learn about, different forms of learning have been conceptualized. May (1992) distinguishes "political learning", where agents learn about how to improve their "strategy for advocating a given policy idea or problem" (May 1992: 336, see also Radaelli 2009: 1151); "instrumental learning", where knowledge about policy instruments and programs is updated; and "social learning", which targets a deeper level of learning about underlying policy problems and updating beliefs on causal links. Hall (1993), following the term's use by Heclo (1974), employs "social learning" in a more generic way, to describe learning from past experience by politicians, officials, experts and other members of issue networks. Depending on how far-reaching lessons learned are, different degrees of policy change can be induced, from incremental changes to policy instruments and paradigms. In Hall's analysis, learning predominantly relies on past experience in the same constituency, while international transfer is not (yet) considered. Learning as leading to changes in core or secondary beliefs is also a central part of Sabatier's advocacy coalition framework (Sabatier 1988, Sabatier & Jenkins-Smith 1993).

The question of the "location of the agency of policy learning" (Bennett & Howlett 1992: 282) is essential, since these agents are also likely to be the ones drawing lessons from foreign policy experience. Hall (1993) explicitly links what is learned, and the consequences of learning, to who learns: his account of changing paradigms in British macroeconomic policies between 1970 and 1989 is in part explained by new agents drawing lessons from past experience, beyond a traditional group of bureaucrats. On the contrary, Heclo (1974) identifies bureaucrats as the most important group of learners. Other agents of learning in the literature include members of epistemic communities, that is to say, "professionals with recognized expertise and competence in a particular domain and an authoritative claim to policy-relevant knowledge within that domain" (Haas

1992: 3). The question of who learns cannot be answered ex-ante, but it is one to be considered when assessing the flow of information about foreign policy ideas within policy networks, especially when learning is also understood as a political process. Objects that can be learned about are then filtered by considerations such as improving one's position or votes or responding to demands from important incumbents and interest groups (Garrett & Jansa 2015). Learning in such a public choice or regulatory capture perspective (Niskanen 1971, Olson 1971, Stigler 1971) is "political" (Radaelli 2009: 1151). It can be argued however that this type of learning actually rather concerns how ideas are *used* than how stakeholders acquire knowledge about them. It is nevertheless undeniable that interests can contribute to shaping actors' individual cognitive heuristics and the filters that determine which models are taken into account.

E3: The process of learning, which is central to the study of policy transfer, is likely to be influenced by individual "cognitive shortcuts" such as the availability of information, and the prominence of the model and sender, as well as by capacities available for the search of information. Motivations for learning can be functional, but in the context of political learning, the usefulness of knowledge for an agent's position or strategy is also relevant.

2.3.3 Transfer driven by externalities: Incentives, competition, and reputation

There is considerable space between the two extremes in Dolowitz and Marsh's (2000) continuum for policy transfer, that is to say, between coercion on the one hand and transfer driven solely by a thirst for knowledge on the other hand. On a middle ground between these extremes, foreign policies can entail externalities that can drive transfer (Heinze 2011: 17). These externalities include incentives attached to policy ideas by senders, advantages that policies grant in competition with the sender or other countries, and externalities linked to the perceived appropriateness of a policy or a norm by the international community, possibly leading to a desire to conform and mimic policies. All these externalities have in common that they render a policy choice relatively more attractive compared to other choices, including non-action.

A traditional way of conceptualizing incentives linked to a policy model, especially in developing countries, is through conditionality. In this case, grants, loans, or other advantages such as access to trade agreements, are conditioned upon the adoption by a recipient country of policies that the sender favors. The role of conditionality has been considered in particular with regard to multilateral financial institutions, the World Bank and the International Monetary Fund, and has been criticized as unethical and resulting in negative policy outcomes (Collier 2000, Dobbin et al. 2007: 455, Molenaers et al. 2015, Svensson 2003). Incentives provided by senders can also be non-monetary, for example, through capacity building or the production of knowledge by the sender for the recipient (King & McGrath 2004). Policy options might be conceived as more feasible and implementable (Tews et al. 2003: 578) through these incentives. Transfer becomes more likely since decisive determinants for transfer (capacities in the recipient country, cost of implementation) are influenced.

A second set of externalities concerns economic competition between recipient and sender or between different recipients. The underlying expectation is that countries experience a tendency to conform to competitive pressure against the background of growing globalization (Drezner 2001, Tews 2005: 67). Competition can lead to transfer when countries compete for market shares or the locational choices made by taxpayers and companies and therefore see the implementation of policies similar to their competitors' as advantageous (Baybeck et al. 2011, Berry & Berry 1990, Holzinger & Knill 2005b: 6). This competitive pressure has long been suspected to lead to a "race to the bottom" toward less stringent regulation (Drezner 2001: 57–58), but in the environmental policy field, evidence is inconclusive. A particularly prominent case of environmental policy transfer by economic competition is the so-called "California effect" induced by "trading up" (Vogel 1997b), where more stringent environmental norms spread from one state to others and to higher levels of government through channels of trade.[18]

The idea of regulatory competition is challenging to apply to policies for renewable energy deployment because they do not directly concern product-related standards (Matisoff & Edwards 2014: 18). Nevertheless, competition can drive renewable energy policy transfer, when it concerns investment. As evidenced by

18 In this logic, exporters conform to requirements and demand from importing states with stricter rules by adjusting their environmental standards themselves. For companies in the exporting state it can be more efficient to produce to a single set of standards and adaptation toward the more stringent norm may hence occur. See also Porter and van der Linde (1995) for the potential firm-level competitive improvements through stricter environmental standards.

Lüthi (2012), investors prefer some renewable energy policy instruments, potentially pushing countries to turn to policies implemented by competitors. Finally, competition may concern aid or access to organizations, especially for developing countries, as described by Brooks (2007: 706): "To the extent that developing-country governments use policy reform as a signal to international audiences [...] then certain policy innovations salient to [their] concerns may become the focus of competition for governments vying for excludable goods".

The "competitive advantages" a policy model confers need not necessarily be expressed in economic terms (Tews 2005: 67). Certain models can be more or less "trendy" on the international scene at a given point in time and their adoption more or less rewarded by improved reputation. The "desire (or need) of domestic actors to conform to internationally widespread norms" (Heinze 2011: 12) can lead to transfer by "emulation" or "mimicry" if countries want to avoid "looking like laggards" (Drezner 2001: 57). This is particularly likely when a perceived and widening gap between domestic policies and international trends puts the legitimacy of existing policies into question (Finnemore & Sikkink 1998, Heinze 2011: 12, Keck & Sikkink 1999). As Dobbin et al. (2007: 452) point out, a constructivist view on policy diffusion and transfer means that "understanding how public policies become socially accepted is the key to understanding why they diffuse". The number of countries that have already adopted a certain policy, an aim for legitimacy on the global stage or policies' trendiness are likely to affect the chances of transfer, potentially leading to "norm bandwagons" (Finnemore & Sikkink 1998: 893, see also Ikenberry & Kupchan 1990). Governments are expected to be "highly sensitive to the number, or proportion, of other countries that have adopted a particular policy stance" (Simmons & Elkins 2004: 174). When an increasing number of countries implement policies "that reflect a normative consensus", the "reputational payoffs associated with policy choic[es]" for follower governments change, and "reputational consequences" might affect governments that do not conform to a widespread consensus (Simmons & Elkins 2004: 173). In summary, international trends in policies and a perception among stakeholders or the general population that an "idea's time has come" (Hugo 1877: Ch. X, Kingdon 2003: 1) can be powerful drivers for policy change, including through transfer.

E4: Policies are more likely to transfer when their adoption is a condition for the access to funds or other advantages for the recipient, when incentives (material and non-material) are attached to these policies, when they confer an advantage over competitors, or when a high number of prior adopters of the policy have set an international trend.

2.3.4 Characteristics of transfer items as determinants for transfer

Transfer is not only shaped by individual and structural filters, but also by inherent features of the policy transfer item (Tews 2005: 72). The characteristics of policy models, programs, and instruments can impact the likelihood and shape of transfer to the same extent that they can determine the process of their selection in policy making process (Böcher 2012, Linder & Peters 1989, Steuwer 2013: 68–69). As Bennett (1997: 229) points out, "Policies have a distinct set of characteristics that influence the diffusion process, and in turn determine whether the evidence enters the domestic policy debate as objective evidence [...], as anecdote [...] or as a set of 'take-it-or-leave-it' guidelines."

An expectation, in particular from the environmental policy transfer literature, is that policies are more likely to transfer if they are informative rather than redistributive in nature, causing less opposition in the recipient context (Kern et al. 2001: 24). Policies that challenge the status quo of regulatory traditions and that would induce fundamental policy change are less likely to diffuse quickly (Tews 2005: 72). Policies that are seen as "solving a problem in a general and potentially universal (and hence, replicable) way" (Rousselin 2012: 478) are likely to diffuse more rapidly. The complexity of policy models is a further determinant for transfer, since learning about and implementing complex models usually requires more capacities and resources (Klingler-Vidra 2014, Nicholson-Crotty 2009).

The availability of information about a model is a crucial determinant for its transfer since "To the extent that policy diffusion is based on awareness and learning from the experiments of others, observability is very important" (Makse & Volden 2011: 111, see also Heiden & Strebel 2012). Observability may also send signals about how successful a model is. A model for which information is widely available suggests "that it may well be applicable to a wide range of nations", and a visible "high-profile innovation can thus create a tendency toward diffusion that is not driven by functional need alone" (Weyland 2004a: 249). As

Elkins and Simmons (2005:44) note, "The most available policy models, perhaps, are those that are reputed to have been successful. Decision makers will understandably be drawn to such a policy, sometimes letting its success bias their evaluation of its effectiveness" (Elkins and Simmons 2005: 44). This type of bias might lead to learning that "can be quite superficial" and "often involves linking a highly salient outcome with a policy innovation without complete information about the causal connections" (Simmons and Elkins 2004: 175) In a field like energy policy, but also in many other policy areas where target conflicts can arise, one needs to clearly define with regard to which objective a certain policy model is successful and who determines success. Hence, "the empirical record of the successes and failures of a given instrument across problem situations becomes far less important than decision-makers' perceptions of that performance" (Linder & Peters 1989: 36).

There are indeed no policies, especially in the field of environmental or energy policy, that are universally beneficial or only able to achieve one objective (Matisoff & Edwards 2014: 7–8). Therefore, when assessing drivers for the adoption of certain foreign policy models or parts of them, it seems crucial to enquire what objectives policy makers and stakeholders in the receiver countries intend to pursue (Klingler-Vidra 2014). Applied to the empirical case of this book, depending on whether, for example, job creation or climate protection goals are to be reached in priority, different energy policy instruments will be given preference, and different aspects of a promoted model might be taken into account (Steinbacher 2015). Weyland also emphasizes the importance of the perceived success of a model, noting that "a new policy scheme attracts attention and finds adherents due to its promise, not its actual accomplishments" (Weyland 2004b: 10). This would be a sign of rationality bounded by representativeness heuristics.

When addressing characteristics of transfer items as potential determinants for transfer outcomes, the multiple "layers" of policy need to be taken into account, since not all aspects of a policy are likely to diffuse with the same speed and the same likelihood (Radaelli 2005). As Howlett and Ramesh (2002: 34) note, internationalization can affect all types of policy change, and not necessarily simultaneously or in the same way, including the "alteration or reformulation of abstract policy goals or ends, [...] the general nature of the types of policy instruments used as the means to implement those goals [and] shifts in program specifications or the calibration of specific policy instrument components". Considering policy layers for transfer therefore means at least distinguishing incremental adjustments to policy instruments, instruments themselves, and overarching policy orientations, similar to Hall's (1993) three orders of change (see also Howlett 2009).

Finally, it appears essentially to take into account how a policy instrument and its features are framed in a specific context, pointing to the political nature of policy instrument selection: "instruments at work are not neutral devices: they produce specific effects, independently of the objective pursued (the aims ascribed to them), which structure public policy according to their own logic" (Lascoumes & Le Galès 2007: 10). A policy's "image", that is to say "how a policy is understood and discussed" (Baumgartner & Jones 1993: 25, see also Baumgartner & Jones 1991, Linder & Peters 1989), can decisively influence whether the instrument is chosen, regardless of its more objective track record (Steuwer 2013: 61). This dimension is also closely linked to the question of a transfer item's fit with regulatory traditions and dominant ideological beliefs.

E5a: The characteristics of policy models influence the likelihood and outcome of transfer. Policies are more likely to be transferred if they are informational rather than distributive in nature; perceived as addressing a universally salient problem; less complex; seen as transferable and successful in the sender constituency; and in line with local objectives.

E5b: Different layers of a policy (calibration, instrument, overarching objectives) need to be taken into account as transfer is expected to differ according to the level concerned.

2.3.5 Similarity and fit between sender and recipient

An important finding in the diffusion literature is the likelihood of geographical clusters of diffusion, facilitated by the similarities neighboring countries or states share. The understanding of how similarities between sender and recipient affect transfer and diffusion has evolved toward a plurality of definitions of similarity.

One channel through which similarities can facilitate transfer is "links of communication" (Weyland 2004a: 258) that can often be observed between countries in close neighborhood or with shared historical ties (e.g., a shared colonial past or language). Apart from the "practical ease of communication", neighbors may have a "natural tendency" to look across the border first for inspiration on foreign policy models (Bennett 1991: 51). These assumptions explain the focus of many studies on diffusion among sub-national entities within one political system (US states and cities, e.g., Berry & Berry 1990, McVoy 1940, Walker 1969, Shipan & Volden 2008; or Swiss cantons, e.g., Gilardi & Foglister 2008; Strebel 2011), on countries in the European Union (e.g., Bulmer 2007) and between countries in one linguistic and cultural group (especially the United Kingdom and the United States, e.g., Wolman 1992). But on the other hand, given the overwhelming focus of the diffusion and transfer literature on these cases (Marsh & Sharman 2009, Nedley 2004) a potential overestimation of the importance of geographical proximity (see Matisoff & Edwards 2014: 19) and of basic cultural proximity cannot be excluded.

Apart from similarities such as language, the level of economic development or even environmental conditions, "ideological compatibility" (Benson & Jordan 2011: 372, Dolowitz & Marsh 1996: 354, Grossback et al. 2004) or shared "ideational systems" (Hall 1993) – such as liberalism or a belief in strong welfare states – are likely to matter. One case in point is Robertson (1991), who demonstrates the importance of ideological fit in the field of labor policies, putting into question the rationality of learning in policy transfer.

On a policy level, similarities between sender and receiver may matter as far as policy objectives are concerned, since "borrowing a policy when problems or goals differ can limit the ability to learn from the experience of the originating country" (Mossberger & Wolman 2003: 433). While Mossberger and Wolman (2003: 437) predict that transfer in the case of differing policy objectives is likely to be limited to "only a few general lessons" rather than a specific "policy design", this type of transfer can still have value for the recipient state and "contribute to prospective evaluation". Transferred policies may be considered even if policy objectives differ, given that many policies are multidimensional and can be fine-tuned to reach different objectives in different countries.

Beyond shared ideological beliefs, policy transfer can be facilitated by the common belonging of sender and recipient to a "cultural reference group", when "lessons are viewed as more relevant the extent to which a foreign case is viewed as analogous" (Simmons & Elkins 2004: 175). Highly visible cultural "markers" such as language or religion (Simmons & Elkins 2004: 176) might increase the likelihood of transfer, but it is important to stress that a country's "cultural reference group" in *specific* fields of policy might be based on very different. For example, shared aims of leadership in clean technologies can become a cultural reference more important than traditional cultural attributes in new policy sectors such as climate change or sustainable energy (Steinbacher 2015). Such a broader view on "culture" would also prevent exaggerated "culturalism" in transfer studies (de Jong 2009) that would be "implying that any attempt to export it [a program] elsewhere would be doomed to failure because each national culture is deemed unique" (Rose 2005: 92). The hypothesis of similarities between sender and recipient decisively shaping transfer needs to be evaluated carefully against competing expectations such as transfer from particularly prominent, powerful states (Majone 1991: 104) or hegemonic domination (Ikenberry & Kupchan 1990: 283, Meseguer 2005: 31). Whether a sender's policy model "fits" with the regulatory or administrative tradition of the recipient country (Héritier 1996; Knill & Lenschow 1997) or would cause major changes to this tradition is in any case a likely determinant for transfer outcomes. The impact of these regulatory and institutional traditions in recipient countries will be discussed in the next section.

E6: Transfer is facilitated by similarities between sender and receiver, such as culture, language, geographic location, and similar ideational systems. This expectation competes with alternative explanations of transfer from large, prominent, and economically successful nations regardless of similarity. Other types of similarities, such as policy objectives, or specific sender characteristics may affect transfer processes and outcomes in specific policy fields.

2.3.6 National determinants for transfer: Recipient structure and capacities

Transnational communication alone, as Busch, Jörgens, and Tews (2005: 150) state, "cannot explain individual cases of policy change. Domestic actors, interests, institutions, capacities, and policy styles all influence the actual decision of any one country to adopt a policy or instrument that is being communicated internationally." Characteristics of the recipient system hence affect transfer processes and transfer outcomes. [19]

For example, governance styles are likely to influence the journey of ideas, including transferred ones, in the policy system (Lahn et al. 2009, Lavenex & Schimmelfennig 2009, Lavenex et al. 2009). For Dolowitz and Marsh (2012: 342), hierarchy-based modes of governance yield top-down transfer processes with a central role for state-level actors in transfer, while networked modes of governance result in transfer processes that associate agents from within and outside government, with multidirectional flows of ideas. Market-based governance could be associated with regulatory competition as a main driving force of transfer (Carafa 2011, Damro 2010, Lavenex & Schimmelfennig 2009). Bulmer and Padgett (2005) find broad evidence to confirm these hypotheses with regard to different governance modes in different policy areas in the European Union and confirm the importance of institutional/structural variables for transfer outcomes.

While this type of inference from governance modes to the shape of transfer is certainly not possible in all circumstances, the argument points to the importance of considering transfer processes individually, within their political, sectorial, and institutional setting. This is particularly relevant given the practical implications of policy transfer studies, where "Recommendations as to how lesson-drawing is to be done can only be given after diligent case studies of various transplantation processes" and by taking into account recipients' policy traditions (de Jong 2009: 148) or "political culture" (Common 2013: 17).

The institutional setup of policymaking and of regulated sectors of the economy in recipient countries not only influences the journey of ideas in a system (Checkel 1997: 478), but also shapes actors' interests and thereby affects the likelihood of diffusion and transfer (Cortell & Davis 2000). The potential consequences of a given, transferred model on policy change in a country depend on which actor groups are influential and what the impact of a transferred policy on them would be (see sections 2.5.2 and 2.5.4). A policy model's impact again is therefore not universal, but depends on the national context. The openness or closed character of "domestic political structures" and the fit with material and

19 Many of these national determinants for transfer are implicitly considered in the last part of this chapter, where the link between transfer and policy making is explored.

normative preferences of key actors are therefore factors that very likely impact the likelihood and outcomes of transfer (Torney 2015: 30–32). Questions of who defines policy alternatives, who is involved in policy debates, the location of decision-making, and possibilities for opposition shape how international sources are reflected in domestic politics, as in Gourevitch's "second image reversed" (Gourevitch 1978: 907).

Institutions embody "regulatory tradition" or "regulatory style" (Héritier 2001: 16) that are considered decisive factors for policy transfer: "path dependencies and administrative styles and traditions, equally set the course for whether a country" adopts diffused policies "and by which means" (Tews 2005: 70). Policy transfer outcomes that would incur "high transaction costs of institutional adjustment" are thus less likely to materialize (Benson & Jordan 2011: 372).

The degree to which a foreign norm "resonates" with the domestic political system and is compatible with dominant beliefs impacts the perceived legitimacy and desirability of international norms (Checkel 1997, Keck & Sikkink 1999: 98, Schimmelfennig & Sedelmeier 2005, Torney 2015: 31–32). This is in line with the expectations regarding the slower pace of diffusion of complex policy instruments that might either be deemed unfeasible with the capacities present in a certain recipient context or too disruptive given regulatory traditions (Tews 2005: 72), and hence considered politically unfeasible (Tews et al. 2003: 578).

While the emphasis on regulatory traditions and institutions as shaping the room for maneuvering and preferences of actors in policy transfer is linked to a historical institutionalist view (see section 2.5.1), "ideational structures", more familiar to a discursive institutionalist view, can be considered highly relevant national filters for transfer as well. These structures can constrain "which ideas are considered politically viable (or even mentionable)" (Carstensen & Schmidt 2016: 320) in a recipient state. The relevance of ideological fit and political culture has for example been found to play a role in the diffusion of renewable energy portfolio standards in the United States (Chandler 2009, Matisoff & Edwards 2014: 811). As pointed out by Carstensen and Schmidt, (2016: 323) it is important to note that actors can engage with these ideational structures – as with other structures – and are not only constrained by, but can also contribute to shaping them (see also Schmidt 2008: 314).

In addition to the structures of policy networks and policy making in recipient states, the capacities of recipients – financial, human, organizational, and others – are important determinants for transfer (Kern et al. 2001: 8, Radaelli 2005: 932, Weidner et al. 2002). A receiver's "absorptive capacities", to borrow a term from the technology diffusion literature (Keller 1996), determine whether agents are able to take foreign models into account, how they look at those models and whether transfer-inspired policies can be implemented at all (Howlett & Joshi-Koop 2011). Active senders may want to enhance capacities in a specific field in order to make the adoption of policies more likely.

E7a: Institutions in the recipient policy system are likely to affect the form and outcome of transfer by determining access to and the roles of groups and actors in policy making. Items that require more institutional adjustment or conflict with dominant groups are less likely to transfer.

E7b: Regulatory traditions and "ideational structures"/dominant belief systems affect the "fit" of transferred policies into the domestic system; the likelihood of transfer of a particular item decreases if it is in opposition with such structures, and thus less politically viable.

2.4 The Role of Senders for Policy Transfer

2.4.1 Who sends policies?

Transfer processes that are exclusively initiated, driven by, and carried out by endogenous demand from within the recipient, without interaction with agents outside domestic policy networks, are unlikely to be the rule (Tews 2002: 18–24). Policy models are communicated and promoted by agents outside the recipient state, as evidenced by works on the role of transnational actors in the promotion of policies (Finnemore 1993, Stone 2000, Stone 2004). The activities of these transfer agents either complement, reinforce, or trigger demand for information from within the recipient state. As Stone points out for the role of think tanks in transfer, transnational actors can "transfer the ideas and ideologies, the rationalizations and legitimations for adopting a particular course of action, and it is part of their endeavor to draw attention to developments overseas" (Stone 2000: 66). Among transnational policy promoters, the roles of the World Bank and the International Monetary Fund have been studied extensively (Brooks 2004, Edwards 1997, Orenstein 2008). Intergovernmental agencies (Strebel 2011) and transnational transfer agents such as international consultants (Prince

2012), knowledge networks (Stone 2004), think tanks (Stone 2000), advocacy networks (Keck & Sikkink 1999), policy research institutes (Ladi 2005), and epistemic communities (Haas 1992, Dunlop 2009) have also been found to play an important role in the communication and promotion of policies globally.

While there is no doubt that diffusion and transfer are particularly promising areas for research into the role of transnational actors, the relative absence of "actively sending" pioneer countries among the range of senders discussed in the transfer and diffusion literature (e.g., Tews 2005: 68–69) is remarkable, given that policies are often invented and implemented at the national scale (Jordan & Huitema 2014: 716). When the role of *pioneering countries* in environmental policy diffusion and transfer is considered, it is generally done along three specific lines: passively, through their provision of a pioneering policy, through their role in "uploading" of policies to higher levels of governance, and, through their role as leaders in international negotiations. Beyond the identification of who pioneers (e.g., Knill et al. 2012, Liefferink et al. 2009), a fundamental question is what makes countries pioneers (Jänicke 2005, Jänicke & Jacob 2007b, Volkery & Jacob 2003: 5–10). Jänicke (2005a) proposes four factors: in addition to sufficient domestic environmental policy capacities and coalitions that enable pioneering policies, issue-specific, situative, and strategic considerations come into play. Pioneers need a combination of "will and skill" (Jänicke 2005: 139), in addition to other characteristics such as reputation and visibility that also play a role in the transfer literature. Jänicke distinguishes between pioneers that are merely "forerunners" and those who push for specific regulations to be adopted at a higher level of governance (Jänicke 2005: 139). Pioneers may wish to "upload" their own policies (Börzel 2002: 196) in order to reduce the cost of conforming to supranational or federal, e.g., European, regulations or to increase the degree of ambition of common policy frameworks (Bulmer & Padgett 2005: 105–106, Liefferink & Andersen 1998, Vogel 1997). By doing so, "multi-level reinforcement" between national frontrunners and the supranational level can occur (Schreurs & Tiberghien 2007). The role of leaders as facilitators of the achievement of more ambitious (climate) negotiation outcomes has been thoroughly explored (Andresen & Agrawala 2002, Malnes 1995, Saul & Seidel 2011). In addition to the effects of leadership on countries' readiness to cooperate (Saul & Seidel 2011, Schwerhoff 2013), the fact that international negotiations are "two-level games" (Putnam 1988) allows countries with pioneering domestic policies to adopt a more ambitious stance in international negotiations.

Despite their crucial role in transfer and diffusion, as providers of pioneering policy models and technology (Jänicke 2013) and of course as recipients (Jänicke 2005: 132–133), the role of nation states as active policy promoters outside institutionalized negotiation frameworks has received very little attention so far (Biedenkopf 2012, Gilardi 2013a). A general aim to "bring the nation state back in" after a focus on non-state and transnational actors is observable in other aspects of environmental policy already (Duit et al. 2016, Jänicke & Jacob 2007b: 35, Jordan & Huitema 2014: 716). One of the few areas where the role of states in actively promoting policies in a transfer perspective has been studied thoroughly is "democracy assistance", in particular by the United States (Scott & Steele 2011). The effectiveness and success of this type of "promoted transfer" through either conditional aid or democracy assistance programs has raised questions (Bridoux 2014). In addition to few examples in the diffusion and transfer literature (e.g., Barabasch et al. 2009, Holden 2009, Lana & Evans 2004), cases of states promoting certain types of policy models can be found in the development aid literature, often with a reference to conditionality (e.g., King & McGrath 2004, McDonald 2012, Molenaers et al. 2015; see Lepenies 2009, 2014, for critical views on knowledge transfer in a development cooperation context). The role of the European Union as a policy promoter in the world approximates the question of states as policy promoters, even though the case of the EU differs in decisive respects (lack of implementation of policies by the EU, membership perspective that can be offered as an incentive, harmonization as a transfer mechanism). Examples of the study of the EU's approach to policy promotion include Lavenex and Schimmelfennig (2009) on the foundations of the EU's external governance outside Europe (see also Börzel & Risse 2009 for the concept of "transformative power Europe"), Knill and Tosun (2009) on environmental policy, Carafa (2010) on energy rule extension to Turkey, Falkner (2007) on biotechnology regulation, Damro (2010) on market policies, Biedenkopf (2011) on hazardous waste policies, and Torney (2015) on climate policy transfer to India and China. Torney's study of the EU's influence on Indian and Chinese climate policy is a particularly relevant example since it puts an emphasis on the "relational dimension of leadership – that is, the relationship between leader and follower" (Torney 2015: 184).

In summary, the source of policy models is taken into account in bilateral transfer research only implicitly. A structured conceptualization of the role of pioneering countries as active senders in international policy transfer is largely absent from the literature. This book therefore relies on a dedicated analytical framework of "leadership through transfer" to explore the role sending states can play in bilateral policy transfer.

2.4.2 Leadership through policy transfer

The term leadership, "a complex phenomenon, ill-defined, poorly understood, and subject to recurrent controversy" (Young 1991: 281) has been used in a variety of ways in the environmental field, spanning individual, political, and economic and business leadership (Gallagher 2012). It has been used to refer to the implementation of an innovative policy by a country (Knill et al. 2012, Liefferink et al. 2009); to describe how countries, individuals or organizations can influence the outcome of environmental negotiations (Andresen & Agrawala 2002, Malnes 1995, Saul & Seidel 2011); and also to define an outcome in which followers effectively adopt similar policy responses (Dobbin et al. 2007: 456). Although the literature on leadership in climate negotiations is of great relevance, the type of leadership addressed in this book is different. Leaders through transfer seek to gain followers to policies or principles they enact domestically (Steinbacher & Pahle 2015: 3), outside the realm of negotiation settings and coercion (Biedenkopf 2012).

A main requirement proposed in the climate leadership literature is that leaders primarily need to aim for the achievement of a collective goal, such as global climate protection (Eckersley 2011, Malnes 1995, Skodvin & Andresen 2006, Underdal 1992). Countries can hardly be expected to act *only* on altruistic grounds, creating a tension between self-interested and collective objectives for leaders (Skodvin & Andresen 2006: 17). Weidner and Mez (2008: 374) therefore propose that leaders may pursue "enlightened, far-sighted self-interest", while also contributing to the achievement of collectively beneficial objectives. Such self-interest for leaders may come from consequences such as an increase in soft power (Nye 2004) and international reputation, the validation of domestic policies through positive reactions from abroad, and technology cost reduction through economies of scale, as well as from "first-mover advantages" and export opportunities as suggested by the environmental lead markets literature (Beise &

Rennings 2005, Jänicke & Jacob 2007a: 8). Understanding how leadership can effectively contribute to the achievement of a collective goal is important because the relevance of unilateral action for global climate protection has been put in question. The provision of a public good such as a reduction of emissions through unilateral policy action has been expected to lead to a "free-rider dilemma" and reduce the need for others to act (Hoel 1991, Olson 1971; also Weimann 2012 on the case of the Energiewende). However, pioneers have been found to reduce uncertainty about the consequences and costs of certain policies (Andresen & Agrawala 2002, Edenhofer et al. 2013: 18, Schwerhoff 2013), possibly rendering their implementation more feasible in the eyes of followers. Saul and Seidel (2011: 902–903), for example, use a game theory approach to illustrate that leaders, by making a first move in a "tit for tat" may reduce perceived risk for other players, and thereby increase the likelihood of cooperation. Unilateral action can also generate a "signaling effect" for potential followers thereby inducing action in other countries (Brandt 2004), especially as policies and norms get adopted by larger numbers of countries (Busch & Jörgens 2012, Fankhauser et al. 2015, Finnemore & Sikkink 1998) and become socially accepted as the appropriate policy solution (Dobbin et al. 2007: 452).

2.4.3 Variants of leadership through transfer

The extensive literature on leadership, in particular on climate policy, can be integrated into three variants of "leadership through transfer" or diffusion (Eckersley 2011, Eckersley 2016, Malnes 1995, Saul & Seidel 2011, Schreurs & Tiberghien 2007, Steinbacher & Pahle 2015, Underdal 1992, Underdal 1994, Young 1991, Wurzel & Connelly 2011a). These three variants – unilateral, active, and effective – are all based on a pioneering policy model and are summarized in Table 1. They are not necessarily conditioned upon each other (i.e., leadership through transfer can in theory occur through a powerful example only) and differ in nature, with effective leadership being an outcome rather than a strategy (Steinbacher & Pahle 2015). The first variant of leadership considered in this book, "unilateral leadership" or "leadership by example" (Underdal 1992: 5), means "demonstrating that a certain 'cure' is indeed feasible and does work" (Underdal 1992: 4). Underdal's definition encompasses both a pioneering model (a potential "cure" to a policy problem) and the willingness of a leader to "demonstrate" the solution to followers by implementing it domestically in the sending constituency. In this book, unilateral leadership equates to the Energiewende as it is implemented in Germany, combined with the country's expressed interest in finding followers (Chapter 5, see also Steinbacher & Pahle 2015: 9).

Rather than relying on the hope that "by virtue of a policy's employment in one locality, other policymakers may hear of [...] a policy innovation" (Klingler-Vidra 2014: 60), senders may reach out actively to potential followers and thereby become active leaders (Steinbacher & Pahle 2015: 5, Steinbacher & Pahle forthcoming). Active leadership means that an aim for leadership is translated into outreach targeted at gaining followers (Young 1991: 287). A range of strategies is summarized under the heading of "active leadership", with three main variants being considered: [20] promoting transfer through communication and information, creating "spheres of leadership" (Eckersley 2011: 9), and offering structural incentives to potential followers.[21] There are striking similarities between these ways of leading and expectations from the transfer literature regarding the availability of information, the importance of institutionalized transfer channels, and the role of externalities for transfer.

The goal of active leadership through communication is to inform about a policy model and thereby render it more "available" in order to trigger an "international spread of policy innovations driven by information flows" (Busch & Jörgens 2005c: 865). Tailored communication and information about the model in recipient countries can have an impact on transfer since "the attraction of using a single model is especially strong when the motivation for adoption is enthusiasm about a site visit [or] a response to promotional efforts by policy entrepreneurs" (Mossberger & Wolman 2003: 434). Another possibility is for leaders to frame their policy model as a preferred way of reaching a collective objective ("issue-framing", Saul & Seidel 2011: 907) or to link the model to problems, issues, and policy objectives of particular relevance to a recipient

20 Other categorizations of leadership activities are possible. Saul and Seidel (2011: 903) distinguish five different modes of leadership (unilateral, structural, intellectual, institutional, problem-solving). Since the transfer processes studied in this thesis concern settings outside institutionalized negotiations, problem-solving leadership through the use of negotiating skills is omitted here. In addition to different modes of leadership, Wurzel and Connelly (2011b: 13) further introduce a distinction between different leadership styles in negotiations, namely transactional and transformational leadership, drawing from Hayward (2008).

21 Germany's efforts to exert leadership are discussed in depth in Chapter 5 of this thesis.

("issue-linking", Flachsland et al. 2012: 263). The production of information about possible solutions and underlying policy problems is a further variant of leadership, and has been termed "intellectual" (Saul & Seidel: 903) or "cognitive" leadership (Wurzel & Connelly 2011b: 13) This form of leadership through the provision of knowledge about policy problems, causal relations, and potential solutions, means that leaders can employ their "cognitive resources – knowledge, weltanschauung (comprehensive world view), ideas or moral convictions – to convince others to do their share in solving a collective problem" (Saul & Seidel 2011: 903).

The existence of channels through which knowledge can spread is a central condition for policy transfer. Leaders can increase the number and quality of such channels and thereby establish "spheres of leadership" (Eckersley 2011: 9). By contributing to the creation or redefinition of clubs, international organizations, partnerships, and dialog formats, leaders exercise "institutional leadership" (Saul & Seidel 2011: 907). The importance of membership in such organizations, in particular for norm diffusion, has been identified (Finnemore & Sikkink 1998, Orenstein 2008). The creation of channels for transfer and of networks in which information can be exchanged is also highly relevant with regard to "socialization" (Börzel & Risse 2009). In this mechanism, stakeholders and decision-makers are exposed to new knowledge and ideas by engaging in regular interactions with counterparts from other countries and "trust as well as normative and political convergence are gradually created" (Heinze 2011: 20). Torney, in his study of EU climate leadership vis-à-vis India and China, finds evidence for attempted socialization "through institutionalized and ad hoc dialogue" through which followers are "encouraged to change their actions to conform with particular expectations concerning appropriate behavior" (Torney 2015: 188), albeit with limited effectiveness.

A third form of active leadership is through incentives. Leaders can attach incentives to particular models and thereby change the relative cost of adopting one policy over another or over non-action. While the idea of leadership through "threats and offers" is contested (Malnes 1995, 98), non-coercive active leadership is taken into account (Young 1991: 288). Active leadership may include financial incentives such as preferential loans, sometimes blurring the line to conditionality, but also includes support that is not directly monetary, such as capacity building measures. Active leadership can thereby change the environment in which transfer takes place and may shift the weights within a policy network.

Table 1: Categories of leadership through transfer. Adapted from Steinbacher and Pahle (2015: 4)

Pioneers		...implement a policy innovation ahead of most other countries.	
Unilateral leaders	Active leaders	Provide a policy model that…	is visible and available.
			is perceived as legitimate.
			is framed to serve / serves a collective goal.
		Show willingness to…	contribute to resolving a collective action problem.
			have others contribute to it as well.
		Take measures to…	engage with recipients.
			create knowledge.
			communicate about policy and underlying issues.
			create channels through which transfer can take place.
			provide incentives to potential followers (material and non-material).
Effective leaders by transfer		Induce action in follower countries that contributes to the achievement of the overarching goal.	

The boundaries between different forms of leadership are likely to be less clear-cut in practice than Table 1 suggests. Thinking about leadership through transfer as a four-tiered approach involving a model, willingness to lead, actions taken, and an outcome is, however, useful in the conceptualizing of the phenomena studied.

E8: Unilateral leadership combines the existence of a pioneering policy model (transfer item) and an aim to gain followers. Active leadership translates this aim into action; through outreach to potential followers, through knowledge-creation or communication, the establishment of transfer channels, or the provision of incentives. Sender-related transfer agents may engage with stakeholders in recipient states and thereby change framework conditions for transfer.

2.4.4 *Soft power, sender characteristics, and legitimacy*

Defined as “the ability to get what you want through attraction rather than coercion or payment” (Nye 2004: x), soft power is a concept akin to leadership and of relevance to the study of senders in policy transfer. For Nye, the potential sources of “attractive power” include culture and values, as well as policies – but only when these “policies are seen as legitimate in the eyes of others, […] soft power is enhanced” Nye (2004: x).

Translated to the realm of policy transfer, the potential for sending countries to effectively promote transfer is increased if their domestic policies are seen as legitimate and as being implemented in a credible manner. Legitimacy distinguishes soft power from coercive power and includes several aspects. Bodansky (1999: 601) sees it as an authority that is either perceived as justified and hence accepted by a “relevant public” or “well founded – […] justified in some objective sense”. Distinguishing both dimensions of legitimacy, one external and determined by validation from a “relevant public”, the other one endogenous and determined by a policy’s or an action’s inherent characteristics, is challenging in practice as different audiences might have differing views on what is legitimate in an “objective sense”. For the study of policy transfer, keeping both dimensions – related to the sender and its model – in mind is nevertheless helpful, as not all policies seen as “well founded” are also considered legitimate choices in the specific context of a recipient country. With regard to climate leadership, Karlsson et al. (2012) find the legitimacy of leaders in international climate negotiations to be essentially defined by leaders’ perceived commitment to a common objective. As Eckersley (2011: 8) points out, leaders may pursue “multiple motivations and purposes” but a requirement for leadership is that “at least one of the leader’s purposes is to provide collective benefits and that this is recognized by a particular constituency of followers as a basis of common action”, since “it is this sharing of purpose that distinguishes legitimate power, and legitimate leadership, from social domination”. Schimmelfennig and Sedelmeier (2005) view the legitimacy of exported rules as being contingent upon domestic resonance, i.e., the degree to which an idea or a policy is perceived to fit into an existing regulatory tradition. Legitimacy and sender reputation are important determinants for transfer. Finnemore and Sikkink (1998: 901) find the adoption of international norms by “crucial states” to trigger “norm cascades”, that is to say, the global diffusion of principles.

Norms promoted by prominent states, that is to say, "states widely viewed as successful and desirable models" (Finnemore & Sikkink 1998: 906, see also Majone 1991: 104) are expected to be more likely diffuse. Jänicke (2005a: 139) emphasizes that reputation and visibility of a country are not necessarily linked to political or economic power, hence a certain attractiveness of leadership roles in environmental policy for smaller countries in Europe. This view is contradicted by what Eckersley (2011: 4) describes as a realist perspective on the "communicative power" of "Great Powers [who] have greater informational, intellectual and human resources at their disposal and [whose] views are more frequently reported in the media".

E9: Soft power, i.e., prominence, perceived success, visibility, and perceived legitimacy of senders likely increases the chances of transfer of their policies. Legitimacy can relate to the sending country (perceived aim of willingness to contribute to a collective objective) as well as to its model (i.e., salience, success, effectiveness).

2.4.5 Public diplomacy and environmental foreign policy

One tool to enhance a country's soft power and at the same time translate soft power into influence is public diplomacy. Defined as governments "engaging with foreign audiences" (Cull 2010: 12), public diplomacy is used to address the public in a foreign country and aims at creating awareness and understanding for policies, culture and values of the sending country. Its goal is to increase the attractiveness of a country in the eyes of civil society in the target country. Motivations for public diplomacy range from merely reinforcing the appeal of a country as a destination for work, leisure, and investment to shaping the preferences of a target country's population and thereby influencing its foreign policy decisions. [22]

From an empirical point of view, the public diplomacy literature has so far largely focused on the United States. A particularly relevant exception is Ostrowski's study of the practice of public diplomacy in German embassies (Ostrowski 2010, see also Zöllner 2009). The growing importance of outreach to

22 The underlying expected mechanism connecting the preferences of a society (i.e., for negotiations over confrontation, for democracy etc.) to foreign policy outcomes is rooted in preferential liberalism (Moravcsik 1997, Ostrowski 2010: 49).

civil society and to multipliers abroad in order to communicate German positions, culture, and policies is recognized by an overwhelming number of respondents in this survey of 158 German representations abroad (Ostrowski 2010: 179). The topic of "sustainable energy, environment, and climate" is identified as the most promising thematic area for public diplomacy (Ostrowski 2010: 88). Although the processes under scrutiny here are not identical with public diplomacy in the traditional sense – they do not address foreign populations directly and do not intend to influence the recipient country's *foreign* policy – this literature provides relevant accounts of the importance of communication to facilitate the transfer of ideas, evidence, and information.

Germany's Energiewende leadership is not only a topic of major interest in public diplomacy efforts, but is increasingly becoming part of the country's environmental foreign policy and climate diplomacy efforts (Li 2016, Tänzler & Wolters 2014), despite criticism that "Energiewende foreign policy" is not pursued vigorously enough (Messner & Morgan 2013). Germany's environmental foreign policy has its roots in the 1980s and was decisively pushed by the election of the Green Party into Parliament in 1983 (Mostert 2011, Schreurs and Tiberghien, 2007). The multiplication of bilateral agreements and informal cooperation fora, in addition to a proactive strategy in international negotiations, and in particular the adoption of a leadership role at the EU level, characterize Germany's efforts in this domain (Mostert 2011: 402-403, Chapter 5).

2.4.6 Lead market strategies

Pioneering policy innovations not only constitute the basis for leadership in policy transfer, but also are often either motivated by or entail economic advantages through the "lead markets" they help create. Lead markets are "countries that first adopt a globally dominant innovation design [...] lead the international diffusion of an innovation and set the global standard" for a given technology, product, or process (Beise & Rennings 2005: 7). While being more connected to the technology diffusion literature, lead market strategies are closely linked to domestic policy innovations and may rely on the simultaneous export of technology and of regulations that create demand for the technology in follower countries (Jänicke 2012a, Jänicke 2013).

Beise and Rennings (2005) identify "transfer advantages" as one of five key pillars of lead markets; transfer advantages occur when leading countries are able to either "increase the perceived benefit of a nationally preferred innovation design for users in other countries" or when "national demand conditions are actively transferred abroad" (Beise & Rennings 2005: 8). The transfer of renewable energy promotion policies can thus enhance "demand conditions" for renewable energy technologies; the promotion of stringent emission standards similar to those in place in the sending country can increase demand for cleaner combustion technologies produced in the lead market etc. By implementing an innovation, lead markets create an international "demonstration effect" (Beise & Gemünden 2004: 93, Beise & Rennings 2005), where the adoption of a technology by the leader is a proof of concept that reduces "uncertainty and therefore risk for subsequent adopters" – a mechanism that has also been identified as key for leadership in climate policy (Schwerhoff 2013). The authors recognize the importance of integrating studies of technology diffusion and "non-technological factors for the diffusion of environmental innovations", including the spread of "regulation […] and the flow of communication" (Beise & Rennings 2005: 15).

Germany has linked technology export strategies in the field of efficiency and renewables and action regarding non-technological factors (Beucker et al. 2014, see also Jänicke 2012a, Rennings & Cleff 2011, Schreurs 2012). Key domestic legislation such as the 2004 amendment of the EEG was justified by the aim of creating a lead market that would subsequently help diffuse the technology globally (Bundestag 2004). Closely linked to the lead market literature is the question of market power as a driving force for the export of regulation, as shown in studies of the California effect (Vogel 1997a). If countries hold market power, they are, "by dint of their market size", able to "alter the beliefs of other actors over the likelihood of possible outcomes. Their standards act as an attractor, causing other actors to converge to their preferences" (Drezner 2008: 32–33).

Outside the field of environmental regulations and products, Holden's study of policy and product exports in the United Kingdom healthcare services sector provides a compelling example of the strong links between technological and policy export promotion (Holden 2009). Holden finds that the UK, despite strong internal criticism of a particular model of health sector private-public partnerships, implemented a "strategy of trying to 'export' the policy itself, despite the fact that the public sector technical capacity needed to make the policy effective may be lacking in the target countries" (Holden 2009: 313). Beyond Holden's critical view on the adaptability of the model, his work also sheds light on the commercial interests that can be drivers for policy transfer. Holden (2009: 322) mentions "programmes for visiting health ministers and other 'inward missions' to see the NHS [National Health Service] approach to PPP/PFI [Public Private Partnership / Private Finance Initiative] and other issues" as well as partnerships, scoping missions, and memoranda of understanding that are similar to German export initiatives for renewable energy and energy efficiency.

2.4.7 Effectiveness and ethics of leadership through transfer

Once the links between leaders and transfer are established, the question remains of when exactly leadership through transfer is effective. From a sender's perspective, the indicator that matters to assess effectiveness is the agenda behind leadership efforts. If the motivation is to inspire a certain course of policies or the adoption of a policy instrument, policy outcomes can be used to assess to what extent leadership through transfer was effective. A particular emphasis was hence put on exploring the perceived agenda of Germany in its activities in Morocco, South Africa, and California. It cannot be automatically assumed that leaders want to export their models exactly as they are implemented domestically, so a leader's agenda needs to be understood to assess the effectiveness of leadership. The effectiveness of leadership through transfer does not necessarily require the leader's policy model to be fully imitated (Biedenkopf 2012: 106, Klingler-Vidra & Schleifer 2014). Leaders can have an impact on how policies and regulatory frameworks are designed, by inspiring general policy directions and even by providing negative lessons that can be used in an evaluative way to design adapted policies, even if some objectives differ (Mossberger & Wolman 2003). The "export" of objectives through emulation or socialization, following a logic of action dependent on what is seen as appropriate (Börzel & Risse 2009, March & Olsen 1989, Strang & Macy 2001) is thus a possible but by no means a necessary outcome of effective leadership through transfer. In general, the question of partial transfer and the transfer of broad orientations without instruments are highly promising areas for further theoretical work and have so far been neglected. In the field of renewable energy and climate policies, for example, policies can be adopted for reasons such as improving security of supply or job creation and still contribute to a collective objective the leader might pursue (Steinbacher 2015). For Saul and Seidel (2011: 902–903), leadership is not conditioned upon the existence of followers at all. By demonstrating willingness to lead and by reaching out to potential followers, leadership can become visible. It is a dynamic process, in which gaining followers is a desired effect rather than an inalienable condition for leadership (Steinbacher & Pahle 2015). It should also be stressed again that an assessment of the effectiveness (e.g. environmental effectiveness) of transferred or pioneering policies once these are implemented in the recipient context (Dolowitz & Marsh 2000) is beyond the scope of this research.

The question of whether it is ethically appropriate or questionable for pioneers to turn to the rest of the world in an attempt to promote their policies or a general policy direction is largely absent from the policy transfer literature (for exceptions, see Bridges 2014). This is in part due to the focus of transfer as a rather mechanical process and also related to the literature's selection bias toward OECD countries, in particular the EU and the United States (Evans 2004c: 12, Nedley 2004). Important impulses for thinking about the ethical dimension of the aim to "find followers" can be found in the development literature, and in particular in the critique of knowledge transfer in development practice (King & McGrath 2004, Lepenies 2014, Molenaers et al. 2015). The controversial debate on the roles of the World Bank and the International Monetary Fund in promoting structural adjustment programs through high-conditionality loans is a prominent example of discussions on the quality of transfer outcomes (Collier 2000, Sachs 1988). As underlined by Lepenies (2009: 34), a focus of development practice on transferring knowledge from developed to developing countries is deeply rooted in a belief that countries converge toward a common point of modernity. This convergence could be understood as the ultimate, all-encompassing diffusion of policies, institutions, and norms, promoted by "know-it-all" [*Besserwisser*] experts who serve a demand they contributed to creating (Lepenies 2014: 229). A main criticism regarding knowledge transfer is the lack of taking into account local objectives, structures, and knowledge and a type of "partner-orientation" that often stops at rhetoric (Fukuda-Parr et al. 2002: 194). Because of the non-transfer of central elements of the Energiewende, the cases selected promise insights into Germany's ability to adjust leadership to local objectives and the content of what is transferred. The inclusion of cases at different levels of economic development also allows for a comparison of Germany's approach and the identification of universal aspects of policy transfer in a policy field where traditional categories of "developing" and "developed" countries are less and less justified (Frankfurt School-UNEP Centre & BNEF 2016).

2.4.8 *Summary: How senders can impact policy transfer*

Figure 2 summarizes the ways in which leadership can impact the likelihood, prospects, speed, and outcome of transfer processes. It is important to acknowledge the interdependencies between different determinants. Leaders' inherent characteristics (size, economic power, soft power) have a potentially

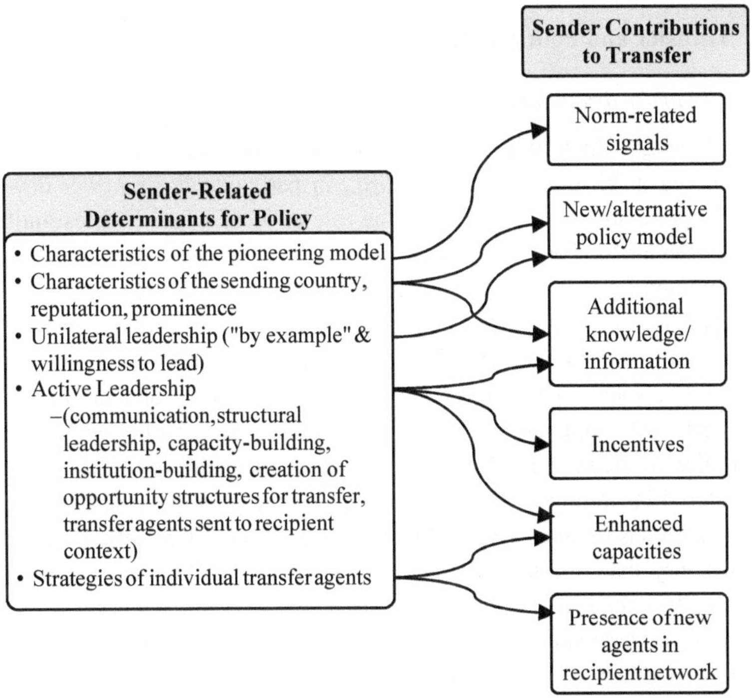

Figure 2: Sender-related determinants for and contributions to transfer. Graph by the author

strong influence on how their model is perceived by receiver countries. Another factor inherent to the sender that can facilitate transfer is similarity with the recipient (cultural, linguistic, colonial past, level of economic development), although this assumption needs to be tested critically given the over-proportionate selection of similar cases for transfer studies (see 2.3.5). Linked to the reputation of the sending country is the question of legitimacy. Here, the properties of the policy model and its implementation in the sending country can increase the prominence of a sender in a particular area by supporting claims about a policy innovation with credible evidence. A crucial impact of senders on the transfer process is the degree to which they make information about their model accessible and thereby increase availability for interested recipients.

2.5 Transfer and Policy Making

2.5.1 Transfer through the lens of historical institutionalism

The lens through which policy processes are considered in recipient countries is largely rooted in historical institutionalism, in particular in its newer developments that put a stronger emphasis on the role of ideas. Historical institutionalism (HI) is understood as a framework in which "Actors are strategic, seeking to realize complex […] goals […] in a context which favors certain strategies over others" and "cannot simply be assumed to have a fixed (and immutable) preference set [and] to be blessed with extensive (often perfect) information" (Hay & Wincott 1998: 954[23]). The "context" corresponds to formal and informal rules (institutions) that shape and can be shaped by the behavior of agents (Steinmo 2008, Thelen & Steinmo 1992). This view also corresponds to the arguments of actor-centered institutionalism, which puts an emphasis on the role of agency in interaction with structures (e.g., Mayntz and Scharpf 1995: 49). Rather than being a theory that formulates hypotheses on how variables are causally and systematically linked, HI is a theoretical framework in which studies driven by the desire to *understand* empirical outcomes (Steinmo 2008) can be embedded. Compared to other "new institutionalisms", rational choice and sociological institutionalism, a main feature of HI according to Hay and Wincott (1998) is its overcoming of the somewhat artificial dichotomy between appropriateness and

23 In Hay (2006: 63-64), the author uses a similar formulation in the context of constructivist institutionalism, emphasizing that actors' "desires, preferences, and motivations are not a contextually given fact" but reflect a "normative (indeed moral, ethical, and political) orientation towards the context in which they will have to be realized".

rationality as determinants of actors' behavior. Although the promise of equal attention to "calculus" and "culture" drivers for action has not always been realized in HI scholarship – and led for further developments, in particular constructivist institutionalism (Hay 2006) – the openness in principle to different and simultaneous behavioral drivers is an attractive feature of HI from a transfer perspective, in which determinants also stretch across categories.

The methods used by many transfer scholars – case studies, thick description, process tracing – and their acknowledgment of the importance of institutions as "opportunity structures for transfer" (Evans 2004a: 1, Evans & Davies 1999: 371), and of the interdependency of structure and agency (e.g., Ladi 2011) point to a proximity of viewpoints between policy transfer studies and HI. It is nevertheless rather rare for transfer studies to be explicitly embedded in HI. Among these exceptions are Ladi's study of Europeanization from an HI perspective (Ladi 2011) and Saint-Martin's study of lobbying in the European Union (Saint-Martin 2010). Historical institutionalism's fundamental interest in the temporal context, the sequence of events and its acknowledgment of past policies or "legacies" (Howlett & Ramesh 2002: 38) shaping future policies and even politics (Steinmo 2008) make it a promising lens through which to study policy transfer. Ideas that are transferred per definition have a source, that is to say, a past, and this source, as well as the moment and point at which the policy idea enters policy debates in recipient countries, matters (Dolowitz & Marsh 2012). As Rose (1991: 21) points out, "we would never expect a programme to transfer from one government to another without history, culture and institutions being taken into account." The transferred policies are hence not "highly plastic and timeless entities" (Saint-Martin 2010: 5), as transfer studies often depict them, but are generally still in implementation, with potentially evolving results, in the sender state.

As underlined by Steinmo (2008), historical institutionalists generally refuse the idea of full independence between variables explaining empirical outcomes. Translated to the transfer context, important interdependencies between different explanatory variables for transfer might occur. Interlinks and mutual influences between transfer activities and institutions in the recipient country can arise, e.g., when policy transfer and institution-building go hand in hand, when institutions are transferred, when the transferred policy would likely change "rules of the game", or when senders attach incentives to transferred policies.

Additions and evolutions at the frontier of HI toward a stronger integration of ideas in the analysis of institutional or policy change allow for an even better integration of transfer studies into this context. In HI, exogenous shocks and the punctuated equilibrium concept[24] (Baumgartner & Jones 1993) traditionally play a dominant role in explaining institutional change in a world that is largely characterized by incremental change and "constrained by past trajectories" (Thelen 1999: 388). While still acknowledging the relevance of path dependencies in the study of policy change, HI has increasingly turned to more idea-centered approaches[25], in which institutions, context, and ideas are seen as "co-evolutionary" (Steinmo 2008).

In the policy field under scrutiny, physical legacies such as grid infrastructure or power plants, in addition to regulatory and institutional path dependencies, can shape the preferences and actions of agents regarding policy objectives and how these can be reached. The combination of an HI perspective, with its acknowledgment of path dependencies as structuring preferences, and newer approaches in the field, which bring in ideas in the form of norms, frames, and discourses (Blyth et al. 2016) therefore create a useful background for the study of transfer within policy making.

2.5.2 Policy networks and policy transfer

The policy networks literature proposes a way to look at the policy making process in a specific issue area through the interdependencies between actors from different backgrounds. Rather than considering agents in isolation, the approach takes into account the relationships between them, their way of organizing, the number of access points, and the distribution of power within the policy network.

---

24 The punctuated equilibrium approach (Baumgartner and Jones 1993) is a traditional view on sources of institutional changes in HI. This view suggests that important exogenous shocks such as crisis, important changes in public opinion, or changes in governmental coalitions are responsible for fundamental change, whereas only incremental change occurs at "normal" times in policy making. The approach has been extended toward a "general punctuation" theory to take into account the role of continuous processing of information in policy making (Jones and Baumgartner 2012).

25 For Schmidt (2008), historical institutionalists (as well as researchers in rational-choice and sociological institutionalist traditions) who emphasize the role of ideas in institutional and policy change actually belong to a newer "new institutionalism", discursive institutionalism, "whether or not they would label themselves as such" (Schmidt 2008: 313).

It also addresses the interests and preferences, goals and norms, of more or less powerful, more or less organized, and interdependent actors (Howlett et al. 2009: 82). Policy networks are useful for the study of transfer because they provide a way to open the black box of recipient countries. But a look at the policy networks literature is also relevant to assess how transfer, changes in policy networks, and policy change are linked (Börzel 1997). Through its taking into account of both endogenous and exogenous factors that can change preferences and shift power in the network, it is hardly surprising that the policy networks approach is used in the policy transfer and diffusion literature (Lee & Strang 2006, Mintrom & Vergari 1998, Stone 2001). Even where links between the policy networks approach and transfer studies are not explicitly stated, views on the policy process tend to overlap. Both the policy networks and the policy transfer literature consider various types of agents – civil servants, politicians, business, scientific communities, advocates, and potentially also foreign agents and look at learning as one main mechanism leading to policy change. The trend toward transnational issue networks (Sikkink 1995) and transnational advocacy coalitions (Keck & Sikkink 1999: 93) allows for a further integration of the policy networks approach and transfer studies.

Although the importance of learning and ideas is recognized, policy networks also depend on the institutions that structure them and determine the relative bargaining power of certain groups or individual actors (Keck & Sikkink 1999: 90). The way political systems are organized influences how differences in preferences and power among different groups play out and how rule-defining players can be influenced. As Gourevitch (1978: 904) puts it, "by setting down the rules of the game, institutions reward or punish specific groups, interests, visions, persons. […] The impact of structures lies […] in the way a given structure at specific historical moments helps one set of opinions prevail over another". Overarching national political systems, with their varying degrees of centralization or fragmentation, shape both the "distribution of power and the type of interaction within policy subsystems" (Adam & Kriesi 2007: 138). The policy network approach is therefore applicable in a variety of contexts and of relevance in comparative works with heterogeneous cases.

The focus on one policy area is another argument in favor of a policy networks framework that recognizes the specificities of different policy areas for studying policy making: "Policies differ according to the incentives and resources they provide for group formation, the expectations they raise among groups or the masses, their visibility salience for mass publics, and the traceability of their effects" (Adam & Kriesi 2007: 140). The concentration and power of interest groups in regulated sectors such as energy is a particularly important aspect to keep in mind, with regard to the potential of "regulatory capture" (Levine & Forrence 1990), where regulatory decisions are shaped by pressure from organized, concentrated groups within a regulated industry. For Gourevitch (1978: 905), scrutinizing "how specific interests use various weapons by fighting through certain institutions to achieve their goals" is therefore crucial to understanding "how the process of getting a policy adopted affects its content". By using a policy network prism, international transfer can be studied in a specific issue area while taking into account the impact of network structures (Howlett & Ramesh 2002: 44).

Policy networks are only one way of looking at policymaking through the links between agents and organizations involved in a specific policy issue subsystem like energy policy. The "advocacy coalition framework" (Sabatier 1988, Sabatier & Jenkins-Smith 1993) is one of the most popular such approaches. It has been extensively used to study policy change through changes in the belief systems of coalitions bound by a common view on problems and causal links to solutions. Energy and environmental policy have been predominant fields of application for this framework (Sabatier & Weible 2007). Compared to the more generic policy network approach, looking at a question through the advocacy coalition lens presupposes stronger assumptions regarding the existence and nature of coordinating coalitions. The policy network approach on the one hand and other coalition-based approaches, such as advocacy or discourse coalitions (Hajer 1993, see also Fischer 2003: 94–114), on the other hand are not mutually exclusive. In particular, the theory of the individual in the advocacy coalition framework can be a meaningful complement to a policy network approach. Individuals in advocacy coalitions can follow logics of appropriateness and of consequentiality (March & Olsen 1989: 38, Sabatier & Weible 2007). Such an overcoming of a priori dichotomies in the micro-foundations of action is useful for policy transfer studies that expect functional, norm- and incentive-based drivers to all play a role (Heinze 2011: 9).

The core beliefs individuals hold translate into coalition views about how a policy issue ought to be dealt with, but do not automatically lead to shared "secondary beliefs" across a coalition about how each specific sub-problem is to be addressed or how an instrument ought to be fine-tuned (Sabatier & Weible 2007: 195). As in the policy networks approach, "policy-oriented learning", that is to say, the accumulation of technical knowledge and evidence (Jenkins-Smith & Sabatier 1993: 42, Sabatier 1987) constitutes the grounds for policy change. Change in the "really important components of a coalition's belief system – its policy core – is likely to require the gradual accumulation of evidence over a long period of time" (Jenkins-Smith & Sabatier 1993: 42). But, according to Jenkins-Smith and Sabatier (1993: 42), this accumulation is not sufficient per se for fundamental changes in public policy, since "such alterations also require changes in the distribution of political resources of subsystem actors arising from shocks exogenous to the subsystem".

The integration of international policy transfer into a policy network approach is depicted in Figure 3. Transferred knowledge can change the policy-domain specific context in which the policy network operates and ideas within the network through network members' "policy-oriented learning". The transfer item itself can generate externalities if adopted (see 2.3.3), impacting the

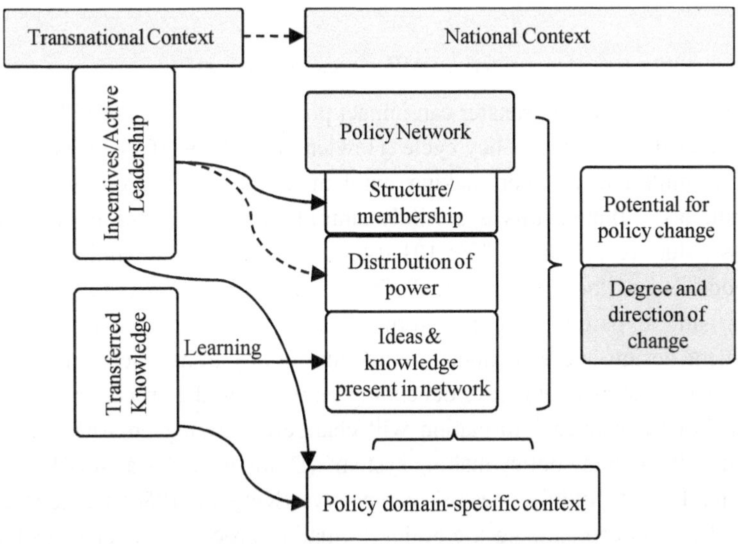

Figure 3: Policy transfer in policy networks. Graph adapted from Adam and Kriesi (2007: 148).

distribution of power and resources in the network. Incentives attached to transferred knowledge and active leadership by senders can also impact this distribution. Agents from the transnational realm may join a policy network, thereby changing its membership and potentially its structures. In summary, international factors "may influence policy network at the national level by redistributing resources, opening up new access points, and creating new venues that allow for reopening matters previously settled at the national level" (Adam & Kriesi 2007: 137). The following expectations can be derived with regard to the interlinks between transfer and policy networks:

E10a: Transfer and sender activities can affect the knowledge and ideas present in a policy network, and also their structural set-up (membership, capacities).

E10b: Transfer is likely to impact different actors in the policy network differently, and positions regarding transferred policies are hence likely to differ as well.

E10c: In conjunction with structural factors (access points, decision power), the relative strength and role of stakeholders in a policy network and their view on transfer is likely to affect transfer processes.

2.5.3 Policy transfer across the policy making process

To study how and when transfer can impact policy making, it is useful to refer to the basic elements of the policy cycle (Howlett et al. 2009: 10) – agenda-setting, policy formulation, decision-making, implementation, evaluation, and possibly termination – although this is a wildly simplified model of the policy making process (Blum & Schubert 2009: 101–140, Jones 1970). The empirical focus in this book is on "policy formation" (Zahariadis 2007: 65), covering agenda-setting and, in particular, policy formulation and decision-making, including instrument design and selection. International policy transfer can interact with policy making in different ways depending on where in the process it occurs and "the use of transferred information will change depending on where an agent […] interacts with the policy making system" (Dolowitz and Marsh 2012: 341).

Kingdon's "multiple streams" perspective (Kingdon 1984; Kingdon 2003) is useful for conceptualizing the policy formation process and for investigating the role of policy transfer therein. The framework's being "sensitive to the way information affects choice", its view that "collective output [is] formulated by the push and pull of several factors" (Zahariadis 2007: 66), and its taking into account of the temporality of policy choices (Zahariadis 2007: 68) are particularly relevant for its integration with a transfer perspective. Policies invented abroad fit well into the picture of streams of policies, politics, and problems in Kingdon's framework: when problems arise, policy solutions may not only be taken out of "garbage cans" (Cohen et al. 1972), but also from sources external to the political system (Kingdon 2011: 17). Contrary to an HI view on transfer (and findings in this book), Kingdon, however, argues that the source of an idea does not matter: "Ideas come from anywhere, actually, and the critical factor that explains the prominence of an item on the agenda is not its source, but instead the climate in government or the receptivity of ideas of a given type, regardless of source" (Kingdon 2011: 72).

Kingdon's conception of the policy process is largely contradicting rational policy making, where problems are addressed through systematic, thorough searches for and analysis of information. In terms of underlying mechanisms, Kingdon's framework is therefore one based on bounded rational learning and "has nothing to do with a rational, linear view of the policy process" (Radaelli 1995: 169). It is important to underline that transfer can not only affect the policy stream in such a framework, but also the problem stream (by bringing global problems to the agenda) and possibly even the political stream, when transferred ideas or the externalities they entail change the distribution of resources and power.

A transferred idea's time may come when a "window of opportunity" or "policy window" opens (Kingdon 2011: 166, also Kingdon 1984). For Bennett, problem pressure in the recipient state then pushes decision-makers to consider new solutions, especially those tested abroad: "as a ready-made solution [...] the program of a foreign counterpart, in whole or in part can act as a timely expedient to mollify political pressure" (Bennett 1991: 35). The notion of timing and opportunity is crucial in this framework, and it is relevant for the analysis of transfer outcomes as well (Dussauge-Laguna 2012). The actors that are most likely to use their resources and access in order to couple streams are "policy entrepreneurs" (Kingdon 2011: 179), who identify windows of opportunity (or even contribute to creating them). Exogenous shocks stemming from economic or technological developments, the international context or shifting attention at the national level may destabilize dominant coalitions within a policy network (Adam & Kriesi 2007: 142, Jenkins-Smith & Sabatier 1993), potentially opening windows for change. The same is true for new knowledge and "changes in policy fashion, ideas or policy frames" that may "disrupt existing policy systems, power relationships, and policies" (Richardson 2000: 1017–1018). This new knowledge can be brought in by actors from outside the recipient country as well as from insiders who turn to foreign examples.

There is no full agreement in the literature as to when in the policy process transfer is more or less likely. While Klingler-Vidra (2014) finds that broad, agenda-setting ideas diffuse more easily, the type of learning associated with paradigm shifts rarely occurs in this phase, according to Hall (1993). Policy entrepreneurs – who may also be transfer agents – are expected to play a role in the phase of agenda-setting, where new information about policy problems and solutions, in combination with exogenous drivers, can rapidly bring issues to the attention of policymakers, since "Evidence that another state has taken action can dispel arguments that this is not a serious issue [...]" (Bennett 1991: 34). For the Bennett, the "quality" of transfer or the depth of learning depend on where in the policy cycle transfer takes place, with transferred evidence in the agenda-setting phase entering "in very vague or anecdotal terms: the Americans/Germans/ Japanese/ Swedes etc. have a program to deal with this, we ought to have one too" (Bennett 1991: 35). It is questionable whether information about one particular foreign policy model suffices to bring an item to the agenda, but international norms and trends may achieve this when "Over time, the accumulation of evidence from abroad can fuel enthusiasm [...] producing a desire to jump on a band-wagon" (Bennett 1991: 34).

A more evaluative consideration of foreign evidence is more likely in the policy formulation phase (Bennett 1991: 35). Policy formulation encompasses the process of debating, designing, and choosing a "course of action" for addressing a policy issue (Howlett et al. 2009: 110). When looking at factors that influence policy instrument choice, the most common types of explanations, functionalist, norm-oriented, and public-choice driven (Böcher 2012, Gawel et al. 2016, Howlett & Ramesh 1993, Howlett 2009, Linder & Peters 1989), can all be accommodated in a transfer framework. Howlett and Ramesh (1993), in their integration of policy learning and policy instrument choice literature, conceptualize instrument selection as "shaped by the characteristics of the instruments, the nature of the problem at hand, past experiences of [other] governments in dealing with the same or similar problems, the subjective preference of the decision-makers and the likely reaction to the choice by affected social groups" (Howlett & Ramesh 1993: 13). Malnes sees instrument selection as being impacted by governments' "beliefs" about "how different measures will affect national interests and values" (Malnes 1995: 102–103).

E11: While policy transfer may occur at any stage of the policy cycle, the expectation is that it is more evaluative and linked to a more in-depth consideration of foreign models in the policy formation phase rather than in agenda-setting. Transferred knowledge may be used to devise policy alternatives, to choose policy instruments, and to decide on their design, especially when a need for action arises from the opening of windows of opportunity.

2.5.4 The politics of policy transfer: Strategic utilization and harnessing of ideas

The view on links between transfer and policy making presented in this chapter imply that transferred knowledge, ideas, and policies are not "value neutral" (Zahariadis 2007: 69), but have a political dimension. This may seem straightforward, but the political dimension of policy transfer remains neglected, and flows of information and knowledge tend to be seen as somewhat neutral (Garrett & Jansa 2015). As Stone emphasizes, "the struggle over ideas" is in fact "the essence of policy making in communities" and "a mode of influence even more powerful than money and votes and guns" (Stone 1988: 7). The central "question of whose interest is served by this [transfer] process" still "goes unasked" (Dolowitz & Marsh 1996: 355) in much of transfer research. As pointed out by Adler and Haas (1992: 370), learning and the use of knowledge in policy networks should rather be regarded "as a process that has to do more with politics than with science, turning the study of political process into a question about who learns what, when, to whose benefit, and why." Even beyond transfer research, the study of the role of knowledge in the policy making process does not systematically include the "political use of policy knowledge" (Daviter 2015: 492).

Dolowitz and Marsh, two of the founding fathers of recent transfer studies remind researchers in a piece on the "future of transfer studies" that "The 'games' that transfer agents engage in will shape what is borrowed, where it comes from, how it is understood, how it is sold, where it is used in the policy cycle and how the information is used (reused) as a policy works its way through the development and implementation processes" (Dolowitz & Marsh 2012: 341). This understanding of transfer means that the information international transfer brings into the policy making process can be harnessed to further and legitimize positions, and that "Knowledge and information should be seen as one more 'resource'" since "Information about public policy is not utilized in a neutral or depoliticized fashion" (Bennett & Howlett 1992: 291). Also, the interests of bureaucrats and other agents in the transfer process can affect transfer outcomes, since "self-interests that go beyond the policy transfer instrument […] may determine the success of policy transfer [and] can impact on the ultimate policy goals and outcomes" (Tambulasi 2013: 82). It also means that transfer, as the "utilization" of the foreign model, "implies a mobilization of that knowledge to further some political aim" (Bennett 1991: 33).

As for knowledge in general (Weiss 1979: 429), not all information about foreign policy models is likely to constitute a resource of similar value to members of a policy network. Different "actors from different coalitions are likely to perceive the same information in very different ways" (Sabatier & Weible 2007: 194), and one and the same foreign policy model can thus be strategically used to support different positions. This subjectivity of policy instruments and the importance of taking context into account have been stressed in the policy instrument selection literature (Linder & Peters 1989: 35, Malnes 1995: 102–103, Steuwer 2013: 66). Information about foreign models is hence used in a "highly selective" manner and with the objective "to reinforce positions and to legitimate decisions already taken", sometimes ignoring "the problems of transferability" (Bennett 1991: 38).

The selective consideration of information is therefore not only a function of cognitive heuristics, but also a deliberate choice that agents make regarding the "timing of the introduction of evidence [and] the nature of the evidence presented" (Bennett 1991: 33). One such example is transnational advocacy networks (Keck & Sikkink 1999: 93), in which local advocates may strategically use transnational contacts to change the course of policies nationally. Problems identified domestically by advocates are "sent" to the transnational sphere, where they may be reframed or be supported by evidence, then bounced back in a "boomerang pattern" to the national level, where the international attention and validation of local positions increases pressure for action (Keck & Sikkink 1999: 89). For Robertson, "advocates of change will tend to invoke foreign lessons in an attempt to place an issue on the political agenda [...]", while in the phase of formulating and adopting policies, "opponents will more forcefully use negative lessons to emphasize the risks of other polities' initiatives, to associate these programs with negative consequences and to highlight the unique features of their political system that make emulation unlikely to succeed" (Robertson 1991: 56).

More than just a battle of ideas, policy transfer includes a political economy aspect. Depending on how transferred ideas or policies affect the "material interests of relevant domestic actors", international policy evidence and norms are more or less "likely to gain traction domestically" since "if new ideas strongly contradict preexisting material interests, it is unlikely that those new ideas will gain traction with domestic actors" (Torney 2015: 31). Testing this assumption is particularly relevant in the policy field that this book deals with. Energy transitions and the policies that further them have potentially disruptive consequences on the material interests of incumbents. How these react to evidence about a foreign policy model that can be considered threatening for incumbents is thus a highly relevant question.

A political view on transfer and diffusion therefore implies that empirical research needs to explore "the motivations underpinning an agent's use of foreign information", with a view to not only to "explain where information is sought and the extent to which the agents learn from this information, but [...] also [to] explain how this information is subsequently used" (Dolowitz and Marsh 2012, 341).

E12: Transferred knowledge can be harnessed strategically to inform or support positions or to provide arguments against other stakeholders' positions. Agents' utilization of transferred ideas can depend on expected material consequences and/or on beliefs about what constitutes appropriate policy in the recipient context.

2.5.5 Summary: What transfer brings to the policy making process

A main criticism of the policy transfer literature is its lack of clarity on how transfer and diffusion are linked to policy change (Braun et al. 2007). As Majone expresses it, "The relevant question, therefore, is not whether policy imitation is possible or desirable [...] but why a particular model becomes influential at a given time" (Majone 1991: 104). Authors such as James and Lodge (2003) deplore the lack of explanatory power of the transfer and diffusion approach. This criticism does not appear justified when transfer is understood as a process and as a dependent variable, since numerous determinants for transfer have been developed in the literature (see section 2.3), leading Bennett (1997: 214) to note that "if anything this literature is over-theorized". The issue becomes more challenging when transfer is (also) seen as an independent variable and hence as a contributing factor to policy outcome and policy change. The literature on transfer therefore needs to be embedded in a broader framework for policy making in the recipient state. Policy making is understood here as an "unfolding of actions, events, and decisions that may culminate in an authoritative decision, which at least temporarily, binds all within the jurisdiction of the governing body" (Schlager 1999: 294). Establishing causal links between policy transfer and policy making requires thinking about what transfer brings to the process compared to policy making that does not involve international transfer, as shown in Figure 4. For Howlett and Ramesh (2002), internationalization potentially affects the prospects of policy change through two main channels. On the one hand, internationalization can introduce new ideas; on the other hand, it can result in the presence of "new actors in the policymaking process" (Howlett & Ramesh 2002: 34), which change the structure and potentially the balance of power in policy networks.

Information can "change the game" in different ways, depending on what underlying logic of action in decision-making processes is assumed. If policy making is rational, the success of the foreign model in reaching objectives relevant for the recipient will be evaluated and policies will be almost automatically implemented if they are judged successful in some rational sense. For James and Lodge (2003), international learning and lesson drawing are thus broadly identical to any form of rational policy making. In such an understanding of the policy making process, successful policies, on which information is available, appear less costly and risky to implement and are thus likely to lead to policy change, in line with expectations from the economic literature on the value of unilateral leadership (Schwerhoff 2013). Negative evaluations of foreign experience could then result in non-adoption. This view on the role of ideas in policy making as an immediate, almost automatic process assumes that any rationally beneficial idea will automatically result in policy change, overlooking the fact that different agents may have different opinions and differing interests in whether a transferred policy is adopted or not (see 2.5.4).

In an attempt to mitigate the, in their view, "shaky" theoretical underpinnings of policy diffusion, Braun et al. (2007) integrate the changes diffused knowledge can bring into a decision-making process in a formalized model, where diffusion can affect two variables for decision-makers, namely payoffs and effectiveness (of policies). This model relies heavily on a public-choice view of actor decisions and appears simplistic for transfer research interested in uncovering the complex interlinks and interdependencies of different transfer mechanisms. International policy transfer increases information about policy alternatives and their effects, but potentially also transmits new norms and discourses that can frame and change actors' behavior and preferences within policy networks and affect what is seen as "appropriate" (Olson 1971, Saint-Martin 2010: 5).

Although ideas are certainly a central contribution international policy transfer brings to policy making, they are not the only one. In addition to updating the knowledge of actors within a system, transfer can also change the relations of power within a policy network. New policy models transferred from abroad might create new "winners and losers", both in material terms and in terms of their importance as thought leaders in policy networks. As Radaelli (1995: 165) points out, the perception of what actors' interests are might shift not only through material incentives, but also through knowledge that can "generate a definition of interests by illuminating certain dimensions of an issue, from which an actor can deduce her/his interests. Interests therefore become a dynamic 'dependent' variable, framed by knowledge" (see also Haas 1992).

Not only can the prospect of transferred policies change the balance in policy networks and information about those policies be harnessed for political purposes, but the transfer of institutions themselves can change the relationship between actors and restructure policy spaces. As shown by Aklin and Urpelainen (2014) and in particular by Busch and Jörgens (2010: 101–136), institutions such as environmental ministries, environmental agencies, and advisory councils have been found to diffuse widely and are likely to restructure policy landscapes in recipient countries.

Policy transfer and the policy it occurs in are interdependent and dynamic. While ideas, including transferred ones, impact policy making, cognitive heuristics, institutional structures, and power relations between members of policy networks need to be taken into account to trace the role of transferred knowledge in policy making. Transferred knowledge can enhance capacities for some actors, empowering them and granting a comparatively bigger importance to their position.

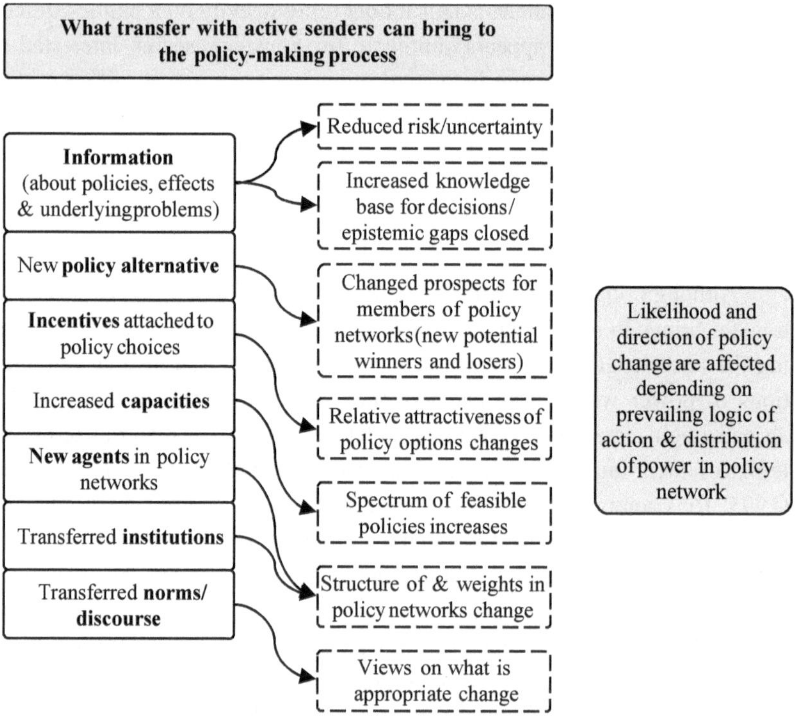

Figure 4: Contributions of transfer to policy making and links to variables affecting policy change. Graph by the author

The expectations E1-E12 presented throughout this chapter feed in to the analytical framework presented in Chapter 3. Following a presentation of how theory is used in this book, hypotheses are derived from the theoretical framework. Transfer is categorized into three variants and operationalized for the empirical cases studied.

The methods employed to explore the consideration, use, and reflection of Energiewende-related knowledge in Morocco, South Africa, and California are discussed in Chapter 4. In Chapter 5, Germany's unilateral and active Energiewende leadership is explored in depth. Empirical findings are then presented in Chapters 6, 7, and 8 on each of the recipient cases studied, and results are discussed in Chapter 9.

3 Analytical Framework and Hypotheses

3.1 An Iterative Approach to Research: Use of Theory

Rather than following a deductive research agenda, where hypotheses from the literature are tested, an iterative, abductive approach was adopted. This means that a general theoretical framework and guiding hypotheses were developed prior to field research, but these were updated and fine-tuned to reflect empirical knowledge gained (Héritier 2008: 65, Reichertz 2007, Reichertz 2010). The choice of an iterative approach – going back and forth between the theoretical framework and data – was guided by several considerations. First, the lack of literature related to the international diffusion of energy transition models in general, and the German Energiewende in particular, asks for an adaptation of existing hypotheses from the environmental policy transfer literature. Secondly, this book addresses neither only the sender nor only the recipient part of transfer and therefore requires combined hypotheses that take into account both sides and their interdependencies. This concerns in particular the question of how Germany as a sender promotes transfer and how its leadership affects classic determinants for transfer. Thirdly, the selection of cases of selective or negative transfer (see Ch. 4) offers opportunities to develop transfer research further and requires a flexible theoretical framework.

The way theory is used is akin to Strauss and Corbin's understanding of "grounded theory" (Corbin & Strauss 1990, Reichertz 2007, Strauss & Corbin 1994). Grounded theory, generally speaking, is the development of specific theory out of empirical data (Charmaz 2014). According to the interpretation developed by Strauss and Corbin of the initial framework conceived by Glaser and Strauss (Glaser & Strauss 1967, Glaser & Strauss 1999), pre-existing theoretical frameworks can be "*elaborated* and modified as incoming data are [...] played against them" (Strauss & Corbin 1994: 273, emphasis in original). In this understanding of grounded theory, the researcher is not expected to carry out empirical research without preexisting knowledge, but develops it further as new categories and patterns emerge from the data.

K. Steinbacher, *Exporting the Energiewende*, Energiepolitik und Klimaschutz.
Energy Policy and Climate Protection, https://doi.org/10.1007/978-3-658-22496-7_3

Several preliminary choices, based on prior case knowledge, empirical interest, and theory, were made. One important choice was to look at renewable energy policy making in Morocco, South Africa, and California through the lens of bilateral policy transfer rather than through, for example, a comparative policy, transition studies, or development policy focus (e.g., Marquardt et al. 2015). Also, a strong emphasis is put on one sender, Germany, although other senders' approaches are considered in contrast on a case-by-case basis. This decision is rooted in prior knowledge of the stated aim of "finding followers" in rhetoric in Germany and the frequent mentioning of the German Renewables Act being an "export hit" (Ch. 1, Ch. 5). Focusing on transfer in the direction from Germany to Morocco, South Africa, and California and concentrating on bilateral transfer processes, rather than multilateral or local ones, are further preliminary decisions taken to narrow down the focus of this book and ensure that in-depth insights into processes could be gained within the resource constraints of the project. Finally, the question of Energiewende leadership could have been addressed through a foreign policy lens, embedding the question into the broader framework of Germany's foreign policy and role in the global arena. Such a perspective might, however, have made it difficult to give due consideration to the inner workings of recipient cases and pay due attention to the specificities of sustainable energy as a policy field, an aspect that emerges as relevant on the basis of empirical findings in this book.

On the basis of these choices, the strands of literature outlined in Chapter 2 served as a guiding framework for field research. The emphasis on particular segments of these literatures, expressed in the form of hypotheses presented in section 3.4, emerged along empirical research. The lines along which transfer and leadership literature could be further extended based on the findings are presented in Chapter 9.

3.2 Integrating Transfer and Leadership for the Energiewende Case

This book ties transfer to both the sending source and the recipient policy environment. Policy knowledge more or less immediately related to the Energiewende makes its way through transfer channels into policy networks and to policy making in the recipient cases, as illustrated in a simplified manner Figure 5. This movement of ideas can be triggered by the Energiewende as a compelling example (unilateral leadership, see section 2.4.3) and through the active search for information by stakeholders in the recipient context. But active Energiewende leadership in its different forms presented in Chapter 5 can also facilitate it through various channels. In addition to ideas, Germany's active leadership can bring new members into policy networks, modify their structures, and add incentives to particular policy options, making them more attractive. Transferred knowledge traverses individual and structural filters as it is used in the policy making process and possibly reflected in policy outcomes.

As Hypotheses I-III reflect, important links exist between each of the different elements of the schematic transfer process shown in Figure 5. The Energiewende, as it is implemented in Germany, is seen in this book as the fundament for unilateral leadership and as a background for active leadership through various channels (Ch. 5). In the transfer process, the Energiewende example (i.e., unilateral leadership) and German policy promotion activities (i.e., active leadership) are hence linked, but are not identical. This has two implications. On the one hand, Germany's Energiewende translates into active leadership but may also directly trigger demand for information from stakeholders in recipient countries. On the other hand, active leadership, such as the presence of German transfer agents, the production of studies or policy advice is not necessarily targeted at promoting "the" Energiewende but may also aim at promoting general principles. Drawing this line between the Energiewende and Germany's activities abroad, i.e., between unilateral and active leadership, is important conceptually, but both are closely linked in practice. Taking into account that not all "Energiewende transfer" is a direct transfer of the Energiewende's instruments is particularly relevant given the cases selected in this book.

Whether its entry into the recipient system is driven by local demand or facilitated by Germany as a sender, individual filters (cognitive heuristics, consequences of the transferred policy on individual groups'/actors' positions, capacities) and structural filters (regulatory traditions, set-up of policy-networks, distribution of power in networks, institutions) affect the use of transferred knowledge. These filters can be shaped by active leadership in the form of communication, the targeted provision of information, capacity building, incentives, or the reinforcement of transfer channels. Importantly, the presence and activities of German transfer agents within policy networks – whether or not they are officially part of German programs, or individuals, experts, and researchers with a background in German energy policy – is to be taken into account in these processes. Finally, transferred knowledge can be reflected in policy outcomes, depending on the distribution of power and structure of a policy network and the view relevant stakeholders hold regarding the transfer item, as explained in more detail below.

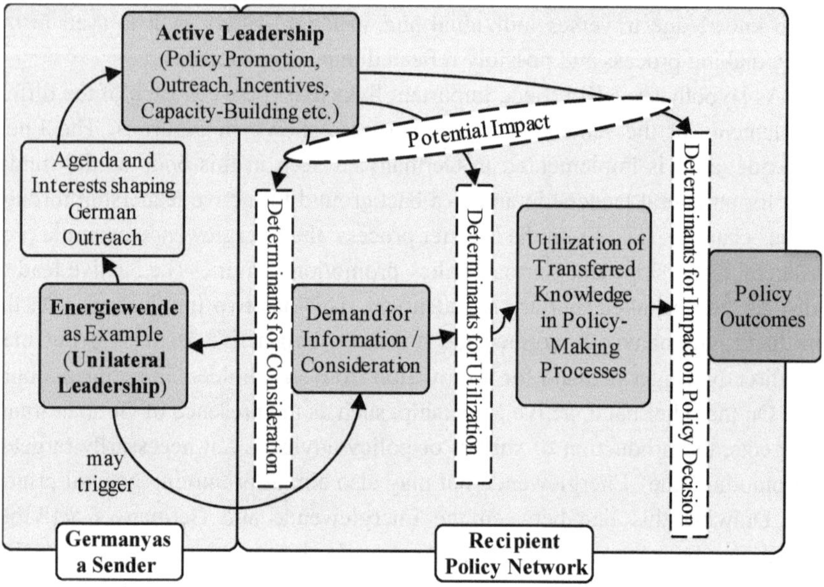

Figure 5: An integrated view on Energiewende leadership and policy transfer. Graph by the author

3.3 Operationalization: When is Transfer Transfer?

There is no standard definition in the literature to decide what is transfer and what it is not. The spectrum goes from calling any use of information "transfer", including from the past within the same organization (Rose 1991: 78), to defining transfer narrowly as the adoption of policies similar to the sender's model, resulting in convergence (Evans 2009b: 244, Wolman 2009: 1). Busch and Jörgens (2005a: 4) suggest that policy diffusion, and, by extension, transfer, can be identified by showing that other countries' experiences were taken into account in the decision-making process. This understanding corresponds to Dolowitz and

Marsh's widely used definition of transfer as a process in which knowledge about foreign policy models "is used in the development of policies [...] in another political system" (Dolowitz & Marsh 2000: 5).

To guide the qualitative analysis of interview material and other sources, three categories of transfer were built. Categorization allows determining "which aspects of the material should be taken into account and then working through the material step-by-step" along these categories (Mayring 2000). Based on the literature, in particular Dolowitz and Marsh (2000) and Bennett (1991, 1997: 213) and insights from empirical research, a threefold categorization of transfer was developed. Transfer is categorized as (1) the *consideration* of information about the Energiewende by members of policy networks, (2) the purposeful *use* of information in recipient policy debates, and (3) the *reflection* of transferred knowledge in policy decisions. These three variants are all transfer outcomes in their own right and can be used to assess the effectiveness of German leadership through transfer in Morocco, South Africa, and California. The categorization deliberately includes the transfer of negative lessons and transfer resulting in non-action or diverging policies as integral parts of the transfer spectrum in this book.

Transfer, in summary, occurs whenever members of the recipient policy network consciously *consider* the German energy transition example or information channeled by German transfer agents and/or *use* this knowledge, with different possibilities on how this use can *impact* policy output, as detailed in Table 2.

Table 2: Categorizing and operationalizing transfer

Category	Operationalization
Considera-tion/ awareness of the German Energie-wende example and infor-mation provided by German transfer agents	Consideration is assumed to have occurred when statements made in interviews confirm that respondents are aware of particular aspects of the Energiewende and German energy policy. This can be confirmed by the provision of details such as mentioning of when interviewees became aware of the example, through which channels, and concerning what elements/characteristics of the policy model. Statements on the degree of knowledge further refine this outcome (degree of details mentioned about instruments used in Germany, policy results, current developments, challenges, and perception of the German experience, importance of the example relative to other energy policy models). The same indicators for "consideration" apply to knowledge produced and promoted by German transfer agents, either directly concerning the Energiewende as an example or principles derived from it. To confirm consideration, interview statements should report personal encounters and cooperation with German transfer agents leading to transfer of information, consultation of documents produced by German transfer agents or participation in workshops, events, training sessions etc. sponsored by German organizations. Interview statements hinting at the consideration of knowledge are triangulated through document analysis. Meeting reports, minutes or public statements that confirm the conscious taking into account of transferred knowledge are used to operationalize this transfer outcome.
Use of transferred knowledge	The transfer outcome is considered to have materialized if stakeholders – including German transfer agents – provide concrete examples of how and when they used transferred knowledge either to build their own position, to underpin an argument, or to counter arguments made by other stakeholders in policy debates within the recipient network. The reasons given by interviewees for the utilization of information and the circumstances in which it took place are taken into account in a more fine-grained analysis of this transfer outcome. Furthermore, stakeholder statements providing concrete examples of when and how other members of the policy network used transferred knowledge or evidence about the Energiewende, either positively or negatively, in debates, discussions, negotiations, media, at events, or in personal interaction are taken into account to triangulate results regarding utilization. As far as they are available, public statements, presentations from workshops, minutes, reports and position papers are scanned for the use of transferred information that either references the German energy policy example or is clearly linked to publications/information provided by German transfer agents.
Reflection of trans-ferred knowledge in policy output	Given that sources of inspiration are generally not stated in regulations or legislative texts, the impact of transferred knowledge on policy outcomes is largely achieved through the triangulation of interview data from different respondents. Impact/reflection is strongly indicated if stakeholders from different institutions agree that a piece of transferred knowledge changed the course of a policy debate, facilitated or accelerated the development of a policy design, reinforced the likelihood of policy change or, to the contrary, inspired a different policy solution or non-action due to negative lesson drawing. The cross-section of perceptions of German transfer agents on the effectiveness of their leadership efforts with the perception of stakeholders from the recipient side allows for the identification of instances of transfer that are likely to have impacted the likelihood or direction of policy change. Policy comparisons of adopted regulations and decisions with German policies put these perceptions into context.

3.4 Hypotheses for Effective German Leadership Through Transfer

When reviewing all transfer mechanisms outlined in Chapter 2, some seem ill adapted to explore policy transfer in the cases studied. Since countries are under no formal obligation to engage in energy transitions, let alone adopt German-style policy instruments, coercion as a driver for transfer can be excluded. Another reason coercion is excluded as a hypothesis is that the "German model" is in competition with other energy policy orientations and different understandings of sustainable energy. Given the absence of a binding international regime for renewable energy and the choice of countries outside the European Union as cases, harmonization was also ruled out as a transfer mechanism. The focus is therefore on voluntary processes, although these might be spurred by incentives and externalities. General expectations from the transfer and leadership literatures regarding when transfer occurs and how senders can shape its outcome were formulated throughout Chapter 2 (expectations E1-E12). By combining these expectations and for cases studied in this book, three hypotheses can be defined for German leadership through sustainable energy policy transfer:

3.4.1 Hypothesis I: Consideration of the Energiewende

Consideration, that is to say, awareness and the "taking into account" of information on the Energiewende by members of the recipient policy network, is considered as a degree of transfer of its own right (Heiden & Strebel 2012: 348). Consideration is a necessary element of transfer since, "Without awareness of what others are doing (and, to some extent, with what effects) it is not even possible to engage in emulation or learning" (Makse & Volden 2011: 112). It is a test in particular for the effectiveness of German "unilateral leadership", raising the question of whether the Energiewende as a pioneering model is sufficiently compelling and visible to trigger consideration.

Hypothesis I: Consideration of the Energiewende in recipient contexts
The Energiewende and related ideas are more likely to be considered by stakeholders in a recipient con-stituency, the more information about them is **available**, the more **capacities** there are in the recipient context to search for and process information, and the more the German example is perceived as promi-nent and **successful** with regard to its implementation in Germany. These factors can be impacted by active leadership through communication (esp. linking transferred information to local objectives/ "issue-framing"), reinforcing capacities, and the creation of transfer channels by Germany.

Different determinants affect the likelihood of a particular model to be at all taken into account by stakeholders in the recipient constituency. First, the availability of capacities (time, staff, access to foreign media, and sources of information) in the recipient case to perform searches for information and process information is likely to affect how deeply the model is considered (see sections 2.3.2, 2.3.6). Germany as an active sender can impact the likelihood of consideration of its policy model by reinforcing those capacities, by creating channels of transfer that facilitate access to information, and by rendering knowledge and evidence on its policy model easily available. The existence of individual "heuristics" or cognitive shortcuts (Meseguer 2005, Weyland 2006) signifies that agents are unlikely to consider all existing policy models and even less so, to the same extent, as a strict interpretation of rational learning would require (Evans 2004a, Meseguer 2005, Stone 1999, Weyland 2006: 5). Giving consideration to a model like the Energiewende is an essential element of transfer because it means that a policy model is integrated into the knowledge base of stakeholders in the recipient policy network. It constitutes a condition for the utilization (Hypothesis II) and potential reflection of transferred knowledge in policy making (Hypothesis III). The availability of information on the Energiewende, its observability (Heiden & Strebel 2012, Makse & Volden 2011: 111), and perceived success in achieving targets in Germany increase the likelihood of consideration (Elkins and Simmons 2005: 44; see section 2.3.4). Consideration can be demand-driven (as in demand-driven learning discussed in section 2.3.2) or can be facilitated by Germany as a sender, through active leadership (section 2.4.3). Germany's prominence as a sender and perceived legitimacy are further factors likely to impact consideration of its Energiewende model (section 2.4.4).

Whether a stakeholder sees the Energiewende as successful and therefore worth considering is likely to depend on individual criteria for success of a policy and the priority attached to different policy objectives by individual stakeholders (Mossberger & Wolman 2003, see section 2.3.5). These differences in the perception of the Energiewende's track record are not only likely to affect interest and awareness of the model, but will also decisively impact its use in policy debates, as highlighted in Hypothesis II.

Drawing a clear line between the consideration of the Energiewende as a specific policy example and of general principles such as "more renewables" that may be communicated by German transfer agents is challenging (see section

9.1.7). Mediating the Energiewende experience by translating its broad orientations into communication to the recipient might at first not appear to equal a consideration of the Energiewende example. But given the very strong links between Germany's domestic model and its outreach activities (Ch.5), the consideration of principles related to the Energiewende by means of interaction with German transfer agents is a form of effective leadership through transfer.

3.4.2 *Hypothesis II: Utilization of transferred knowledge*

However essential, Germany's aim to find followers is unlikely to be satisfied by the mere consideration of the Energiewende model and its principles by stakeholders abroad. Understanding whether and how Energiewende-related transferred knowledge is *used* in the policy making process is therefore essential. Hypothesis II formulates expectations as to when knowledge about the Energiewende or its principles is likely to be used by stakeholders in the recipient transfer network.

Hypothesis II: Utilization of transferred knowledge
Transferred information is more likely to be used by members of the policy network when it is perceived as serving **objectives supported** and as **feasible** in the local context, when it can **underpin a personal or institutional position** and when it **closes epistemic gaps** that impede change. **Active leadership** can affect utilization through the entry of new agents to the network, incentives, the targeted provision of infor-mation to specific stakeholders ("empowerment"), and by increasing the spectrum of what policies appear feasible.

Whether stakeholders will make use of transferred knowledge in policy debates and their own position – a key aspect of transfer (Bennett 1991: 33, Dolowitz & Marsh 2000: 5) – likely depends on several factors. Whether the transferred model is perceived as addressing objectives of relevance to the stakeholder or their institution will determine whether and how evidence about the foreign model is employed to support a position or contradict others' arguments, as a political view on the use of knowledge suggests (Daviter 2015). An important nuance is that the extent and way in which transferred knowledge is used – positively or negatively, to underpin or contradict and argument etc. – will depend on whether members of the transfer network would gain or lose if the transferred policy were implemented (see section 2.5.4). Gains and losses could be material, but could also concern thought leadership or the relative position of an institution. Using a foreign policy model that fits objectives or enhances one's standing in the policy network can be influenced by beliefs of what any element of the German energy model is able to achieve and whether its implementation is feasible and appropriate in the recipient context. When objectives between the sender's model and the recipient's coincide, the German model can be used as a reference to legitimate and support choices and to "mollify political pressure" (Bennett 1991: 35). It should be underlined that Germany as a sender country may play a role in shaping the receiver's definition of a policy problem and even policy objectives to be reached (Saul & Seidel 2011: 903).

Active leadership cannot only produce information that may close important epistemic gaps in the recipient context, but also influence the ways in which information is used. Information can be made available, but it can also be tailored to the specific needs of certain actors, and it can be framed as to address local objectives (Saul & Seidel 2011: 907). Whether stakeholders see transferred knowledge as potentially beneficial or detrimental to their position can also vary along incentives attached to a policy or measures taken by the sender that make the policy appear more feasible in the eyes of opponents. Leadership through transfer is effective if information is predominantly used in a way that supports a move into the direction of principles supported by the sender, hence generally a positive (but possibly selective) use of the sender's model. The eventual impact of whether and how transferred knowledge is used will vary, among other aspects, with the relative strength of different members of policy networks and their respective views on transferred policies (Biedenkopf 2012: 109). *Who* uses information affects how it is used in the policy making process, with the presence of German transfer agents as members of policy-networks being a factor to take into account.

3.4.3 Hypothesis III: Reflection of transfer in policy outcomes

Hypothesis III extrapolates from Hypothesis II to assess the aggregated effect of the use of transferred knowledge on policy outcomes. This book is interested in transfer as a process but also considers its impact as a variable in policy making. This is relevant for the question of effectiveness since leadership is eventually interested in the adoption of policies that further aims/collective objectives also supported by the leader.

Hypothesis III: Effects of Transfer on Policy Outcomes
The likelihood of Energiewende-related transfer impacting policy outcomes (in the sense of leading to a specific policy design or policy change) depends on the relative **power of stakeholders in the policy network** and their **respective position with regard to transferred policies**. Whether transferred knowledge affects policy outcomes also depends on the **presence of windows of opportunity** that ask for a politi-cal/regulatory action. Active leadership can impact these factors by attaching incentives to policy choices (mitigating losses, enhancing gains for particular groups or agents), and contributing to the recognition of windows of opportunity.

For transfer outcomes, not only the way information is used matters, but also who uses it, to what end, and what the role of these agents in the policy network is. How knowledge impacts policy action and its direction depends on the structure and the balance of power (in terms of authoritative power, regulatory capture or thought leadership) within a policy network (Adam & Kriesi 2007: 137, Howlett et al. 2009: 82). The structure of the network needs to be taken into account along agency, since "institutions provide differential empowerment of actors" (Radaelli 2005: 935). This assumption corresponds to expectations formulated by Jenkins-Smith and Sabatier (1993: 42) with regard to the impact of policy learning, where the accumulation of new knowledge is a necessary but insufficient element to explain public policy change. Political resources, authoritative power, and the structure of decision-making, in conjunction with different actors' positions regarding transfer, thus need to be taken into account to explain how transferred knowledge is reflected in policy outcomes. Leadership can impact the relative strength of actors, e.g., by tailoring information provided to the arguments of certain stakeholders or by attempting to change dominant views in a segment of the policy network through information, e.g., through study tours and training sessions. It can also alter the very structures of policy networks, by promoting the establishment of new agencies, the reinforcement of existing capacities in some institutions over others, and the integration of new (transfer) agents to recipient networks.

The very nature of the policy model from which knowledge is transferred affects who wins and loses from implementing a similar policy and is therefore of relevance to analyzing the prospects of transfer in a policy field. If strong incumbents (or incumbent beliefs) are challenged by the basic principles linked to a policy model (such as democratizing electricity generation to a large number of players or replacing one source of electricity by the other), transfer is likely to hit obstacles. Incentives, and hence externalities (Heinze 2011: 12), can be attached to a policy model to mitigate losses. This can increase a model's attractiveness compared to others and decrease the cost of adjusting to it. Finally, the feat of transferred knowledge in a policy network is also dependent on a temporal dimension and transfer is more likely to occur at some points in time than at others. Decision-making in policy networks can be incremental, i.e., when a regulation is scheduled to be updated, but more fundamental change is likely to be linked to the opening of windows of opportunity (Kingdon 2011: 166), and pressure for political action (Jenkins-Smith & Sabatier 1993: 42). In both situations, transfer can play a role.

In summary, the effectiveness of leadership through transfer can be confirmed – albeit to differing degrees – when Germany's model is taken into account, and/or used, and/or reflected in policy outcomes in a way that furthers the achievement of objectives the sender supports with its leadership strategy. In the case of Germany and its Energiewende leadership, transfer that facilitates moving into the direction of more renewables in the electricity mix, can be considered as a requirement for leadership through transfer to be effective. It is also important to emphasize that a final evaluation of the effectiveness of leadership through transfer in reaching overarching objectives, such as climate protection, requires an assessment of policy impacts (e.g., on emissions) and implementation that goes beyond the scope of this research.

Following a presentation of methods in Chapter 4, Germany's unilateral and active leadership are discussed in Chapter 5, providing a more in-depth view into what the Energiewende model is and how leadership is built on it. The ways in which leadership by example and active German leadership affect policy transfer – from consideration to utilization and reflection – in Morocco, South Africa, and California, are then explored in Chapters 6-8. These empirical findings are discussed in comparison and with regard to the hypotheses just outlined, in Chapter 9.

4 Research Design and Methods

4.1 Methodological Positioning and Research Process

4.1.1 A nuanced interpretivist approach

Research presented in this book is rooted in a nuanced interpretivist epistemology that recognizes the intrinsic links between objects and their interpretation by "meaningful actors" and calls for scholars to "aim at discovering the meanings that motivate […] actions rather than relying on universal laws external to the actors" (Della Porta & Keating 2008: 24). Studying events, from an interpretivist perspective, is not useful "without looking at the perception individuals have of the world outside", and their "imperfect knowledge and complex motivations" (Della Porta & Keating 2008: 25). The subjectivity of agents whose actions are studied, but also of the researcher, is acknowledged, and the possibility of establishing firm and universal laws of causation in social science research is met with skepticism. However, the interpretivist approach is nuanced, as in many works in the policy learning literature, which Münch (2016: 31) considers in fact to be located outside the interpretivist tradition. Ideas and knowledge are central to the analysis of transfer and act as filters and as resources (Daviter 2015): They are, "on the one hand […] cognitive filters, and on the other hand they can be considered as main components of action" (Steuwer 2013: 56). Although they can change eventually if ideas and policies change, structural factors (institutions, rules) are seen as "the banks and riverbeds that channel and shape participant behavior" (Mucciaroni 1992: 466) at a given time. This view is contrary to a strictly constructivist approach, which denies "the distinction between facts and values altogether" (Della Porta & Keating 2008: 31). The understanding of ideas is hence in line with much of the policy learning literature, where policy ideas as well as the duality of agency and structure are considered (Jenkins-Smith & Sabatier 1993, Kisby 2007, Münch 2016: 31).

The epistemological positioning in this book translates into an iterative approach to research, where conceptual frameworks and empirical research "are interlinked with continuous feedback" (Della Porta & Keating 2008: 29). An interpretivist approach to research aims at *understanding* (Charmaz 2006: 126)

K. Steinbacher, *Exporting the Energiewende*, Energiepolitik und Klimaschutz. Energy Policy and Climate Protection, https://doi.org/10.1007/978-3-658-22496-7_4

and looks at cases as "interdependent wholes" (Della Porta & Keating 2008: 27). The inner workings of cases can be explored by using theoretical frameworks as guidelines and developing them further going back and forth between data and overarching concepts (McKeown 2004: 141, see also section 3.1).

This not only has implications for the use of theory, but also impacts the choice of methods. The present study takes inspiration from grounded theory since categories and codes to structure material are developed *from* data, rather than being defined ex-ante by theory only. Nevertheless, theoretical and empirical knowledge acquired prior to field research feeds in to the process of data collection and treatment (Corbin & Strauss 1990, Reichertz 2007: 215, Strauss & Corbin 1994, Yin 2009: 71). Process tracing appears as a particularly well-adapted methodological approach to this end, since it "provides a way to learn and evaluate empirically the preferences and perceptions of actors, their purposes, their goals, their values and their specifications of the situations that face them" (Vennesson 2008: 233). Process tracing not only allows for chains of events to be identified, but most importantly for their context to be understood, and provides opportunities for "empirical applications where *both* agents and structure matter" (Checkel 2008: 114, emphasis in original).

4.1.2 Overview of the research process

Figure 6 provides an overview of the research process and individual steps are described in more detail in the remainder of the chapter. Starting with empirical knowledge about the German energy transition and public statements regarding its leadership aim, a literature review was carried out and research gaps were identified. Early on in the research process, background conversations (i.e., exploratory interviews, see rationale in Rathbun 2008: 696) with experts and stakeholders in Germany were conducted and extensive document analysis was carried out to fine-tune the researcher's understanding of Germany's activities as an Energiewende leader. Findings from this step in the research process are presented in Chapter 5.

Following case selection, field research was prepared. A first, exploratory, field trip to Morocco and the attendance of a high-level conference in November 2013 in Casablanca led to the establishment of contacts and the preparation of the subsequent research stay of close to two months in February and March 2014. In Morocco, the researcher was affiliated with the Secretariat of the

Moroccan-German energy partnership operated by GIZ and located in the Moroccan energy ministry, allowing for first-hand insights into energy cooperation between Morocco and Germany. In South Africa, field research was carried out from October to December 2014, mainly in Pretoria, Johannesburg and Cape Town. In California, during two months of field research in March and April 2015, the researcher was informally affiliated with the University of California in Berkeley's Environmental Science, Policy, and Management department. In all cases, energy sector events were attended with an aim to understand the context of energy policy making in each recipient state (see Table 43 in the Appendix).

Prior to field research, case knowledge was acquired regarding structures, important agents, energy policy developments and cooperation channels with Germany, in order to focus as much as possible in the interviews on what cannot be observed from the outside. Throughout field research, memos were written and interview transcription was started. Initial codes were developed during field research and were refined in the follow-up to each research stay.

Draft chapters for each case were written following each period of field research. Another set of interviews at the very end of empirical research, in summer 2015, targeted evolutions in Germany's leadership approach and was used as a plausibility check for preliminary findings. These plausibility checks were also carried out with the help of experts, some of them former interview partners, in Morocco, South Africa, and California. The aim of these checks was to further the thought process, refine arguments and conclusions and in some cases update information regarding processes that were not yet concluded at the time of field research. Finally, the case reports were harmonized in structure and the cross-case discussion presented in Chapter 9 was drafted.

The following sections present the use of case studies as an overarching structure; of process tracing as a framework for analysis (section 4.2); the selection of Morocco, South Africa, and California as relevant cases for the research question this book addresses (4.2.2); and data collection and treatment through interviews, document analysis, rankings, transcription, and coding (section 4.3). In addition to the discussion of biases and caveats in section 4.3.7, the appropriateness of methods used is discussed in more detail in Chapter 9.

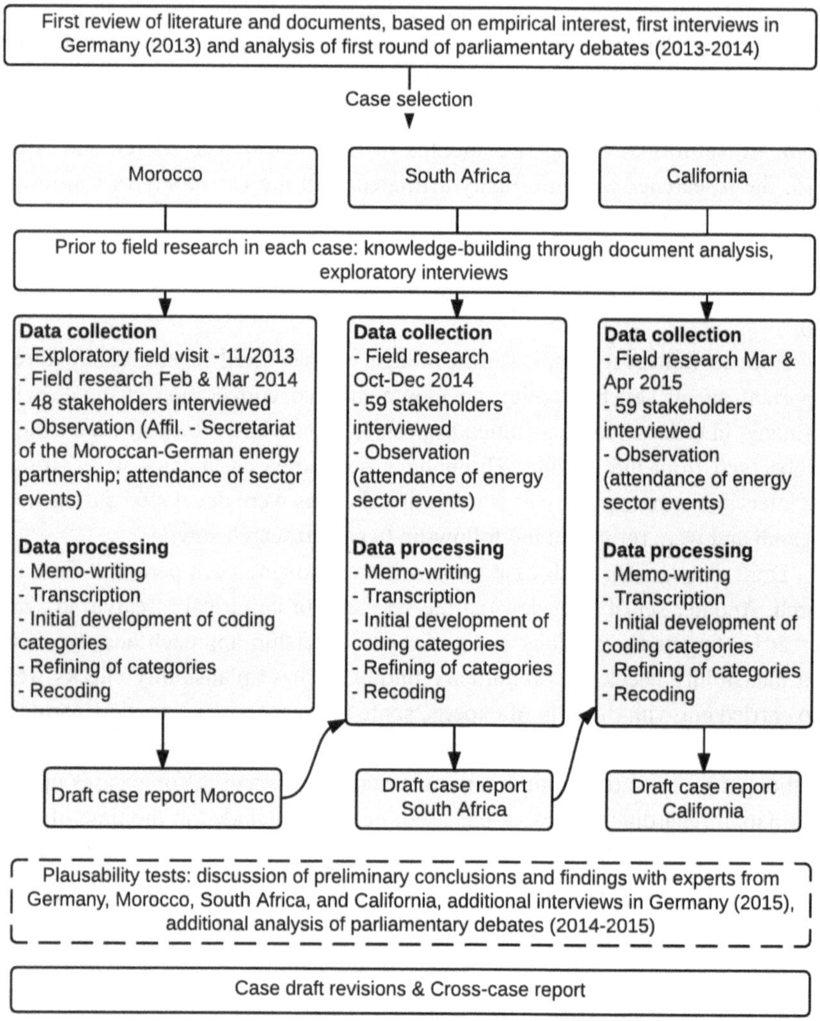

First review of literature and documents, based on empirical interest, first interviews in Germany (2013) and analysis of first round of parliamentary debates (2013-2014)
Case selection
Morocco
South Africa
California
Prior to field research in each case: knowledge-building through document analysis, exploratory interviews
Data collection
- Exploratory field visit - 11/2013
- Field research Feb & Mar 2014
- 48 stakeholders interviewed
- Observation (Affil. - Secretariat of the Moroccan-German energy partnership; attendance of sector events)
Data processing
- Memo-writing
- Transcription
- Initial development of coding categories
- Refining of categories
- Recoding
Data collection
- Field research Oct-Dec 2014
- 59 stakeholders interviewed
- Observation (attendance of energy sector events)
Data processing
- Memo-writing
- Transcription
- Initial development of coding categories
- Refining of categories
- Recoding
Data collection
- Field research Mar & Apr 2015
- 59 stakeholders interviewed
- Observation (attendance of energy sector events)
Data processing
- Memo-writing
- Transcription
- Initial development of coding categories
- Refining of categories
- Recoding
Draft case report Morocco
Draft case report South Africa
Draft case report California
Plausability tests: discussion of preliminary conclusions and findings with experts from Germany, Morocco, South Africa, and California, additional interviews in Germany (2015), additional analysis of parliamentary debates (2014-2015)
Case draft revisions & Cross-case report

Figure 6: Overview of the research process

4.2 Case-Centered Research

4.2.1 Choice of case studies as a framework and cross-case comparison

Case studies, defined by Gerring (2004: 341) as "an intensive study of a single unit with an aim to generalize across a larger set of units", provide the framework for empirical research. The use of single cases or small-n comparative case study designs is widespread in the transfer literature (e.g., Dussauge-Laguna 2013, Evans 2004b, Holden 2009, Tambulasi 2013, Weyland 2006). As underlined by Busch & Jörgens (2005a: 4), country case studies complement quantitative research designs present in the diffusion tradition (e.g., Berry & Berry 1990, Gilardi & Foglister 2008, Gilardi 2014, Holzinger et al. 2009, see also Marsh & Sharman 2009).

By "zooming in" on a single or few entities, case studies allow the researcher to "peer into the box of causality" and, ideally, "see X and Y interact" (Gerring 2004: 348). They are particularly relevant for the study of "contemporary phenomena" in their "real-life context"(Yin 1981: 98) and for issues with a high number of potential explanatory factors and complex causal relationships, as can be expected in transfer research (Marsh & Sharman 2009: 273). Yin particularly recommends a case-study approach for the study of "knowledge utilization", since "if one is desirous of answering 'how' and 'why' questions instead of or in addition to questions of frequency, case studies are the more appropriate strategy" (Yin 1981: 100, see also Gerring 2004: 347). As pointed out by Vennesson (2008: 225–226), case studies are not linked to one particular epistemology. They are not a method per se, but a means to frame the collection, processing, and analysis of empirical data. The cases are the "political, social, institutional, or individual entities or phenomena about which information is collected and inferences are made" and the context within which variables and their interaction can be understood (Brady & Collier 2004: 275). Even though Morocco, South Africa, and California are referred to as cases, Ragin's discussion of what a case actually is (Ragin 1997: 4–5) needs to be considered. To be precise, the cases this book is interested in are not the entire states of Morocco, South Africa, and California, but sub-segments, namely the policy networks dealing with (renewable) energy policy debates. For the sake of conciseness, these cases are referred to as "Morocco", "South Africa", and "California". Germany is studied in depth with regard to its energy policy model and its policy promotion strategies, but since it constitutes a potential explanatory factor and a "common denominator", it is not a case properly speaking in the framework of this book.

Given the lack of prior, in-depth studies of Energiewende transfer, and of the role of pioneering countries in transfer in general, a comparative approach and heterogeneous cases were chosen to prepare the ground for further generalization. Due to the novel field it studies, the aim of generalizing is limited in this book to identifying initial patterns regarding the role of senders in policy transfer and of Germany in particular. It is obvious, however, that additional case studies are needed to solidify any attempts for generalization of the international effects of the Energiewende. A comparative approach to the study of transfer is also called for by authors such as Evans (2009b: 254), who argues that "All studies of policy transfer should adopt a comparative methodology but few do." Comparative policy transfer research is also of interest from a policy standpoint, since it allows senders to "better understand the challenges of cross-national transfer and to offer guidance for practice" (Mossberger & Wolman 2003: 348). The role of the Energiewende and Germany's leadership is the common element across cases. These are put in contrast to identify commonalities and differences in Chapter 9. This book thereby follows a "multiple-case" and "embedded" case design, in which several units of analysis (the events and processes traced) are studied within the framework of each case, and cases are studied separately and then are later compared (Yin 2009: 48). A "replication approach" (Yin 2009: 56) is adopted, where a theoretical framework, although not an unmodifiable one, is built, methods are chosen and individual case studies are then carried out. Individual case reports (Chapters 6-8) are written that feed into the cross-case conclusions, and adaptations to the theory, and possibly lead to policy implications, as suggested by Yin (2009: 57).

Finding a balance between the number of cases to be studied and the need to generate in-depth insights was a central concern. As Bennett (1997: 214) notes, "The need to discover that policymakers in country A are aware of policies in country B and moreover that evidence is *utilized* within the domestic policy process severely limits the number of cases that may be analyzed within the comparative framework". By selecting a smaller number of cases but delving deeply into each one of them, "richer, more contextualized and more insightful empirical research" (Bennett 1997: 214) can be produced – an objective that is at the core of this book and motivated the choice of a small-n design with three cases in the limits of this project.

4.2.2 Case selection: Heterogeneous cases of apparent non-transfer

Applying a random sampling logic, as advocated for by positivist social science researchers (King et al. 1994), to the study of single or a small number of complex entities is often "misplaced" (Yin 2009: 56). Alternative case selection strategies are available to qualitative researchers (Gerring 2008, Seawright & Gerring 2008, Starke 2013). Several strategies were combined, with an aim of exploring a new empirical research area while addressing gaps in the literature regarding the consideration of complex transfer outcomes and to shed light on Germany's sender strategies in different contexts.

Narrowing down the "universe of recipient cases" and selecting the three cases was done in several steps. First, European countries were excluded from the pool of potential follower cases to be considered. This is because intra-European transfer is not only better explored already, but also often driven by a powerful and particular transfer mechanism – harmonization (Holzinger et al. 2008: 556). The inference of general conclusions from a European set of recipients to any other potential followers would likely have been limited. This is an issue of concern given the policy relevance of understanding bilateral policy transfer outside institutionalized negotiation settings and top-down processes. Another aspect taken into account is a strong lack of studies that do not focus on either EU member states or US states in particular and the OECD in general (Evans 2004b, Stone 2001: 19). As highlighted already, the invitation Marsh and Sharman extend to transfer researchers to look beyond the scope of similar and neighboring industrialized countries is gladly accepted (Marsh & Sharman 2009: 281).

In a second step, countries that are particularly targeted by German cooperation efforts in the renewable energy field were identified through interviews conducted in Germany and the presence of countries in different initiatives. The goal here was to identify "plausible" (Seawright & Gerring 2008: 306) or "most likely" cases (Gerring 2007: 232), insofar as German Energiewende leadership could likely materialize there. As far as developing countries are concerned, the identification of this group of particularly targeted countries was possible by taking into account host countries of International Climate Initiative (IKI) projects with an renewable energy focus, countries with which an energy partnership existed or was planned at the time of case selection, and members of the Renewables Club initiated by Germany in 2013. This group was validated by interviews with German ministerial officials (Interviews D4, D6, D7, D9). With industrialized countries, institutionalized transfer channels such as the Transatlantic Climate Bridge or bilateral fora are more rare and bilateral delegation visits or exchange between advocates serve as vectors for transfer. Here again, the view of interview partners was taken into account to identify particularly "targeted" countries.

Among these "desired followers", several cases stood out as not having adopted on a large scale the core policy instrument of Germany's approach to renewable energy promotion, feed-in tariffs. This is not to assume that every country in contact with Germany would automatically adopt its policies, but it is still an intriguing fact given the large number of countries and states that have adopted feed-in tariffs globally (REN21 Secretariat 2015, see also Ch.1). Exploring the reality and mechanisms of transfer to these recipients with very close ties to Germany but who have taken an alternative policy route is therefore highly promising in terms of developing transfer research further. The lack of "negative" cases has been deplored in policy analysis in general and in transfer and diffusion research in particular (Dobbin et al. 2007: 463, Evans 2010a: 159, Solingen 2012). The selection of "negative" or selective cases, where an outcome would have been very much possible – following Mahoney and Goertz' (2004) "Possibility Principle" – is therefore essential to further theory building. In particular, understanding these cases and particular domestic factors is important to explain variance in transfer outcomes, an insight large-n diffusion research cannot provide (Klingler-Vidra 2014: 18) These cases are also crucial from an empirical point of view, since they shed light on what Energiewende diffusion and transfer can mean beyond the commonly used metric of the global diffusion of feed-in tariffs. At the time of case selection, Morocco and South Africa were the only cases with whom all forms of cooperation in the energy sector with Germany were established, but where feed-in tariffs were not implemented. A similar conclusion can be drawn for the third case studied, California. Although it is of course a federal state and not a country, California, the eighth largest economy worldwide (Sisney & Garosi 2015), enjoys wide-reaching independence in the formulation of its energy policy (Elliott 2013, Litz 2008), putting it on par with the nation states selected. As outlined in Chapter 8, the ties between California and Germany in the energy field have traditionally been very close and at eye level, given their shared aim of leading in sustainable energy (Galiteva & Moss 2014). In California, feed-in tariffs were only implemented in small programs in some municipalities and in a very limited and strongly modified form at the state level. Of course, the Energiewende in the renewable energy sector – as shown in the three case studies – means more than feed-in tariffs, but the shared feature of not adopting this iconic policy instrument makes Morocco, South Africa, and California crucial cases to study what "Energiewende transfer" means and how (effectively) Germany exerts leadership in various settings.

To achieve this variance, a final concern in case selection was diversity with regard to classic determinants for transfer, in order to achieve "useful variation on [...] dimensions of theoretical interest" (Seawright & Gerring 2008: 296). The selection of South Africa, Morocco, and California, is akin to a "most different" systems design. In this case selection approach, the "researcher tries to identify cases where just one independent variable" – in this case, "German Energiewende leadership" – "as well as the dependent variable" – here, the non-adoption of the German model of feed-in tariff – "covary, and all other plausible independent variables show different values" (Seawright & Gerring 2008: 306). The intuition here is to see the "independent variable of interest", the German Energiewende and German outreach, play out in diverse contexts (Seawright & Gerring 2008: 300).

Choosing Morocco, South Africa and California as cases makes it possible to cover a broad range of transfer channels and potential causal relationships regarding transfer from Germany, from the most institutionalized channels to informal interaction. The cases are also heterogeneous with regard to their political systems, levels of economic development, geographic location, capacities, and accumulated regulatory tradition in the renewable energy field. Thereby, "noise" is reduced and the common denominator across the cases – Energiewende transfer and Germany's leadership – can be studied in a more focused way. Case selection is thus motivated by an aim to achieve "good variation on diffusion mechanisms instead of trying hopelessly to [...] squeeze [cases] acrobatically in [...] categories" (Gilardi 2012: 15).

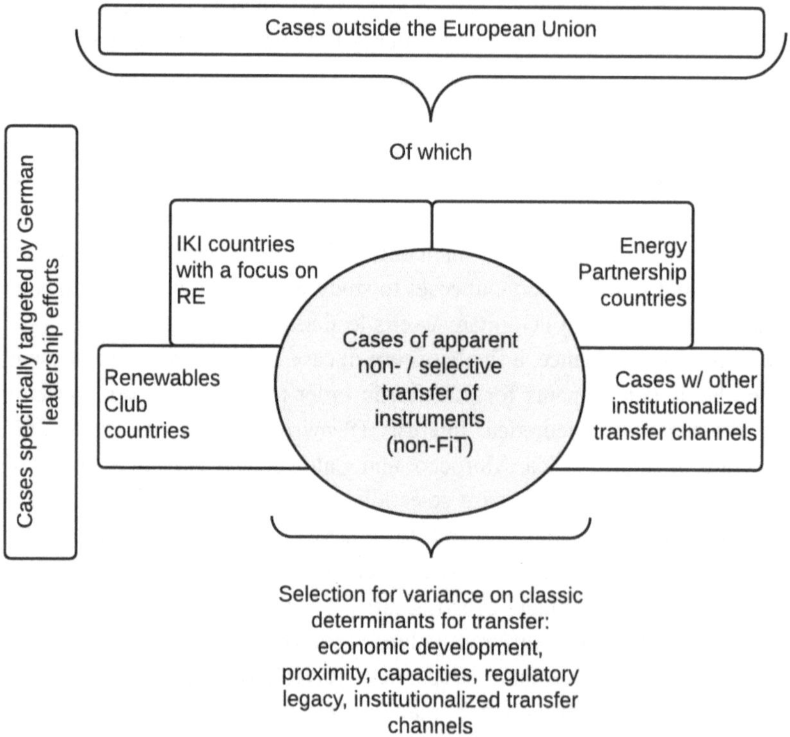

Figure 7: Case selection process and universe of cases. Note: Countries may belong to several groups

Figure 8 illustrates the variance between Morocco, South Africa, and California in terms of determinants deemed important for transfer based on expectations from the literature.

The choice of a "most different systems design" (Seawright & Gerring 2008: 298) was also motivated by a desire to contribute to furthering the expansion of transfer theory beyond its traditional focus on similar cases (Evans 2004b, Stone 2001: 19). Taking into account cases of different levels of economic development also makes it possible to critically assess to what extent policy transfer to "developing" and "developed" countries really differs. Two competing assumptions are generally made about the specificities of transfer between industrialized and developing countries. On the one hand, as underlined by

Brooks (2007: 706), limited resources and capacities in developing countries might make it particularly attractive to take other governments' experiences into account as a means to gather information before enacting policy change. On the other hand, transfer is expected to be more likely between similar countries, including in terms of economic development. The question is raised in particular given the novelty of renewable energy technologies and the potential for developing countries to leapfrog into sustainable energy systems (Haselip et al. 2011), blurring the categories of "developing" and "developed" countries with regard to energy system transformation.

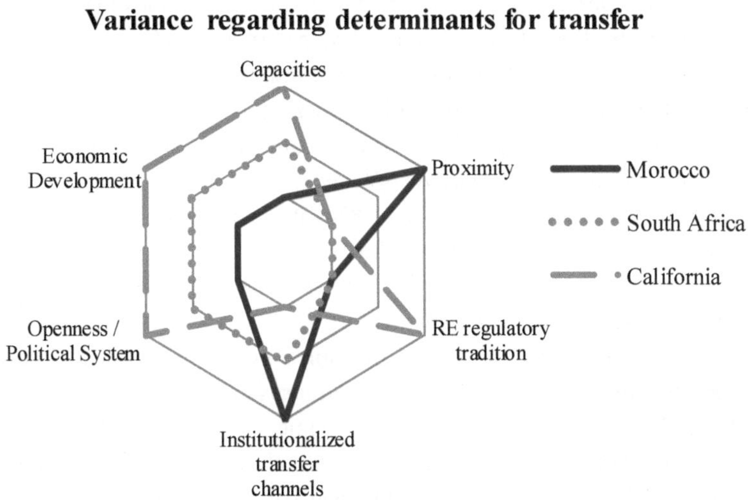

Figure 8: Overview of case variance on classic determinants for transfer.

4.2.3 *Process-tracing*

As Evans and Davies (1999, 381) point out, "proving" how transfer unfolds, i.e., establishing causal relationships is a challenging endeavor. Complex causalities can be expected to occur when policy evidence travels, and "Different causal pathways may lead to the same result (multiple causation), small chance events may cause major changes (nonlinear causation), and individual mechanisms may only produce changes when engaged in tandem with others (conjunctural causation)" (Marsh & Sharman 2009: 273). Rather than refraining from making any attempts to identify causal mechanisms, these should be accompanied by "the best and most honest estimate of the uncertainty of that [causal] inference" (King et al. 1994: 76). Process tracing is a well-adapted framework to address this issue, since it allows for the identification of a "repertoire of causal paths [...] and the conditions under which they occur" (George & Bennett 2005: 207), while providing enough information to put causal claims about transfer in context.

The basic principle of process tracing as a method is to study "the unfolding of events or situations over time" (Collier 2011: 824) through detailed description. As opposed to the "pattern-finding" approach of diffusion, process tracing "follows the spread of a policy or practice from one location to another" and "permits inspection of the role played by external models, and inquiry into why and how a concrete instance of learning or mimicry occurs" (Lee & Strang 2006: 886). Although it is based on "thick description" (Evans 2009b: 254), process tracing does not correspond to the accumulation of all available data, but is a focused and structured endeavor (Vennesson 2008: 235), targeted at "finding diagnostic evidence" (Collier 2011: 824) through different sources. It is hence not a way to collect and treat data, but to put it in context to reconstruct processes and events, with an aim to understand "what" happened, but also how and why processes unfolded as they did.

Process tracing is a method of choice in transfer studies (Lee & Strang 2006), to explore the "detailed twists and turns of policy development" (Bennett 1997: 225). Betsill and Corell, for example, employ process tracing to study NGO influence in international environmental negotiations, by "building a logical chain of evidence linking NGO transmission of information, actors' use/non-use of that information, and the effects/non-effects of that information" (Betsill & Corell 2001: 71). The authors hereby refer to very similar categories of transfer (consideration, use, effects) as the ones used in this book. By collecting and assembling data on "what NGOs were doing (activities and resources), how they tried to transmit knowledge and information to negotiators (access), whether delegates responded to that information, and whether those responses were consistent with the NGO position (goal attainment)", Betsill and Corell (2001: 72) also integrate senders – in this case NGOs – into the transfer processes traced.

Tracing processes eventually requires selecting which events that these processes lead up to should be considered. The selection of policy making processes to be considered more closely in each of the cases was done through empirical research. A background understanding of policy milestones and events was acquired prior to field research, but the identification of key processes in (renewable) energy policy according to stakeholders was also part of the objective of interviews conducted. The timelines presented in Chapters 6-8 situate the processes traced in each of the cases in their temporal context.

4.3 Data Collection and Treatment

4.3.1 Qualitative interviews

Case studies and process tracing are frameworks rather than methods to collect and process data. To fill the case studies with substance, this book relies heavily on qualitative interviews and coding, while triangulating findings through document and content analysis and the use of ranking exercises. For political science researchers in a positivist tradition and/or with a rational choice view on agency, little is to be gained from interviewing agents whose actions are determined by objectively identifiable interests (King et al. 1994). However, although "interviewing has flaws", it is, "on pragmatic grounds [...] often the only means to obtain particular kinds of information" (Rathbun 2008: 690). The type of information this book is interested in is generally not obtainable without engaging in conversations with stakeholders involved in policy processes. Interviewees can report their and others' use of information, express their views on a foreign policy model, and share perceptions on how it affected policy outcomes. Since traces of these thought processes are not generally present in public records, especially in relatively closed political systems, semi-structured interviews were strongly indicated as the main method for data collection. Interestingly, and likely for the practical reasons stressed by Rathbun (2008), the use of in-depth interviews for anything else but adding "a little color to otherwise stiff accounts" (Rathbun 2008: 685) is not generalized in transfer research (Mossberger & Wolman 2003: 435, Wolman & Page 2002; notable exceptions include Biedenkopf 2011, Crow 2012, Mossberger 2000, Ogden et al. 2003, Orenstein 2008, Weyland 2004c). By relying heavily on interviews that are triangulated with written sources, this book stands on the shoulders of larger-n studies that use techniques such as focused, structured comparison (Busch & Jörgens 2010), event-history analysis (Berry & Berry 1990) or "dyadic approaches" (Gilardi & Foglister 2008) to identify patterns of diffusion but are often ill adapted to identifying the mechanisms at play in particular instances of transfer (Howlett & Rayner 2008, Starke 2013: 565).

4.3.2 Interviewee selection and key interview data

Finding the evidence that a lesson has been drawn demands "excellent access to key informants in informal decision-making processes" (Evans 2009b: 249). Efforts were therefore concentrated in this book on gaining access to stakeholders in Germany, Morocco, South Africa, and California who were closely involved in policy making processes and in cooperation between Germany and recipients. Interview partners were selected for their direct involvement in (renewable) energy policy debates and policy making, with an aim to cover all main groups of actors present in the policy network (see sector affiliation in Figure 9). Approximately one-third to one-half of the interview partners in each case were identified prior to field research, through exploratory interviews and recommendations, as well as through their frequent appearance on lists of participants and as speakers at energy sector events such as industry conferences, public hearings, etc. The remaining interviewees were identified using the "snowball method" (Bleich & Pekkanen 2013: 87) based on referrals from other interview partners, until recurrent recommendations of the same names indicated saturation. A common theme across the cases was a very high rate of positive responses to interview requests, including among top officials, with fewer than five negative responses or contacts that were eventually not followed up by an interview for practical reasons in California and Morocco each. The rate of positive responses was equally very high in South Africa, with the single exception of ministry representatives other than those affiliated with the office implementing the renewables program. The high response rate and stakeholders' readiness to discuss the issue of policy transfer from Germany is in contrast to the lack of public references to policy transfer in all cases. This finding confirms the usefulness of qualitative interviews with agents directly involved in the policy process to identify transfer, in particular where policy outcomes are not indicative of convergence.

In total, 181 individual respondents were interviewed in 168 interview situations for the purpose of this book. In addition, 16 background conversations were led. In these background conversations, interlocutors were aware of the purpose of gathering information for the research project, but only specific questions were asked, feedback was sought, or initial findings from research were discussed for plausibility checks. Background conversations were not used as direct quotes in the cases, but rather helped establish knowledge to put findings in context. The main groups of actors covered and the total number of interviewees from each group across all interviews are presented in Figure 9. More details concerning interviewee affiliation and selection are contained in each of the case studies in Chapter 6-8 and in Chapter 5 for Germany.

Of the 181 interviewees, 15 were interviewed in Germany (plus ten background conversations), 59 in California (plus one background), 59 in South Africa (plus three background), and 48 in Morocco (plus two background). Interviews lasted 53 minutes on average and were digitally recorded, except where respondents explicitly asked for only handwritten notes to be taken.

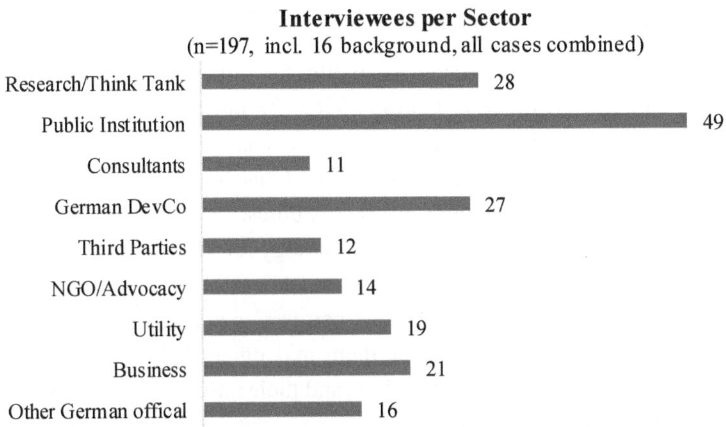

Figure 9: Number of interviewees per sector (all cases combined)

Interview partners were granted anonymity, based on the experience of numerous early interviews, especially in Morocco, but also in Germany, where interview partners insisted that their names not be cited. As Berry emphasizes, interviewees "are talking about their work, and, as such, justifying what they do. That's no small matter" (Berry 2002: 680). Interviewees in all cases provided deep insights into the politics of how ideas and knowledge are used, thereby confirming that transfer is not a value-neutral, mechanical process. These insights in some instances were also a critique of intra-institutional politics and decisions, increasing the need for interview partners to remain anonymous in order to be able to share inside information. This is all the more true as top-level officials and heads of institutions, departments, and units, accepted to be interviewed, but on the condition of anonymity. Energy policy is a contested field, rich with political and economic interests, and policy networks in the renewable energy space were small and closely knit, increasing the concern of interviewees not to be identified by name in some cases. Since the primary goal of empirical research in this book was to generate deep of insights into how transfer plays out in policy making, increasing the chance of getting this type of information by granting anonymity was privileged over disclosure of interviewees' names. General information regarding the sector affiliation of interview partners is given in each of the case chapters, while non-anonymized transcripts and interview notes remain with the author.

4.3.3 Questionnaire design and interviewing

In line with Strauss and Corbin's recommendation to start an enquiry by asking "What is going on here?" (Strauss & Corbin 1998: 77), one part of the interview questions was targeted at filling gaps in the understanding of policy making processes that published material often leaves open (Busch & Jörgens 2010: 66). With process tracing as the backbone of case studies, clearly identifying the order of events, actors involved, and the chronology of who learned about a certain piece of information at what time was crucial. A second layer of questions was then targeted at assessing interview partners' perceptions of the Energiewende as an example, on motives for them and others to use information related to it, and on German outreach activities and their impact. The sequence of questions went from general ones regarding policy events and facts to questions focusing on transfer from Germany and interaction with German transfer actors and questions of perception. Contrary to Yin's recommendation, the interviewer did not try to "appear genuinely naïve about the topic" (Yin 2009: 107), but built questions based on prior knowledge and also discussed findings from earlier interviews and other sources to triangulate views. The focus of interview questions was "not only on what happened, but also on how it happened" in order to "use process tracing to examine the reasons that actors give for their actions and behaviour" (Vennesson 2008: 233). The development of questionnaires was done on the basis of through publicly available material and background conversations in all cases. Acquiring prior knowledge was essential, since "Each case study investigator must understand the theoretical or policy issues because analytical judgments have to be made throughout the data collection process" (Yin 2009: 71).

Interview questionnaires, based on the guiding questions presented in Table 3, were adapted for each interview partner, and were tailored to the respondent's role, past positions, opinions expressed publicly, and the decision-making processes they were involved in. Questionnaires were followed as an orientation by the researcher, but were adapted to new information in the course of interviews if needed, with an aim to carry out interviews in a conversational style (Berry 2002: 679) and to follow leads into unchartered territory of the transfer processes traced. All questions were open-ended, i.e., no questions asking for standardized responses were posed. In line with the expectation that case study investigators are to deplete "their analytical energy" at the end of each day of field research and "create a rich dialogue with the evidence" (Yin 2009: 69), evidence was reviewed and initial codes were developed during field research. Examples of questionnaires for each case can be found in the Appendix (Tables 32-35). The categories in Table 3 were covered across all interviews in recipient states and were complemented with respondent-specific questions:

Table 3: Overview of questions addressed in interviews

Overarching Interview Questions – Adapted for each interviewee
▪ What were major (renewable) energy policy developments, debates, and milestones?
▪ Who are/were opponents and proponents to renewable energy development/ specific regulations?
▪ When did the respondent first become aware of the German energy transition example and/or enter into contact with German stakeholders regarding energy issues?
▪ How did the respondent's perception of the example and elements of it evolve over time? Why?
▪ What elements of the model, which policy elements were considered?
▪ Has the respondent used the German example in debates, and if so, at what point in time and to what end (concrete examples)?
▪ What were reactions from other members of the policy network to this use?
▪ Who else used the example (which elements and arguments) and how?
▪ (How/when) did the respondent interact with German stakeholders / transfer agents (frequency, concrete framework, person involved, role)?
▪ Were German stakeholders present at meetings, conferences, in policy making processes? When exactly and why, according to the respondent?
▪ How is German outreach perceived? What are the assumed driv-ers/agenda?
▪ Are there examples of how transferred knowledge and/or activities by German stakeholders changed the debate or were reflected in policy outcomes?
▪ Ranking: [What are objectives pursued through renewable energy in the case?]

The decision to use a semi-structured interview questionnaire was necessary due to the novelty of the subject studied, uncertainty as to what transfer processes were to be discovered, and in particular, what elements of the German energy transition example had been taken into account. However, it could be appropriate in future similar transfer studies (or in cases where a second round of interviews is feasible) to add a set of standardized questions. These could, for example, capture the perception of general success or failure of the German energy transition example on a scale. More standardized questions could also allow for the creation of in-depth, systematic network analysis and lead to an integration of quantitative and qualitative methods, i.e., the integration of diffusion and transfer research in larger research programs (Starke 2013: 562, see also discussion in Ch. 9).

4.3.4 Coding and processing interview data

The recorded interview material was transcribed and then coded in several iterations using the software MaxQDA®, as outlined in Figure 10. In parallel to conducting interviews, "memos" on coding categories were written in personal research journals for each of the case studies, informing the development of categories and analysis. Memos, an element that is also central to grounded theory, are short notes about categories and patterns that emerged in interviews. They are "narrated records" of a researcher's "analytical conversations with him/herself about the research data" (Lempert 2007: 248).

Initial codings were carried out on part of the transcripts (all transcripts in the case of Morocco, and about one-fifth of the interviews in the other cases). This was done as a preparation to refine the coding categories (Charmaz 2006: 11, Mayring 1983: 77). The decision was taken to refrain from "word by word" or "line-by-line" coding in this initial coding phase (Strauss & Corbin 1998: 57); instead, incident-to-incident coding of sentences and paragraphs was performed (Charmaz 2006: 53, Strauss & Corbin 1998: 120). In an effort to reduce complexity, "axial coding" Strauss & Corbin 1998: 123 reduced the number of codes in the second coding phase. Prior theoretical and empirical knowledge and the specific research interest informed the process, but codes were developed from the data in a bottom-up process (Kelle 2007: 201, Strauss & Corbin 1998: 48–52).

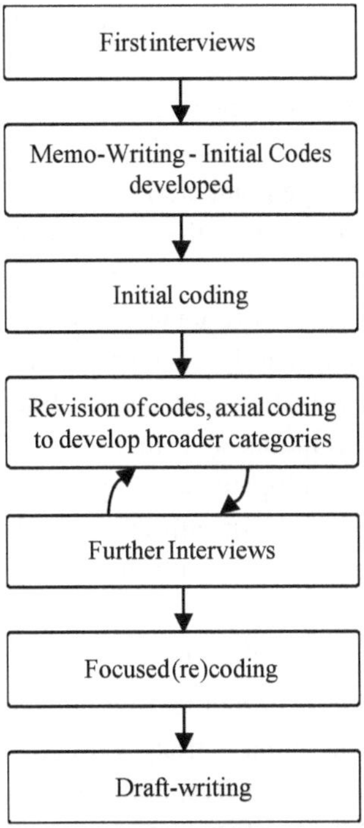

Figure 10: Overview of the coding process. Adapted from Charmaz (2006: 11)

Across the three cases, the final codes applied to the transcribed interview material fall into two categories, one case-specific and factual and one analytical and targeted at theory building. The first category of codes permitted to structure the wealth of interview data around events and processes, such as specific policy debates, legislative processes etc. Interviewee statements were hence labeled with the policy or topic (i.e., "smart inverter regulation", "Renewables Law 13-09") to which they referred to ensure that reports of transfer could be placed into the right context. The second, more analytical set of codes aimed at capturing elements such as motivations, perceptions, mechanisms of transfer, and strategic use of information or beliefs about policy instruments and personal and policy objectives. Text segments could be coded for several codes in parallel, such as an event-code, a use-related code, and a code referring to a theoretical concept or development. In total, more than 5,300 codings were generated and used for the identification of patterns, stories, processes, and trends, which are narrated in the cases in Chapters 6-8.

After due consideration, the decision was eventually taken not to represent coding outcomes in quantitative terms (i.e., number of respondents agreeing, percent of segments indicating a point etc.). This decision is motivated in part by the lack of standardized questions in the questionnaires, limiting comparability between interviewees. An even more important reason is that the "contextual grounds of meaning" are essential to understand motivations and processes and are lost when statements are merely counted (Mishler 1986: 5). Codes are an interpretation of material gathered, and are necessarily influenced by the researcher's frames and assumptions (Mishler 1986: 4). While coding still appeared as the best strategy to mitigate biases, qualitative interview data, although coded, can hardly be rigorously quantified. Interview material and verbatim citations are used throughout the text to illustrate and corroborate accounts.

4.3.5 Exploring objectives: Ranking exercises

Similar policy objectives are expected to be a relevant determinant for transfer. Interviews in Morocco and South Africa were therefore complemented with ranking exercises of interviewees' perceptions of objectives of renewable energy policies in the countries.[26] In California, due to a greater amount of secondary literature and public information about objectives, questions regarding policy objectives for renewables were integrated to the interviews.

The ranking exercises were modeled on the approach used by Joas et al. (2014, 2016) in their study of objectives of the German energy transition. Types of "card sorting" or "q-methods" have been employed in research fields such as information architecture, psychology and the social sciences (Müller & Kals 2004). Interview partners were asked to rank small paper cards at the end of the interview, to respond to the question of "What are the objectives Morocco/South Africa pursues through its renewable energy strategy?" Interviewees were asked not to express what these objectives should be from their personal point of view, but what they perceived to be the actual goals pursued through the new Moroccan energy strategy and the South African renewables program, respectively. The

26 For Morocco, see Steinbacher 2015 for a more detailed discussion of ranking results.

objectives proposed were the following: energy independency, security of supply, job creation, creating an industry, electricity exports (Morocco), societal project, climate protection, local environmental protection (South Africa), decentralization, increasing competition, avoiding/replacing nuclear energy, expectation from the international community, society's expectation (South Africa), country leadership, attracting investors, and affordable electricity prices (South Africa).

Interviewees were free to exclude one or several cards and ask for objectives to be added. Full results for rankings in Morocco and South Africa are shown in additional online material related to this book[27]. The selection of objectives for the Moroccan case draws upon official statements such as a speech held by Morocco's Minister for Energy, Abdelkader Amara, on the objectives of Morocco's energy transition in Casablanca in November 2013 (MEMEE 2013a). For South Africa, to ensure comparability, largely the same objectives were proposed and interview partners were encouraged to suggest objectives to be added.

Some caveats need to be stressed as far as the analysis of ranking outcomes is concerned. Interviewees were free to rank several objectives at the same level of importance, which could have warranted a recalculation of ranks. For example, when three cards were ranked as top priorities, each of them could have been attributed (1+2+3)/3 = 2 points as an average rank. The subsequent card, ranked as a second-level priority by the respondent, would then have received rank 4. The choice has been made not to use average ranks but to consider multiple cards ranked as top priorities as equally important as a single card ranked first in a strictly hierarchical ranking. Results provided in the case study sections therefore need to be considered within the context of the interviews.

The ranking frequently sparked lively discussions and often revealed important insights into the change of policy objectives over time. They therefore were crucial tools for understanding the broader energy transition within which policy transfer occurred. Dedicating more time exclusively to this point and carrying out the same type of ranking exercises on a large scale and across more countries is a promising avenue for further research.

27 To access the book's appendix, please follow the URL given in the imprint or visit www.springer.com and search for the author's name.

4.3.6 Triangulation: Document analysis and observation

While coded qualitative interview transcripts constitute a main source of data used, other sources were used to corroborate findings and put them into perspective, in an effort to "triangulate" data and thereby solidify results (Yin 2009: 103, Checkel 2008: 119). This was mainly achieved through the qualitative analysis of documents and reports produced by institutions covered through interviews, policy documents, and press statements, media articles, meeting reports and minutes. Secondary literature was used where possible, e.g., to clarify policy processes, but was rarely explicit concerning the specific question of transfer. The extent to which written documents could be taken into account varied from case to case, with California providing the most comprehensive set of minutes from meetings, regulatory proceedings, public statements, public comments to consultations, and press releases allowing for a tracing of the use of the German example in the debate. Data availability in South Africa was far less comprehensive, but secondary literature, media articles, and public statements – in addition to two major renewable energy conferences attended by the author – provided context for the interview data. Morocco's political system does not offer the same degree of transparency and open debate. This lack of written sources was compensated to the extent possible by the high number of interviews compared to the size of the energy policy network in Morocco and the opportunity to observe cooperation while being hosted by the Secretariat of the Moroccan-German Energy Partnership.

Two approaches to the selection and search of documents were used. Prior to field research, cursory, exploratory searches were undertaken to grasp the field, identify positions of different groups regarding sustainable energy policy and get a sense of the intensity and framework of interaction with German counterparts. From one document to the other, a snowball technique was used to explore policy debates and at least attempt to "exhaust [...] secondary sources and publicly available primary sources before beginning" (Rathbun 2008: 695) and thus use interviews to gain "valuable (and often invaluable) information" (Rathbun 2008: 695) only they can provide. A second approach to document selection, used during field research and during the drafting phase of case reports, was "purposeful sampling" (Morse 2007: 237) or strategic search for evidence that could confirm or contradict interviewee statements, in particular where a particular position appeared to be isolated.

The reliance on written documents was important regarding Germany and the assessment of its general leadership efforts. In addition to interviews carried out in Germany, 101 parliamentary debates were coded and analyzed[28] to assess the presence of leadership aims related to the Energiewende. The debates took place between the beginning of the 18th legislative term (November 2013) and the end of field research for this book (May 2015). The documents were scanned for terms such as "Energiewende", "Energie" (energy), and "Erneuerbare" (renewables), and in a second step for connections to terms like "Vorreiter" (front-runner, given the lack of a viable alternative term in German), "Modell" (model), "Pionier" pioneer), and "Vorbild" (exemplary/model). Speeches identified as relevant were then reviewed and relevant text segments pertaining to the international dimension of the Energiewende and an aim for leadership were identified. The 150 relevant text passages are presented in online material related to this book, together with speakers' names and party affiliation. Insights from this analysis are presented section 5.4.1. Publicly available, written sources and information obtained through requests to ministries' information services were used for the background chapter (Chapter 5). Table 4 summarizes which methods were used for data collection regarding main intervening variables of importance for Hypotheses I-III.

28 Part of this analysis, based on a subset of the debates used for this thesis, was presented in Steinbacher and Pahle (2015: 9–10)

Table 4: Methods and type of data used to operationalize key intervening variables

Intervening Variables	Methods and sources used to operationalize variable
Policy objectives in recipient context	Policy objectives for renewable energy strategies ranked by interviewees in ranking exercises in Morocco and South Africa and objectives stated in interviews by stakeholders in California. Triangulated with objectives present in policy documents, statements made at events attended.
Relative power of members of transfer network	Contacts and influence perceived by interviewees; presence in debates; analysis of information formal decision-making processes and institutional competences/prerogatives/mission, organigrams and publicly available information about different decision-making processes (minutes, proceedings).
German leadership	Document analysis of descriptions for programs and projects sponsored by Germany available online or in writing and obtained from agencies, incl. content, budgets, success indicators, sponsoring and implementing agency, local partners, duration, and evaluations. Analysis of reports of training sessions, delegation visits and study tours organized, including lists of participants, agendas, preparatory documents made available by organizing agencies. Analysis of data regarding loans and official development assistance (ODA) available from the OECD and German ministries. Analysis of *Bundestag* debates.
Communication of Energiewende-related knowledge	Document analysis of information material produced by, sponsored by or communicated by German agents, incl. studies produced upon demand of partners in recipient countries (e.g., renewable energy potential studies, reports regarding necessary policy adjustments, feasibility studios, energy scenarios). Interviewee statements regarding the context in which these studies were produced and how they were used or perceived. Interviewee statements regarding the appearance of German agents at conferences, participation in meetings, and information communicated in this context. Triangulation with lists of participants or media reports of events.
Availability of information	Self-assessment by interview partners of knowledge about the Energiewende; interview partners reporting information searches; awareness of German example; contact with German policy promotion activities (participation).
Perception of the model (feasibility, prominence)	Stated views in interviews on the German model or on information promoted by German transfer agents; positions in minutes and policy documents; public statements.

4.3.7 Reflection on research design and biases

As Charmaz (2006: 15) points out, "Just as the methods we choose influence what we see, what we bring to the study also influences what we can see". Awareness and reflection on how personal views on a topic as widely discussed as the Energiewende can influence study design, questions asked, and the interpretation of material collected cannot eliminate subjectivity (Checkel 2008: 124). The researcher's generally positive view of the German Energiewende creates a risk of looking at policy transfer related to the model as desirable *per se*, potentially obscuring critical aspects of transfer. The researcher's affiliation with GIZ in Morocco could be seen as another potential source of biases. However, GIZ counterparts were in no way involved in the design of the research, questionnaires, choice of interview partners, interview scheduling, or the analysis of findings. The researcher did not present herself to interview partners as being affiliated with GIZ to avoid any confusion on whose behalf the interviews were conducted.

The most important strategy to mitigate some of the biases every qualitative (and quantitative) researcher may be subject to was to transcribe and code interviews as closely as possible to the data. By working with full transcripts, whenever possible, and by developing codes that fit the data rather than forcing data into pre-defined codes, the researcher had to reflect on the data more intensively than if interviews had only been used as background matter (Rathbun 2008: 685). A potential concern regarding this research was linked to reactivity (Mishler 1986: 114). Since the researcher was affiliated with a German university throughout the research presented here, one could imagine that respondents may have been led (for reasons of social desirability, politeness, or concern for working relationships with German counterparts) to report positively on interaction with German counterparts. Two arguments can be put forward to relativize this concern. First, the researcher made an effort to clarify her origin (Austrian) whenever interviews partners used phrases like "you in Germany", "for Germans like you" etc. But the most important indicator that relativizes the impact of this bias were the nuanced views on the Energiewende's implementation, concerns regarding its transferability and a sometimes skeptical view on German outreach activities shared with the author. German leadership was critically discussed even in a case like Morocco, where the sheer amount spent by Germany on development cooperation in the energy sector could have led interview partners to feeling obliged to provide positive answers, strongly relativizing the likely impact of this particular bias.

Validity and reliability are concerns in much of qualitative, and especially interview-based, research, due to the uniqueness of each interaction between the researcher and a respondent (Berry 2002). Several strategies were adopted to mitigate these justified concerns. As proposed by Berry (2002: 681), the number of interviewees was maximized within what could be achieved in the framework of each research stay and diversity in terms of backgrounds (industry/public sector/NGOs, etc.) was aimed for. Acquiring case-knowledge prior to field research made a more critical view on interview statements possible than if the interviews had been the first source of information. Furthermore, interviewees were asked to contrast their position, view, and strategy with other stakeholders' approaches and were asked to provide concrete examples to underpin their statements.

Methodological choices, overarching results, and avenues for future transfer research will be discussed in more detail in Chapter 9. The next chapter presents Germany's unilateral and active leadership related to the Energiewende (Chapter 5), and recipient case studies are then presented in Chapters 6 (Morocco), 7 (South Africa), and 8 (California).

5 Energiewende Leadership: Internal and External Dimensions

5.1 The Energiewende: A Longstanding Energy System Transformation

5.1.1 Chapter overview

On 23 August 2015, renewable energy generation in Germany hit a new record: between 1 pm and 2 pm, more than 83% of electricity consumed in Germany came from wind, solar, hydro and biomass[29] (Agora Energiewende 2016a: 10). Over the course of 2015, the share of renewables in the German electricity mix exceeded 30% for the first time (Agora Energiewende 2016a: 8). The Energiewende, with its threefold aim of decarbonization, reducing primary energy consumption, and phasing out nuclear power is well under way and is one of the world's most prominent and controversially discussed plans of transforming an energy system (Morgan & Weischer 2013). From its very first steps, Germany's energy transition had an international ambition, especially with regard to facilitating the deployment of renewable energy in the electricity sector. Making renewable energy technologies accessible to other countries by introducing them at scale in Germany, and thereby contributing to global climate protection efforts, has been a core theme of Germany's Energiewende since its beginnings (Röhrkasten 2015: 165; Bundesregierung 1988).

The history of Germany's energy transition dates back to well before Chancellor Angela Merkel's decision in spring 2011 to reduce the operating times of the country's nuclear power plants, after initially prolonging them in late 2010. The term "Energiewende" is sometimes associated with this specific decision (Dehmer 2013) and international commentators readily use the Germanism *Energiewende* (The Economist 2012) to refer to it. The Energiewende's roots lie in anti-nuclear protests in Germany since the 1970s and in the pioneering efforts of

29 This record might have been broken on 8 May 2016, with more than 95% of electricity demand at 11am being covered by renewables in Germany. Final generation data was yet unavailable at the time of writing to confirm this new record. See Enkhardt (2016).

K. Steinbacher, *Exporting the Energiewende*, Energiepolitik und Klimaschutz. Energy Policy and Climate Protection, https://doi.org/10.1007/978-3-658-22496-7_5

communities in Germany to increase the share of renewables since the 1980s (section 5.1.2 and 5.1.3). The term "Energiewende" is used in a broad sense in this book, encompassing the transition since its beginning, with a focus on the period since 2000, when both nuclear phase-out and more ambitious renewable energy policies were introduced at the national level. Although "Energiewende" translates rather as "energy turnaround" than as "transition", it is impossible to identify a single turning point in German energy policy. The term was introduced through a report entitled *"Energiewende – Wachstum und Wohlstand ohne Erdöl und Uran"* (Energiewende – Growth and Prosperity without Petroleum and Uran) published in 1982 by the research institute *Öko-Institut Freiburg* (Krause et al. 1980). From the beginning, for its defenders, the Energiewende meant ending nuclear energy in Germany while also reducing emissions as well as cutting back primary energy consumption.

Germany's policy model and international efforts, i.e., its unilateral and active leadership, are intrinsically interwoven. This chapter first describes the dimension of unilateral leadership, that is to say, the Energiewende as it is implemented in Germany (section 5.1). Taking into account Germany's domestic policy model is relevant for several reasons: the Energiewende is a pool from which German transfer agents abroad can draw, it is a driver behind leadership efforts, and an example that triggers interest and leads to lesson drawing. The chapter also addresses the particular challenges of dealing with a transfer object as complex and evolving as the Energiewende. The second aspect of leadership taken into account is active leadership. To increase the chance of spillover effects from the Energiewende to other countries, Germany has been highly active in international renewable energy governance (Röhrkasten 2015) and has created a range of initiatives and programs to promote the uptake of renewable energy abroad (see also Steinbacher & Pahle 2015: 18–21, and online material related to this book[30]). These active leadership efforts are described in section 5.2, while concrete projects are described in each of the case studies in Chapters 6-8.

In addition to qualitative document analysis of primary and secondary literature, this chapter draws from 15 interviews and 9 background conversations with officials and experts involved in shaping Germany's external Energiewende

30 To access the book's appendix, please follow the URL given in the imprint or visit www.springer.com and search for the author's name.

strategy, including heads of units from governmental departments involved in international Energiewende activities.

In addition, the understanding presented in this chapter draws from numerous informal conversations with staff involved in international energy cooperation activities within GIZ and ministries. Interviews were carried out in two series, one in the fall of 2013 and one in the spring and summer of 2015, in order to assess potential changes in outreach activities. Figure 11 shows the number of interviewees per sector. The following organizations were covered through interviews in Germany: Agora Energiewende; BMUB; BMWi; BMZ; Energy Watch Group; Federal Foreign Office; GIZ; IASS; Konrad-Adenauer-Stiftung; RENAC.

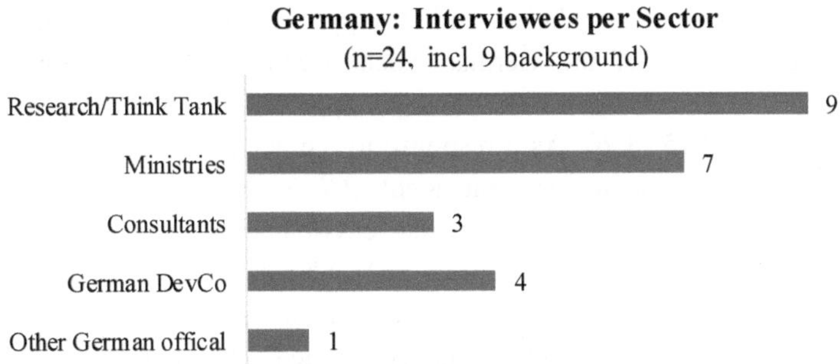

Figure 11: Number of interviewees per sector (Germany)

5.1.2 *Roots of the Energiewende: Opposition to nuclear energy*

The genesis of Germany's Energiewende is inconceivable without taking into account two closely interlinked factors: the long history of the German anti-nuclear movement as well as the involvement of communities, villages, and cities in promoting renewables, which have led to the strong bottom-up characteristics of the transition. Unlike in other countries, there has never been enthusiastic popular support for nuclear energy in Germany (Jahn & Korolczuk 2012, Schreurs 2013: 88; for a comprehensive history of environmentalism in Germany, see Uekötter 2014). From the beginnings of the country's nuclear program, communities in the immediate vicinity of planned nuclear plants opposed their construction. Over the course of the 1970s, protests, some of them violent in nature, took place against nuclear power plants in Breisbach, Whyl, Brokdorf, and Grohnde as well as in big German cities, drawing together hundreds of thousands of participants (Hake et al. 2015: 535, Jahn 1992, Schreurs 2013). Despite growing public opposition to the technology, the nuclear energy program was publicly subsidized and supported by all political parties (Hake et al. 2015: 535). Out of the public movement against nuclear power in Germany emerged the German Green party (*Die Grünen,* later: *Bündnis 90/Die Grünen)*, which entered the *Bundestag* for the first time in 1983. This marks a turning point for the green movement in Germany and the institutionalization of calls for nuclear-phase out (Jahn 1997, Schreurs & Papadakis 2007: 101-103). Just prior to the Chernobyl nuclear disaster in 1986, half of the German population already opposed nuclear energy and only 10% supported it after the accident (Hake et al. 2015: 536, Jahn 1992: 396). Chernobyl, perceived as an immediate and considerable threat to the health of the German population and heavily mediatized in Germany, also changed the position of the German Social-Democratic Party (*Sozialdemokratische Partei Deutschlands,* SPD) toward supporting a gradual nuclear phase-out (Hake et al. 2015: 536). As a response to the accident, and to demonstrate political action, Chancellor Helmut Kohl (*Christlich-Demokratische Union,* CDU, 1982-1998) established the German environment ministry in 1986, "almost overnight" (Weidner & Mez 2008: 362). Shortly after Chernobyl, another driving motive behind the Energiewende entered the political agenda: in 1987, Chancellor Kohl recognized climate change as one of the greatest threats to the global environment and asked a parliamentary enquiry commission, the "Enquete Commission on Preventive Measures to Protect the Earth's Atmosphere" (*Enquete-Kommission zum Thema "Vorsorge zum Schutz der Erdatmosphäre"*), to assess by how much CO_2 emissions in Germany could be reduced (Lauber & Mez 2004: 600, Weidner & Mez 2008: 362). In 1990, on the basis of the commission's final report, the German government adopted a goal of reducing CO_2 emissions in West Germany by 25% by 2005 compared to 1987 levels (Weidner & Mez 2008: 363).

When the Greens and SPD formed a coalition government for the first time in 1998 under the leadership of Chancellor Gerhard Schröder (SPD, 1998-2005), the goal of nuclear phase-out figured in the coalition agreement (SPD & Bündnis 90/Die Grünen 1998). In 2000, a nuclear consensus was negotiated with owners of nuclear assets in Germany and a gradual phase-out until approximately 2022 was decided (Jahn & Korolczuk 2012: 160). Understanding the roots of the Germany energy transition as a consequence of a very longstanding movement against nuclear puts the 2011 Energiewende decisions by Chancellor Merkel (CDU, 2005-present) in context. It also explains why a broader understanding of the Energiewende is used in this book.

5.1.3 Beginnings of renewable energy promotion in Germany

From 1974 onwards, in response to the oil crisis, Germany started to invest in research and development of renewable energies, but budgets were reduced after the entry into office of a conservative-liberal government with CDU/CSU-FDP in 1982 (Hirschl 2008: 128, Lauber & Mez 2004: 599–600). An agreement, which had been in place between electricity utilities and government since 1979 (Lauber & Mez 2004: 600), allowed producers of renewable electricity to feed in into the grid and be compensated at the level of "avoided costs" (i.e., what utilities have to pay for electricity from other sources, not taking into account externalities). The calculation of avoided costs in Germany was less favorable for renewable energy producers than in a similar system in place at the time in the United States. There, and in California in particular, a favorable interpretation of the Public Utility Regulatory Policy Act (PURPA) established in 1978 led to an early boom in renewable energies (Hirschl 2008: 129, Jacobs & Mez 2012: 259).

Reference was made to the California example in several debates in the German Bundestag regarding the design of a renewable energy feed-in law prior to 1990 (see extracts in online material related to this book[31]). While members of the Green party referred to California and the PURPA as a positive example, former research minister Riesenhuber stated that the increase in cogeneration and wind power in California came at the expense of "extraordinarily high public subsidies" and that he was opposed to "permanent subsidies that block markets instead of opening them"[32] (Bundestag 1989: 9470)

To respond to growing calls for a more effective feed-in law, and following a first unsuccessful attempt by two conservative parliamentarians to propose

31 To access the book's appendix, please follow the URL given in the imprint or visit www.springer.com and search for the author's name.

32 "Ich möchte auf das Beispiel von Dänemark und von Kalifornien — Sie sprachen von den Vereinigten Staaten — hinweisen. Bekannt ist, daß mit außerordentlich hohen staatlichen Subventionen — Steuererleicherungen und Direktzuschüssen — dort Windenergie für die Stromerzeugung in den Markt gebracht worden ist. [...] Ich halte nichts von Dauersubventionen, die die Märkte versperren und nicht öffnen."

such a law in 1988, a small market-creation program of 1,000 solar rooftops, and 100 MW and later 250 MW of wind energy was launched in 1989 (Hirschl 2008: 130, Lauber & Pesendorfer 2004: 138, Lauber & Mez 2004: 601). Neither this program nor the earlier feed-in agreement with utilities led to a noteworthy increase in renewable energy capacities over the course of the 1980s (Lauber & Mez 2004: 600). Even more widespread anti-nuclear sentiments since Chernobyl, as well as a growing awareness of climate change, reinforced interest in alternative sources of energy in the late 1980s.

In 1989, an informal coalition of Green MPs and conservative members of the *Bundestag*, who were sensitive to the demands of the first German wind energy technology firms and the potential of renewable energy for farmers, pushed a legislative proposal for an electricity feed-in law (Hirschl 2008: 131). Following an unsuccessful last attempt to reach a voluntary agreement with energy utilities, the electricity feed-in law (*Stromeinspeisungsgesetz*, StrEG) was adopted in December 1990. Utilities only weakly opposed its adoption, amidst the tumultuous period of the German reunification, and likely because the significance of the proposal was underestimated. This "accidentally adopted" law was going to be a milestone in the making of Germany's energy transition (Jacobs 2014). In combination with the 1,000 solar rooftops program and the 100/250 MW wind incentive program, as well as local programs, the StrEG led wind energy electricity production to be multiplied by 78 and solar electricity production to be multiplied by 42 in Germany between 1990 and 1999, albeit from low levels (Lauber & Jacobsson 2016: 149). In absolute numbers and in terms of the share of renewables in the German electricity mix, the success of the law was however limited, especially with regard to the still very expensive solar technologies. This is mainly due to the insufficient incentives the law provided for renewable energy technologies, compensating producers with up to 90% of the retail rates paid in a utility's service area (Hirschl 2008: 181). Such tariffs would appear generous today, but were far from covering the cost of installation and operation for most plants at the time, in particular in the solar PV sector.

Change toward a more effective framework for renewable energy promotion was to come from below: the pioneering actions of cities and towns across Germany during the 1990s were decisive for the further development of the Energiewende. From the beginning of the 1990s, neighborhoods, villages and cities started to make use of a rule from the 1989 federal electricity tariff act. The provision allowed utilities to go beyond the StrEG and instead compensate renewable energy producers at higher rates, allowing them to actually recover their costs (Lauber & Mez 2004: 604). One of the most prominent examples for such a municipal frontrunner was Aachen, a medium-sized city close to the Dutch and Belgian borders. Following pressure from citizens, organized since 1986 in the Solar Energy Promotion Association (*Solarenergie-Förderverein*, SFV[33]), the municipal council of Aachen decided in 1992 to oblige the municipal utility to introduce a cost-covering feed-in tariff for renewables, which entered into force in 1995 following approval by the regional authorities (Jung 2015). Experience with the so-called "Aachen Model" was shared with other cities and villages across Germany and even abroad (Fabeck 2006). Interview partners in California were aware of the example and mentioned it spontaneously (C27, C49). International benchmarking reports also refer to the crucial, pioneering efforts of German municipalities in overcoming a "value-based" with a "cost-covering" approach to renewable energy promotion (Couture et al. 2010: 9). In other pioneering towns, like Freising and Hammelburg in Bavaria, citizens also agreed on the basic "cost-covering" model (*kostendeckende Vergütung*) that was going to lay the ground for the effective 2000 EEG, (Fabeck 2006). In the Aachen model, electricity consumers pay a surcharge on their electricity bills to finance the compensation of renewable energy producers at a level allowing the latter to cover their cost. This meant that, for solar PV, a stunning 2 DM (about 3$ at the time) per kWh were paid to the first installation operating in Aachen in 1995 (Jung 2015: 8). At real values, the feed-in tariff paid to PV plants in Germany in 2015 is 15 times less (9 euro cents on average), pointing to the tremendous decrease in the cost of technology and the learning that occurred since the early days of the German energy transition (Agora Energiewende 2015a: 19).

Pioneers like Aachen also leveled the path by developing frameworks regarding the ownership, compensation, insurance, power-purchasing agreements, and integration of renewable energy (Fabeck 2006, Jung 2015, Lauber & Mez 2004: 605). The pioneering experience of these villages and cities was "upload-

33 SFV still operates in 2016, 30 years after its inception. The association repeatedly raised concerns over reforms of the EEG, in particular the shift from a feed-in tariff to an auction-based model and "target corridos" for how much capacity could be added yearly; these departures from the initial Aachen-inspired model were seen as putting into question the association's overarching goal of a 100% renewable electricity system (Jung 2015).

ed" to the national level and translated to Germany's arguably most important Energiewende law, the EEG.

5.1.4 *The EEG: Germany's iconic renewable energy law*

The 2000 EEG fundamentally changed the logic of compensation for producers of renewable electricity in Germany at the national level, leading to an unexpectedly rapid, fundamental shift toward renewables and decentralized electricity generation. While prior developments were essential in leading up to the EEG, the adoption of the law very much marks the beginning of the acceleration of the Energiewende at the national level (for more comprehensive accounts of the EEG's genesis, see Bechberger 2000, Hake et al. 2015, Hirschl 2008, Lauber & Mez 2004; Lauber & Pesendorfer 2004). The EEG fundamentally changed the logic at a national level of how renewable energy producers are compensated for the electricity they generate. Following the cost-recovery approach pioneering cities had introduced, the EEG ensures producers receive a fixed tariff for each kWh of renewable electricity produced, over a 20-year period. This feed-in tariff (*Einspeisevergütung*)[34] is calculated to cover the cost of building and operating a renewable energy installation of a given size and technology at a given location, plus a reasonable rate of return, and digresses for new installations along with declines in technology and other cost[35]. Guaranteed, cost-covering compensation is only but one feature of the German feed-in tariff. As the StrEG, the EEG guarantees the purchase of any amount of electricity produced by renewable energy plants, guarantees grid connection, and priority access to the market, meaning

34 Particular policy design features traditionally characterize German feed-in tariffs: the absence of a limit on how much renewable energy capacity can be added per year (no program cap), tariffs calculated to ensure cost recovery, differentiation depending technology and location, long guaranteed payment periods of 20 years, priority access for renewables to the grid, and priority sale (dispatch) on the market have defined the "German way" of promoting renewables (Agora Energiewende 2015b, Lauber and Jacobsson 2015). With the 2014 EEG reform in particular, the feed-in tariff system underwent changes such as adjustments depending on the amount of capacity added ("breathing caps" / "atmender Deckel") and the obligation for owners of larger installations to market their electricity directly (BMWi 2015c) Further changes are to be implemented from 2016 onwards, in particular the shift toward an auction-based system for larger wind and solar plants (Appunn 2016, BMWi 2015a, BMWi 2016f).

35 Tariff digression lagged behind decreases in technology prices at the end oft he 2000s and the beginning oft he 2010, leading to a virtual explosion of PV capacity in Germany.

that renewable energy has to be dispatched before all other sources of electricity by system operators (Bundestag 2000). The importance of combining these elements in the German feed-in tariff as the basis for the outstanding boom in renewables cannot be overstated (Jacobs 2014).

Albeit being particularly prominent, Germany's feed-in tariff law was not the first such legislation globally nor in Europe (Jacobs & Mez 2012: 260–261). In the United States, California's interpretation of the 1978 PURPA Act, led to an unexpected increase in requests from renewable energy operators to be connected to the grid. The California Public Utilities Commission (CPUC) asked utilities to offer so-called "standard offer contracts" to qualifying facilities (QFs), to ease the administrative burden for the often small-scale renewables plants. Starting 1982, standard offer contracts were signed between renewable energy producers and utilities (Jacobs 2014: 761). One of contract templates, implemented from 1983, linked prices paid to renewable energy producers to the expected increase of oil and gas prices and guaranteed prices for a 15-30 year period, creating attractive framework conditions, especially for wind power operators (Jacobs 2014: 761). In the two years following the introduction of standard offer contracts, Californian utilities were confronted with an unexpectedly high amount of new capacity requesting connection. Responding to backlash from the utilities, the CPUC revised its interpretation of PURPA only three years later, effectively ending the rapid increase in renewables in the state (Jacobs 2014: 762). Prior to the adoption of the StrEG in Germany, Portugal and Denmark introduced feed-in tariffs. These were based on the avoided cost from oil-based electricity generation in Portugal and on a share of retail tariffs in Denmark (Jacobs & Mez 2012: 260–261, Jacobs 2014: 763). In particular the positive experience of the Danish tariffs for wind energy development was taken into account in Germany as well (Jacobs & Mez 2012: 262). Spain introduced a feed-in tariff in 1997 with higher tariffs than in the German StrEG, but none of these earlier laws led to a comparable increase in renewable capacity as the German EEG and its unique mix of generous compensation, absence of caps and security for investors (Jacobs 2014: 765–766).

The introduction of the EEG was accompanied by Germany's "100,000 solar rooftops" program, in which homeowners had access to low-interest loans with long payback times through KfW (BMU 2003). The program ran from 1999 to 2003, and ensured the financing of 300 MW of PV capacity in small-scale installations. At the end of the program, an amendment to the EEG in 2004 further improved support for small PV and other small-scale renewables. This led to a tremendous increase in renewables, and in particular solar PV, in the second half of the 2000s and the beginnings of the 2010s. PV capacities in Germany increased by double-digit percentages annually, as shown in Table 5, while other renewable energies experienced a more linear increase.

Table 5: Annual increase in PV capacity in Germany, 2005 to 2015. Data source: BMWi (2016g)

Years	Annual Increase	Years	Annual Increase
2005 to 2006	41%	2010 to 2011	42 %
2006 to 2007	44 %	2011 to 2012	30 %
2007 to 2008	47 %	2012 to 2013	10 %
2008 to 2009	73 %	2013 to 2014	5 %
2009 to 2010	70 %	2014 to 2015	4 %

Although targets set in the preambles of the 2004 and 2008 EEG amendments were not binding, their increase from 20% renewable energy by 2020 (2004 EEG) to 30% in 2020 (2008 EEG) is indicative of a speeding up of the transition in the 2000s. Following the "solar amendment" in 2004, the next version of the EEG in 2008 was already developed in the framework of growing pressure from parts of the conservative and liberal members of parliament (CDU-CSU and FDP) to accelerate the digression of tariffs and reign in renewable energy expansion (Lauber & Jacobsson 2016: 152). Yearly PV capacity additions have slowed down under the impact of 2012 and 2014 amendments to the EEG, a development that has been harshly criticized (Bündnis 90/Die Grünen 2016, Fell & Ehring 2014), also by pointing the role of the EEG as an international model that needs to be safeguarded (Bundesverband Erneuerbare Energien e.V. 2015: 2).

Whereas many of the provisions that contribute to or frame Germany's energy system transformation – the EEG, but also efficiency measures, climate policies, and emissions trading within the EU emissions trading scheme – had been in place for years, 2010 and 2011 saw a wider debate about the direction Germany's energy policy should take and popularized the term Energiewende. Figure 12 illustrates the policy history and main milestones of renewable energy legislation in Germany. Rather than constituting a turning point, the 2010 Energy Concept and the 2011 decision to go back to a nuclear phase-out by the 2020s, are points of consolidation of a path Germany had entered two decades earlier.

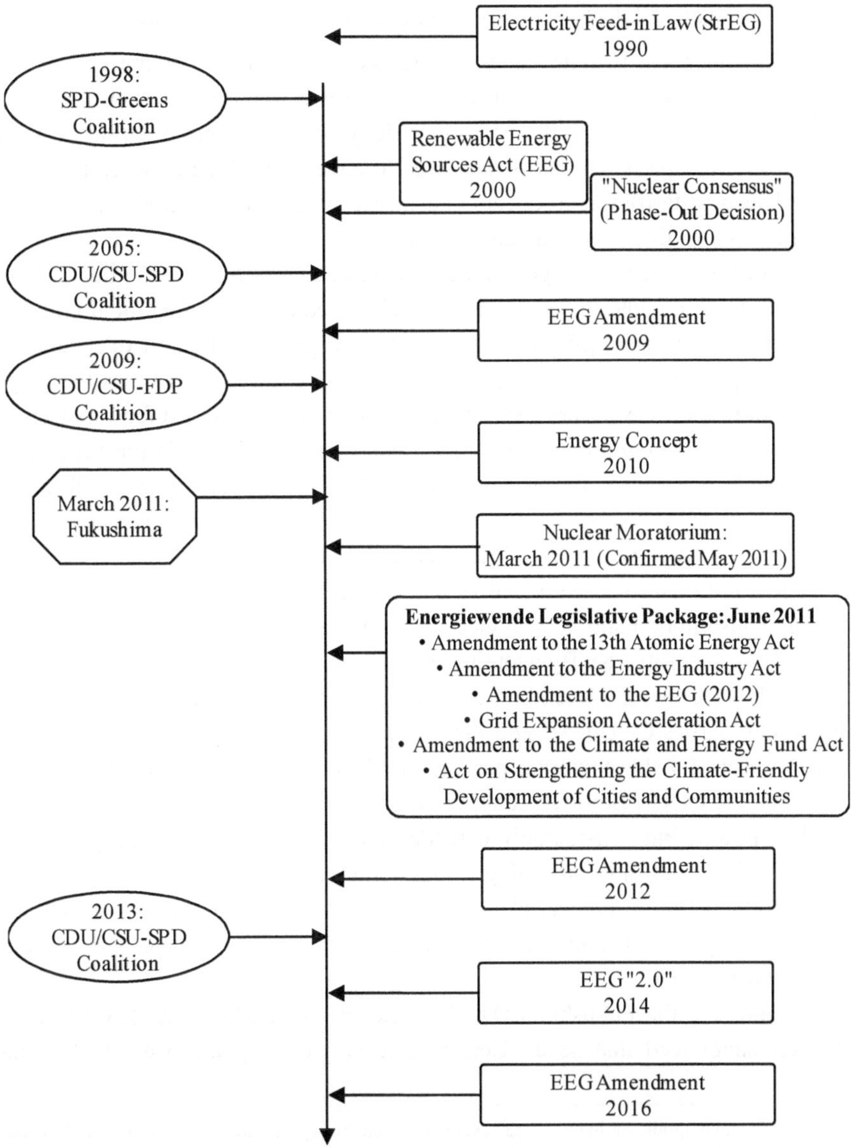

Figure 12: Milestones of renewable energy legislation in Germany. Graph by the author

5.1.5 The 2010 Energy Concept and 2011 Energiewende package

The rapid increase in renewables, and especially solar PV, at the end of the 2000s also led to criticism because of the rise it caused in the renewables surcharge (*EEG Umlage*). In the EEG, electricity consumers pay a top-up to their electricity rates per kWh that is then used to cover the spread between the guaranteed feed-in tariff for producers and market prices for electricity in a given month[36] (Agora Energiewende 2015b; Hake et al. 2015).

When a new coalition government between CDU-CSU and the liberal FDP entered office in 2009, the coalition partners agreed on prolonging the lifetime of nuclear plants as a bridging technology while reigning in the cost of renewable energy deployment. The FDP had been the most vocal (and at times the only) opponent to the EEG in the German parliament and now pressed for a new energy concept. The government's long term energy goals defined in the September 2010 "Energy Concept for an environmentally friendly, reliable and affordable energy supply" (*"Energiekonzept für eine umweltschonende, zuverlässige und bezahlbare Energieversorgung")* foresee a reduction of CO_2 emissions by 80 to 95% in 2050 (vis-à-vis 1990 levels), and a share of 60% renewables in primary energy consumption and 80% in electricity consumption by 2050 (Bundesregierung 2010a, see Table 8). By adopting the 11th Act to Amend the Atomic Energy Act, published on 13 December 2010, (*Elftes Atomgesetz*, BGBl. I: 1814), the governing parties CDU/CSU and FDP decided to grant extensions of between 8 and 14 years for the operating permits of 17 nuclear power plants. In its 2010 decision, the CDU/CSU-FDP coalition did not overthrow the idea of phasing out nuclear power, but considerably extended the lifespans of the country's plants while at the same time providing a long-term perspective for renewables. The 2010 prolongation of nuclear phase-out was met with important popular opposition, bringing tens of thousands of protesters to the streets in major German cities (Reuters 2010).

Following the Fukushima Daiichi accident on 11 March 2011, Chancellor Merkel announced that seven German nuclear power plants were to be shut

36 The exemption of many large, export-oriented businesses from the EEG surcharge has caused continuous debate in Germany as well as heavy criticism from the European Commission, which had always been highly skeptical about the German feed-in tariff approach (Fischer and Geden 2011), Solorio et al. 2014).

down temporarily with immediate effect and one plant to be closed definitively (Schreurs 2013). These immediate shutdowns were made permanent in May of the same year. The moratorium, decided on 14 March 2011, was swiftly followed on March 22 by the establishment by Chancellor Merkel of an "Ethics Commission for a Safe Energy Supply" (*Ethik-Kommission Sichere Energieversorgung*). The commission brought together representatives of civil society, scientists, church and business representatives with the notable exception of representatives from the nuclear industry. Its final report, published on 30 May 2011, clearly states that "the ethics commission is firmly convinced that the phase-out of the use of nuclear power can be finalized within one decade (…). Society should firmly commit to this target and to the necessary measures [to reach it]" (Ethik-Kommission Sichere Energieversorgung 2011: 9).

The findings backed Chancellor Merkel's decision to announce the speeding-up of the country's nuclear phase-out on 30 May 2011. In a press conference, she announced all nuclear power plants would be shut down by 2022, but underlined this was "speeding up" the implementation of the government's 2010 Energy Concept rather than a U-turn in energy policy (Bundesregierung 2011). In addition to the disaster of Fukushima Daiichi, the results of the March 2011 elections in the Land of Baden-Württemberg might have been an additional impetus for Chancellor Merkel to speed up the Energiewende. In these elections, the Green party received an additional 12.5% of the votes compared to 2006 while the CDU lost 5.2%, and with it had to relinquish governing power to a coalition of the Green party and SPD and entered opposition after 58 years in power (ZEIT ONLINE 2011). The decision to permanently shut down the eight nuclear power plants concerned by the moratorium and to end operation for the remaining nine plants between 2015 and 2022 is what fundamentally differentiates the German government's energy concepts of September 2010 and of June 2011. Other targets, summarized in Table 8, such as the share of electricity from renewable energy (80% by 2050) and the reduction of GHG (80% by 2050 compared to 1990 levels) were not modified. Using the term Energiewende only for the decisions taken in response to the Fukushima-Daiichi disaster would thus be misleading.

5.1.6 The Energiewende as a complex transfer item

The 2010 Energy Concept – with the notable exception of nuclear energy operating times – remained in place after the Fukushima disaster, but its implementation was accelerated through a series of amendments to existing legislation (Bundesregierung 2011). Although the Energiewende is referred to as a "model" in this and other pieces of research and writing, it is in fact a collection of political objectives (Joas et al. 2016), targets, laws and ordinances (see Figure 13) that shape the German energy transition. Among the pillars of the Energiewende – renewables, efficiency, decarbonization – progress is most visible today in the area of renewables in the power sector. The fact that the case studies of transfer in Morocco, South Africa, and California revealed a strong focus on renewables (followed by efficiency), in the consideration of the Energiewende and in active German leadership efforts, comes as no surprise given Germany's long-term and frontrunner role in this sector.

Apart from the sheer increase in renewables in the power sector, the ownership structure of renewable energy installations is a specificity of the German approach. The specific structure of the EEG compensation scheme, in particular the security it provides to investors, resulted in low-cost financing and a tremendously high share of private households, farmers and cooperatives entering the business of producing renewable electricity. In 2012, private households or farmers owned 47% of renewable energy capacity, and Germany's big utilities only owned 5% (Agentur für Erneuerbare Energien 2014b).

Linked to this tremendous share of small-scale installations is strong public support for the Energiewende. Polls have indicated for years that more than 90% of Germans favor a further increase in renewables in the electricity mix, with 66% saying such an increase is very or extraordinarily important and 23% important; only 6% of respondents in a representative survey found an increase in renewables to be not important (Agentur für Erneuerbare Energien 2015: 1). Another figure related to acceptance points to a cultural specificity of Germans' high willingness to pay for environmentally sustainable electricity (Mattes 2012: 7). About two thirds of respondents in the above-mentioned survey indicated the EEG renewables surcharge on their electricity bills was either appropriate (57%) or even too low (6%), with less than a third finding the 6.24 eurocents surcharge per kWh (in 2015) to be too high (Agentur für Erneuerbare Energien 2015: 6).

Several features of the Energiewende make it a particularly challenging item for policy transfer and diffusion. According to theory (see section 2.3.4), transfer and diffusion are more likely to occur when the object of transfer is easily identifiable, prominent, and understandable, brings incremental rather than disruptive change, and is universal in nature. The Energiewende fulfills only some of these criteria, such as a high degree of prominence and its origin in a generally prominent, large industrialized country. The Energiewende is a complex set of policies targeting different actors, sectors and mixing different policy instruments, each including dozens of additional provisions per technology or location of renewables sites (see Figure 13). It is a special or even unique policy experiment with limited transferability to other contexts. It is also highly distributive in nature, both in terms of shifting economic power from owners of fossil-fuel plants to renewable energy facilities, and in terms of democratizing energy production by turning consumers into producers of electricity or "pro-sumers". Even within Germany, the question of what political objectives are to be achieved in priority through the Energiewende is subject to discussion (Joas et al. 2016). A lack of clarity about what a model actually consists of is likely to influence transfer processes (Klingler-Vidra 2014).

5.1.7 *Targets, achievements, and challenges of the Energiewende*

Table 6 provides an overview of Energiewende targets and the status quo at the time of writing. While Germany is on track to meet its renewable energy targets, the Energiewende's climate record has been subject to close scrutiny (Löschel et al. 2015). This is the case in particular because of the so-called "climate paradox", a temporary increase in GHG emissions in 2011, 2012, and 2013 compared to previous years (Graichen & Redl 2014). A decrease in 2014 was again followed a slight increase in emissions in 2015 (Agora Energiewende 2016a), despite steadily rising shares of renewables. Several mechanisms explain this increase despite growing shares of renewables in the power mix. Electricity in Germany is dispatched following the so-called "merit-order", that is to say, a ranking of different power plants according to the price they ask for for a marginal unit of electricity produced. Renewables benefit from priority dispatch and have very low marginal costs (since they do not need to pay for fuel such as coal or gas), meaning that they shift the merit order to the right and push out more

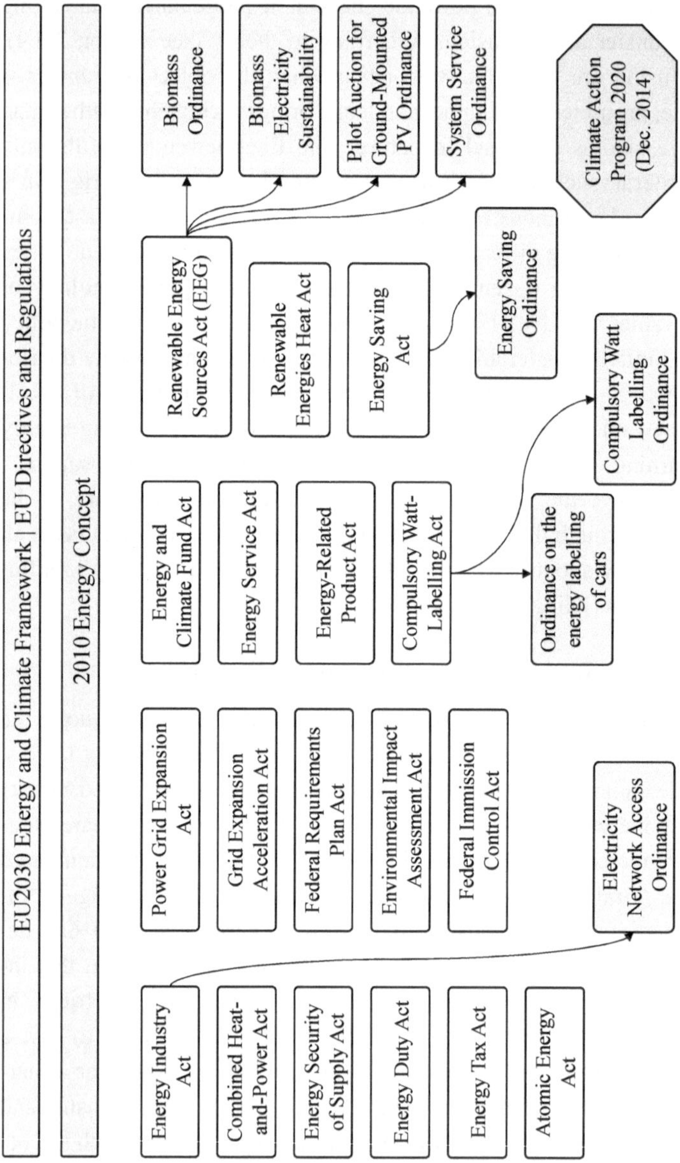

Figure 13: Legislative map of acts and ordinances for the Energiewende. Graph by the author, adapted from (BMWi 2015d). White: acts, grey: ordinances

expensive plants. The "merit order effect" of renewables entering the market has led to dramatic price decreases on the German wholesale electricity market (Agora Energiewende 2015b: 29–30, Cludius et al. 2014). The downward pressure on wholesale electricity market prices, combined with very low CO_2 prices in the European emissions trading system and low global coal prices meant gas was more expensive relative to coal and could no longer compete in the market (Graichen & Redl 2014). Given this combination of circumstances, an increased reliance on coal-fired power plants, especially within Germany's steadily increasing electricity exports, led to momentarily rising CO_2 emissions despite growing shares of renewables and an overall positive tendency of CO_2 reductions since 1990 (BMWi 2015b, Graichen & Redl 2014). The Energiewende's "climate paradox" drew international attention (Karnitschnig 2015, Le Blond 2015, McCown 2013, Nimgaonkar 2015, Rhys 2013: 2, Wambach 2014) and was mentioned in interviews in South Africa and California in particular.

Table 6: Status Quo and Energiewende targets. Data: Agora (Energiewende 2016a), BMWi (2015b)

Area	**Goal**	**Status (2015)**	**2020**	**2025**	**2030**	**2035**	**2040**	**2050**
CO_2 (1990 baseline)	Reduction of CO_2 in all sectors (1990)	-26%	-40%		-55%		-70%	-80- 95%
Nuclear	Shut-down by 2022	8 remaining units to be shut down as of 2015						
RES	Share in final energy consumption	13.5%	18%		30%		45%	>60%
	Share in gross electricity consumption	32.5%	35%	40-45%	50%	55-60%	65%	>80%
Efficiency (2008 baseline)	Reduction of primary energy consumption	-9.1% (2014)	-20%					-50%
	Reduction of gross electricity consumption	-4.8% (2014)	-10%					-25%

To strengthen chances of meeting its 2020 and 2030 GHG reduction targets, the federal government adopted a climate action plan in December 2014 (BMUB

2014a). A new roadmap for an integrated climate policy is expected in the course of 2016 (Agora Energiewende 2016b: 16). Most importantly, the need to phase-out Germany's coal fleet if the energy transition is to achieve its long-term decarbonization goals in a weak European climate framework, entered political discussions prominently in 2015. Controversial discussions took place on whether utilities should be compensated for the phasing out of old coal powered plants and whether a "coal consensus" for a phase-out in the 2035 - 2045 time range should be aimed for (Agora Energiewende 2016b). Germany's electricity mix in 2015 still relied heavily on coal, with hard coal making up 18.2% and lignite 24% of the electricity mix, slightly down from 18.9% and 24.8% in 2014 (Agora Energiewende 2016a: 12). At the end of 2015, Energy and Economics Minister Sigmar Gabriel announced the conclusion of an agreement with utilities RWE, Vattenfall and Mibrag on putting 2.7 GW of coal-fired capacity in a security-of-supply reserve for seven years before retiring the plants definitively (Agora Energiewende 2016b: 9). During the reserve phase, owners will be compensated for the turning off of five plants with €230 million per year until 2020 (Bundesregierung 2016a: 2). In addition to additional efforts in terms of emissions reductions on the electricity supply side, annual monitoring reports on the Energiewende identify considerable challenges in the transport and heat sectors and call for a reinforcement of efforts in the energy efficiency space, in integrating the heat and transport sectors to the Energiewende, and a push for ambitious reforms to the European emissions trading scheme (BMWi 2015b, Löschel et al. 2014, Löschel et al. 2015).

5.1.8 Economic dimensions: Jobs, exports, technology

The Energiewende's track record is not only assessed with regard to an increase in the share of renewables or CO_2 reductions. Economic indicators, from job creation to macro-economic effects to impacts on technology prices are closely scrutinized in Germany and abroad (e.g., Sopher 2014, Zimmermann 2015). As shown by Joas et al. (2014, 2016), stakeholders perceive climate protection as the central, but by far not the only driver behind the Energiewende, and most interviewees in the survey also indicate that the Energiewende "would make sense" even in the absence of climate change. Taking a look at the economic dimensions of the Energiewende is relevant also because of the objectives that drive renewable energy deployment in other countries, as evidenced by the ranking exercises carried out in Morocco and South Africa.

One of the most frequently cited achievements of the Energiewende is its impact on driving down the price of renewable energy technologies, and solar energy in particular (Morris 2016, Pahle et al. 2012). This impact is a direct consequence of the creation of a substantial market for renewables in Germany, which was the single largest market for PV for years (see also Jänicke 2011, Subtil Lacerda & van Den Bergh 2014: 8239). The price of PV systems in Germany has been divided by more than 7 since the mid-1990s and the start of renewable energy promotion in Germany (Morris 2016). Interview partners have also recognized this link as a major contribution of Germany's unilateral leadership in the uptake of renewables worldwide.

Another indicator used to underline the Energiewende's track record is its ability to create employment and to enable companies, by creating a strong domestic market, to become export leaders in segments of the renewable energy and energy efficiency field. The logic behind the link of creating policy-driven demand domestically to capture first-mover advantages and then to move abroad is theorized in the lead market literature (Beise 2004, Beise & Rennings 2005; Jänicke & Jacob 2004; Beise & Rennings 2003). Importantly for this study, creating favorable regulatory conditions abroad can support capturing the export-related advantages of lead market strategies. The number of jobs related to the Energiewende in Germany is around 300,000 or more (370,000 in 2013, up from 160,000 in 20014, see Lehr et al. 2015: 92), a number that was well known to interview partners in Morocco, South Africa[37], and California. While most figures related to the job creation potential of the Energiewende since 1995 or 2010 conclude gross employment effects of several hundreds of thousands of jobs, the net job creation effects are also estimated as clearly positive, with up to 232,000 *additional* jobs to be created by 2050 through the Energiewende (BMWi 2015e).

Much of the Energiewende's estimated positive effects on employment stems from the export strength of companies in the renewable energy and energy efficiency sector. Germany has been the world leader for environmental protec-

37 Interviewees in South Africa were undecided whether South Africa's 300,000 "green job" creation target was inspired by the German job creation figures. In any case, German experts had advised the drafting of the South African strategy.

tion goods, including renewable energy technology, for well over a decade (Gehrke & Schasse 2015: 54–55). Roughly €82 billion of exports were realized in this sector in 2013 (Eckermann 2016: 5). The sum of exports had increased to an all-time high of €85 billion in 2011, before decreasing because of a downturn in Germany's solar industry. In 2012 and 2013, the production of solar PV cells in Germany decreased by over 50% and by 20% for solar PV products in general (Gehrke & Schasse 2015: 43–44). This dramatic decrease is due to the increasingly important role of Chinese solar PV manufacturers on the global market, coupled with the slow-down of solar PV expansion in Germany due to reforms to the EEG. Moroccan interview partners saw the downturn as one driver behind Germany's involvement in promoting policies and in particular concentrated solar power technology (CSP) in Morocco.

5.1.9 A transition in the transition: Evolving Energiewende policies

The decrease in solar manufacturing in Germany was not the only challenge Germany's Energiewende started to face since 2011. A particular concern for critics of the Energiewende have been steeply rising household electricity prices in Germany starting in the mid-2000s, which are only partly due to the rising EEG surcharge, but received much attention, including in international media (Carlyle 2013, Karnitschnig 2014, McCown 2013, The Economist 2014). Average household electricity prices increased by about 10 euro cents between 2006 and 2016, to approximately 29 eurocents per kWh for an average German household[38]. During the same period of time, the EEG surcharge, which is a part of the electricity price consumers pay, increased from less than 0.88 eurocents per kWh in 2006 to 6.35 eurocents per kWh in 2016 (BDEW 2016: 7), due to a combination of several factors. First, the very nature of the EEG surcharge leads to an increase in its level when wholesale electricity prices decrease. As explained above, the EEG surcharge covers the spread between market prices electricity producers receive and the feed-in tariff they are guaranteed. The entry of renewables to the market, in combination with other factors such as overcapacities and low fossil fuel prices, led to a decrease of 100% in the price of peak load 1-year future contracts on the German electricity exchange between 2008 and 2015,

38 The average yearly electricity consumption of a German household is around 3,500 kWh.

while base load prices decreased by about one-third (Agora Energiewende 2015b: 30). The reduction in market prices constitutes not only a problem for operators of conventional power plants, but also leads to an automatic increase in the EEG surcharge. The reduction of electricity generation cost has not been fully passed through to consumers, but the combination of the generation cost of electricity and the EEG surcharge for consumers is expected to remain stable over the coming years before decreasing when expensive older feed-in tariff contracts come to an end (Agora Energiewende 2015b: 30).

In the light of the continued rapid increase in renewable energy and especially in solar installations, fundamental reforms to the EEG were aimed for by the CDU/CSU-FDP coalition, which entered office in 2009. A stringent cap on renewable energy capacity additions and cuts to feed-in tariffs were sought for in the framework of an amendment to the EEG, but were met with protest from a majority of *Länder* in the *Bundesrat*, the upper chamber of the German parliament, leading to a less radical reform of the EEG in 2012 (Lauber & Jacobsson 2016: 152-153). The reform however introduced major changes such as an obligation for new, larger-scale installations to market electricity directly to consumers and exposing them to a market risk so far unknown to producers of renewable energy in Germany. The reform did not halt discussions on the cost of the country's renewable energy policy. Despite continuous popular support for the Energiewende, calls for a slow-down and cost control of the transition dominated political discussions over much of 2012 and 2013, through to 2014. A point of crystallization of this debate was Environment Minister Peter Altmaier's (CDU, 2012-2013) estimate of the cost of the Energiewende being about one trillion euros for the whole project (Mihm & Steltzner 2013) and his resulting call for a "cost-brake" ("*Kostenbremse*"). Although the figure of one trillion euros itself was widely criticized as being too high and disregarding of the benefits of the Energiewende (Reuter 2013), it set the scene for a more thorough reform of the EEG.

In response to price increases and under protests from environmental organizations, citizen initiatives, and the renewable energy industry, the CDU/CSU-SPD coalition in office since 2013 proceeded to push through a major reform of the EEG, with the aim of introducing an "EEG 2.0". Space in this chapter does not allow for a comprehensive account of the reform (see BMWi 2015c; Graichen 2014; Lauber & Jacobsson 2016: 153), but several lines of reform merit to be highlighted. The underlying logic was a move toward a stronger market integration of renewables and the shift toward more market-based policy instruments. Relevant measures include the compulsory direct marketing of electricity from larger installations, with a phasing-in of similar requirements for smaller (new) installations. Self-producers of electricity are required to pay 40% of the EEG surcharge on the electricity they produce and consume on-site. The reform also introduces "target corridors" of how much capacity of onshore and offshore wind, as well as solar should be added year-by-year until the targets shown in Table 6 are reached (Graichen 2014). If the capacity added exceeds the target corridor, the tariff offered decreases more than the automatic digression built into the EEG; if capacity lags behind, the tariff may increase slightly. The most fundamental shift introduced is certainly the introduction of pilot auctions to test a new mechanism to promote renewable energy. First pilot auctions were held over the course of 2015 for ground-mounted PV. Different auction design schemes were tested in the four bidding rounds held between April 2015 and April 2016 and yielded average successful bid prices of 9.2 (15 April 2015), 8.5 (1 September 2015), 8.0 (1 December 2015) and 7.41 eurocents (1 April 2016), which is below the level of feed-in tariffs in place at the time of the auctions (BMWi 2016f). A reform of the EEG in 2016 is set to introduce the auction scheme permanently for large installations and extend it to wind energy (Appunn 2016, BMWi 2015a). The paradigmatic shift from administratively set feed-in tariffs to determining market premiums via auctions has been met with skepticism from environmental groups, citizen energy cooperatives, and the renewable energy industry (BUND 2015, Bundesverband Erneuerbare Energien e.V. 2015: 3, Bündnis Bürgerenergie e.V. 2015). A particular point of concern is the potential loss of "actor diversity" in the Energiewende, i.e., the particularly high share of citizen- and cooperative-owned renewable energy installations in Germany (Jacobs et al. 2014).

5.2 Internationalizing the Energiewende

5.2.1 An early global dimension

In parallel to the beginning of efforts in research and development in the area of renewable energy in Germany, international cooperation efforts were also stepped up. Projects targeted at the promotion of "non-exhaustible energy resources" were sponsored by BMZ from 1975 onwards, in response to growing demand from developing countries, which had also been hit by the oil price shocks of 1973 (Bundesregierung 1981). In 1979, the projects, most of them focused on biomass and solar, were brought together in the Special Energy Program (*Sonderprogramm zur Nutzung nicht-erschöpflicher Energiequellen,* SEP), which also played an important role in Morocco (Bundesregierung 1981). In 1981, the United Nations Conference on New and Renewable Sources of Energy in Nairobi was the first conference of its kind focusing exclusively on renewable energy (see also Hirschl 2008: 480; Röhrkasten 2015: 80). The German federal government considered the action program defined in Nairobi a "guideline" for its own energy cooperation activities (Bundesregierung 1982: 1). A year prior to the conference, the Independent Commission on International Development led by former German Chancellor Willy Brandt (SPD, 1969-1974) had published its final report and called for an "orderly transition from high dependence on increasingly scarce non-renewable energy sources" (Independent Commission on International Development Issues 1980: 171) as well as for the establishment of a "global energy research centre under UN auspices" (p.169) – an idea that translated to the vision of an international renewable energy agency (Hirschl 2008: 440, Röhrkasten 2015: 79).

German legislators were conscious about the importance of Germany's energy policy choices for other nations in the discussions preceding the introduction of the 1990 StrEG. In a response to a request by the liberal FDP group in parliament, the federal government underlined the responsibility of industrialized countries in developing renewables, also with a view to their use in countries of the global South: "The use of renewable energies is part of future energy supply strategies worldwide. (…) Renewable energies can play a bigger role in the countries of the Third World and can contribute to alleviating their economic and social problems. The highly developed industrial countries therefore have to develop and deploy technological systems for the use of renewable energy. Thereby, a contribution is also made to the North-South dialogue" (Bundesregierung 1988: 2). The development of RES is considered from a global perspective, clearly stating that their worldwide spread constitutes an "important contribution to [Germany's] security of supply" and their "great importance becomes visible [only] from a worldwide perspective" (Bundesregierung 1988: 8). The document emphasizes Germany's responsibility in promoting the spread of renewable energy technologies abroad by stating that "the supply with energy of a dramatically growing population, especially in the world's sun-belt requires the use of all environmentally friendly sources of energy and obliges the Federal Republic of Germany, as a technologically highly developed industrial nation, to make a particular effort to promote renewable energies" (Bundesregierung 1988: 9)[39]. Several forms of international cooperation are listed in the document. Apart from cooperation within the European Community (demonstration plants and research), exchange of information and joint ventures through the International Energy Agency (IEA), German-Spanish and German-Greek research projects in the field of solar and wind energy are mentioned (Bundesregierung 1988) as early examples of active German leadership. The role of German companies active in the development and realization of wind and photovoltaic energy installations was deemed particularly important and cooperation with them in developing countries was a declared goal of the federal government (Bundesregierung 1988: 37).

Over the course of the 1990s, progress on renewable energy governance at the international level remained slow, and the topic was virtually absent from the 1992 Rio Conference on Sustainable Development (Hirschl 2008: 475, Hirschl 2009: 4409, Röhrkasten 2015: 80). Germany's focus at the time remained on renewable energy cooperation with developing countries, an area for which 3 billion DM had been spent by 1996 (Bundesregierung 1996: 15). Two of the main proponents of EEG in the *Bundestag*, Hans-Josef Fell (B90/Greens) and Hermann Scheer (SPD)[40], used the international dimension of Germany's renewable energy policy as an argument in debates preceding the adoption of the game-changing act (D13). In a December 1999 debate in the *Bundestag* (three months prior to the adoption of the law), Scheer once again underlined that the law was without equal worldwide and therefore could be used as an example globally (Bundestag 1999: 7372). Fell pointed to international competition in the

39 Interestingly, the document also makes reference to renewable energy policies abroad, casting doubt on the sustainable success of wind energy in those countries. California and Denmark, who are "often depicted as models to be followed with regard to the use of wind energy" (Bundesregierung 1988: 17) are seen as struggling to maintain wind energy development.

40 Both "founding fathers" of the EEG also played a role as transfer agents in South Africa and California, in addition to their general involvement in international renewable energy cooperation and, in particular for Hermann Scheer, the creation of IRENA.

field of renewable energy, from countries such as the United States and China (Bundestag 1999: 7257).

5.2.2 *Renewables 2004: A milestone in active Energiewende leadership*

As shown by Röhrkasten (2015: 118), Germany entered the international stage as a driving force behind the global promotion of renewables at the World Summit on Sustainable Development (WSSD) in Johannesburg in 2002. The summit focused on renewable energies, but did not lead to the agreement on concrete targets for renewables Germany and other pioneering countries had hoped for (Hirschl 2009: 4411). Emphasizing industrialized countries' historic responsibility in addressing climate change through sustainable energy and his country's pioneering efforts in the field, Chancellor Gerhard Schröder (SPD, 1998-2005) announced at the summit that Germany would host an international conference on renewables. He also pledged that Germany would spend €500 million on the promotion of renewables and €500 million for energy efficiency in developing countries within five years (Schröder 2002). In the year preceding the WSSD, a group of renewable energy advocates led by Hermann Scheer founded the World Council on Renewable Energy (WCRE). WCRE joined forces with another organization led by Hermann Scheer – EUROSOLAR[41] – to push for an international agency for renewable energy. These efforts eventually resulted in the creation of the International Renewable Energy Agency (IRENA) in 2009 (EUROSOLAR & WCRE 2009 (2001), Land 2009).

Chancellor Schröder's announcement resulted in the organization of the International Conference for Renewable Energy Bonn – "*renewables 2004*" – from 1 to 4 June 2004. Renewables 2004 was a major milestone, not only for Germany's renewable energy leadership, but also on the way toward institutionalization of renewable energy cooperation in the international arena (Hirschl 2008: 474). The conference, with more than 3,600 delegates from 154 countries, resulted in a political declaration stressing the importance of stable regulatory frameworks for renewable energy and in the adoption of an international action program (BMZ 2010, Röhrkasten & Westphal 2013: 6–7). The action program assembled voluntary pledges and announcements by countries worldwide on their plans for in-

41 EUROSOLAR was founded as the European Association for Renewable Energy by Hermann Scheer in 1988 (EUROSOLAR & WCRE 2009(2001)).

creasing renewable energy. Germany as a host country stood out by pledging an additional €500 million (in addition to the one billion euros pledged in Johannesburg two years earlier) to be made available from 2005 onwards through KfW for sustainable energy worldwide (BMZ 2010). Due to enormous interest in the financing program, its budget was exhausted three years later and a new special facility, the Initiative for Climate and Environmental Protection (*Initiative für Klima und Umweltschutz*, IKLU) was created (Meyer 2015). According to interview partners closely involved in German international energy policy at least since the preparations for the Bonn conference, *renewables 2004* clearly marks the stepping up of Germany's own, visible active leadership efforts in the field of renewable energy beyond development cooperation (D6, D10). Renewables 2004 was not an isolated conference, but was preceded and followed by a number of bilateral and multilateral meetings including the Middle East North Africa Renewable Energy Conference (MENAREC[42]) in April 2004 and led to a number of follow-up events including renewable energy conferences in Beijing (Beijing International Renewable Energy Conference, BIREC) in 2005, Washington (WIREC) in 2008, Delhi (DIREC) in 2010, Abu Dhabi (ADIREC) in 2013 and South Africa (SAIREC) in 2015 (BMZ 2010).

Just prior to *renewables 2004*, the Bundestag hosted the International Parliamentary Forum on Renewable Energy. Chaired by Hermann Scheer, the forum reiterated the call for an international renewable energy agency (EUROSOLAR & WCRE 2009 (2001): 70). The German government, reluctant to pursue this idea of an international renewable energy agency outside the UN framework in Bonn, instead initiated the creation of REN21 (Hirschl 2008: 485, Röhrkasten & Westphal 2013: 7). REN21 is a global network linking governments, NGOs, businesses, and other initiatives in an endeavor to facilitate dialogue on renewable energy. REN21's secretariat is co-hosted by the United Nations Environment Programme (UNEP) in Paris and the German GIZ provides its staff (Röhrkasten 2015: 110, Wienges 2009: 160). Due to its light institutional setup and its mission of connecting and facilitating existing initiatives rather taking on governance functions and centralizing efforts, REN21 was far away from the interna-

42 A series of MENAREC conference have since taken place, up to MENAREC 6 in Kuwait in 2016.

tional renewable energy agency Hermann Scheer and his colleagues were pushing for.

5.2.3 Germany's role in the creation of IRENA

Initiating the creation of an international renewable energy agency was part of the coalition agreements of German federal governments in 2002 (SPD-B90/Greens) and 2005 (CDU/CSU-SPD), but the hope for a process within the UN framework continued to be an obstacle to reinforcing German efforts in the direction of IRENA (Hirschl 2008: 481–482). Several factors shifted the balance in favor of German support for an agency outside the UN framework that was to become IRENA. In 2006 and 2007, two additional sustainable development conferences (Commission on Sustainable Development sessions 14 and 15) were held but failed to define global renewable energy targets with a clear time horizon, as Germany had advocated for for years (Röhrkasten 2015: 85, Röhrkasten & Westphal 2013: 4). Within Germany, the change in government in 2005 with the CDU/CSU-SPD replacing SPD-B90/Greens led to a constellation where the Development Ministry, the Federal Foreign Office, and the Environment Ministry were held by Hermann Scheer's party colleagues from SPD (Röhrkasten & Westphal 2013: 7). Thirdly, the dramatic increase in fossil fuel prices in 2008 and growing attention to environmental protection in the run-up to the 2009 Copenhagen climate conference COP15 increased attention to renewable energy ad the need for a dedicated international organization. This made for a particularly favorable environment for Germany to finally push forcefully for the creation of IRENA through a range of preparatory meetings around the globe. After a number of meetings, the creation of IRENA was announced in Berlin in 2009, with 75 countries signing the founding declaration (Röhrkasten 2015: 133–134). The number of IRENA members has since increased to 147, with 29 more having applied for membership (IRENA 2015a).

Hermann Scheer's idea for IRENA can be linked back to concepts of leadership and policy transfer: "IRENA will have the task to advise governments in the development of policy approaches for renewable energies. It's not necessary for everyone to start their learning curve at zero, one can learn from available experiences" (Land 2009: 71). The establishment of IRENA as the first international agency dedicated to renewable energy is very clearly due to relentless efforts by its "inventor", Hermann Scheer, and to German leadership. Scheer's role as one of the main sponsors of the EEG as in initiating the IRENA idea, clearly show how intrinsically linked Germany's own Energiewende and its international leadership activities are. The fact that Germany was not successful in bringing the IRENA secretariat to Germany nor in nominating a German to head the agency, can be seen as a drawback given the crucial role the country played in establishing global governance for renewables. On the other hand, as pointed out by Röhrkasten and Westphal (2013: 10), German foreign policy pursued other priorities at the time and the establishment of IRENA in Abu Dhabi as the first international organization with headquarters in the Arab world could be seen as an important signal to the region.

5.2.4 Cooperation focused on the promotion of feed-in tariffs

Among the activities initiated by Germany in the domain of international cooperation on renewable energy after the *renewables 2004* conference, the International Feed-In Cooperation (IFIC) stands out as being entirely focused on feed-in tariffs. IFIC was founded at the sidelines of the Bonn renewables conference on the initiative of Germany, co-founded by Spain and joined later by Slovenia and Greece in 2007 and 2012 respectively (International Feed-In Cooperation n.d.-b). Other countries occasionally joined in its activities over the years (International Feed-In Cooperation n.d.-a).

IFIC's founding declaration mentions assistance for third countries wishing to "establish and implement a similar model" like the German and Spanish feed-in tariff as a specific goal of IFIC, but also considers "possibilities for harmonising different national feed-in systems" (BMU 2005: 2). According to the declaration, the promotion of feed-in tariffs as a policy for the development of renewable energy worldwide encompasses an "exchange of experiences", the definition of success criteria for feed-in systems, and "demonstrating the advantages of feed-in tariffs" (BMU 2005: 2). In addition to an aim of facilitating the uptake of renewables worldwide, the creation of IFIC can be understood as a response to tensions between the European Union institutions and member states' heterogeneous choices of renewable energy promotion schemes. This tension is particularly pronounced for the issue of feed-in tariffs (Solorio et al. 2014). The European Commission's notoriously skeptical view of feed-in tariffs as a policy instrument is expressed, for example in communication COM(2005)627 from 7 December 2005 (European Commission 2005). The communication points to the need for harmonization across Europe, the risk of overfunding and doubts regarding the compatibility of feed-in tariffs with EU internal energy market and state aid rules (European Commission 2005: 4) The creation of IFIC can thus be understood as one of many signals of Germany's preference for feed-in tariffs, not only nationally, but also on the European and international levels.

IFIC's activities have largely been limited to workshops held once to twice a year between 2005 and 2013 and the publication of studies on the situation of feed-in tariff systems in EU member states (International Feed-In Cooperation n.d.-a). Background conversations and interview findings confirm that visible IFIC activities have somewhat slowed down, but studies were still carried out for in the framework of IFIC (D6). The German Advisory Council on Global Change (*Wissenschaftlicher Beirat der Bundesregierung Globale Umweltveränderungen*, WBGU) called for a reactivation of the IFIC as a platform from which Germany could assume its responsibility for the international transfer of ideas and knowledge on energy system transformations (WBGU 2012). The advisory council itself also takes a position in support of feed-in tariffs as the most promising tool to promote renewables (WBGU 2012: 4) and recommends Germany, "as a pioneer in the field of feed-in tariffs", to launch "an international initiative to promote the spread of feed-in tariffs all over the world" and create "a mechanism for financing feed-in tariffs in developing countries" (WBGU 2012: 23). Such a mechanism is the Global Energy Transfer Feed-in Tariffs program (GET FiT). The initiative, co-financed by the governments of Norway, the UK, and Germany and implemented in cooperation with Deutsche Bank and KfW uses international funds to pay surcharges on the electricity tariff that are needed to finance a feed-in tariff for producers of power from renewables (DB Climate Change Advisors 2011).

In a speech at the occasion of the foundation of the World Renewable Energy Council (WREC) in 2001, Development Minister Heidemarie Wieczorek-Zeul (SPD, 1998-2009) drew a clear connection between Germany's feed-in tariff-based approach in the EEG and her priorities in development cooperation: "Here in Germany we have drawn the appropriate conclusions [...] with the Renewable Energy Sources Act [...]. I consider this route in particular to be of interest for the developing countries as well. [...] I see here [in the Renewable Energy Sources Act] a crucial advantage over other promotional models, such as the quota systems, which the USA for example is pursuing. This is also our starting point for bringing our influence to bear on the World Bank, an institution that in other instances has shown a tendency to follow the American example" (Wieczorek-Zeul 2009 (2001): 34). Clashes between the German, feed-in tariff-friendly, approach and clear opposition to such a model by the World Bank reportedly occurred in closed meetings in Morocco, according to stakeholders interviewed who had attended these meetings (M1, M13). A federal government response to a parliamentary inquiry from 2009 also clearly points in the direction of feed-in tariff policy promotion and the use of international governance channels: "The Federal Government advises countries regarding the increase in renewable energies […] In this framework, the Federal Government promotes the EEG with priority feed-in and guaranteed minimum compensation as an instrument. It [the Federal Government] has thereby contributed to the introduction of similar rules in 20 EU countries and about 50 countries globally. [….] Furthermore, IRENA, which was founded in January 2009, is an appropriate forum to make instruments, such as the instrument of priority feed-in with guaranteed minimum compensation for electricity from renewable energies, better known" (Bundesregierung 2009a: 55). As the South Africa case shows, both Germany's active outreach and its domestic policy model have changed and alternative approaches to the promotion of renewable energy have entered the toolbox of Germany's international renewable energy cooperation.

The question of whether Germany's external Energiewende strategy was and is targeted specifically at the promotion of its own policy model, in particular feed-in tariffs, was fundamental to the case studies in this book. Many elements hint at such a focus, although interviewees saw a change and a more diverse approach – also simply due to the increase in policy experiences for renewable energy globally.

5.2.5 Bilateral energy partnerships

In addition to Germany's decisive role in the establishment of an international governance framework for renewable energy (Röhrkasten & Westphal 2013), the country has established a range of institutionalized, bilateral channels of cooperation, and transfer. The conclusion of bilateral energy partnerships between Germany and partner countries is one of the cornerstones of Germany's external energy policy. Energy partnerships are political, high-level dialogues coupled with working group meetings bringing together government officials and, where appropriate, private sector actors (Müller 2015). They do not supplement bilateral cooperation projects in the energy sector, but rather act as an enabling "umbrella" and a means to "politically accompany" bilateral energy cooperation (D4, D7). Given their initial focus on energy security, most of these partnerships fall under the responsibility of BMWi, but the partnerships agreed with Nigeria in 2007 and with Norway in 2006 are led by the Foreign Office (personal communication with Foreign Office information services, 15 April 2016).

Public information about the concrete structure and goals of many of those energy partnerships is scarce, which is in some cases linked to a low level of activity of the partnership. This is the case for the bilateral energy partnership with Nigeria, which Chancellor Merkel recognized had been neglected for years (Merkel 2011). Traditionally, the focus of energy partnerships has been on sealing privileged relationships with important energy producing or large energy consuming countries, in order to enhance global security of energy supply (D4). As shown in Table 7, the scope of partnerships has gradually increased to place a stronger emphasis on renewable energy.

In particular, the agreements underpinning the energy partnerships with Morocco (September 2012), Tunisia (January 2013), South Africa (August 2013) and Algeria (March 2015) show a stronger focus on renewable energy and energy efficiency, including from a security of supply and export perspective (BMWi 2016c). The agreements with Morocco and Tunisia were, for example, established with a view to accompany the DESERTEC initiative politically and thereby contribute to the security of supply with "green electricity" both in Northern Africa and in Europe (BMWi 2012a, BMWi 2012b). The shift of responsibilities for renewable energy on the domestic level in 2013 to BMWi, which leads on most energy partnerships, and the acceleration of Germany's Energiewende, as well as growing economic interests of export-oriented German companies in the renewable energy sector, also reinforced the aspect of renewable energy in the work of energy partnerships (D4, D7).

Table 7: Overview of bilateral energy partnerships between Germany and partner countries, adapted and translated from Müller (2015: 3–4), BMWi (2012a), BMWi (2012b), BMWi (2013), and BMWi (2016c)

Country	Created	Areas of cooperation (selected)
Ukraine	2005	German-Ukrainian High Level Group for Economy includes energy cooperation as a part of economic cooperation (e.g., working group on coal)
India	2006	Energy security; increasing efficiency in thermal power plants; demand-side energy efficiency and low-carbon growth strategies; renewable energies; investments in energy projects, common research & development.
Norway	2006	Security of supply with electricity and gas, CCS.
Nigeria	2008	Reinforcement of the electricity sector; construction projects in the energy sector; electricity production/ distribution projects; contribution of Nigerian gas to German gas supply.
Brazil	2009	Energy efficiency in the electricity, heat and building sector; other energy-related cooperation projects.
Russia	2010	Various areas of bilateral cooperation in the energy sector, incl. energy efficiency in the electricity, heat and building sector;
China	2012	Increasing energy efficiency; energy strategy, energy policy and energy law; Conventional and renewable energy generation capacities; energy distribution and transmission; coal incl. transformation of "coal cities"; environmental protection in the energy sector.
Kazakh-stan	2012	Various areas of bilateral cooperation in the energy sector, incl. energy efficiency in the electricity, heat and building sector; Renewable energies.
Turkey	2012	Renewable energies; Energy efficiency; conventional power plants, modernization of power plants, lignite mining; electricity distribution and transmission grids; regulation of electricity and gas markets; energy and electricity exchange; consumer issues.
Tunisia	2012	Reduction of energy consumption, increasing energy efficiency, use of renewable energy sources, greenhouse gas mitigation; cooperation in the framework of the Tunisian Solar Plan, the Mediterranean Solar Plan and the DESERTEC-Initiative; energy markets and regulation, capacity building.
Morocco	2012	Comprehensive energy partnership, including security of supply and environmental protection; technologies for electricity generation from renewable sources and grid management; private sector investments and creation of framework conditions for an increased use of renewable energies, including the implementation of the Moroccan Solar and Wind Plans, flexible cooperation mechanisms (esp. Article 9); energy cooperation in the Mediterranean, including DESERTEC.
South Africa	2013	Renewable energies; energy efficiency; CCS and nuclear safety; capacity building and skills development in the green technology sector.
Algeria	2015	General energy policy, including security of supply and stability of export markets as well as climate protection; exploration and production of carbon-hydrates; energy research, energy efficiency; regulatory aspects of access to electricity markets, in particular for renewable energy; regional grid integration.

5.2.6 *A "Club of Energiewende Countries"*

Largely missing from the group of energy partnerships were alliances with other strong frontrunners in the renewable energy field. The idea of such "clubs" of leaders has been considered a promising option by researchers for the international climate regime in general and an option for German leadership in particular (Keohane & Victor 2010, Messner et al. 2014, Weischer & Morgan 2013). June 2013 marked one of the clearest signs of Germany acknowledging the international dimension of its own energy transition and the potential of shared leadership among frontrunners. Based on an initiative by Environment Minister Peter Altmaier (CDU, 2012-2013), nine countries, in addition to Germany and the director general of IRENA, were invited to sign a communication (BMU 2013) announcing the creation of the "Renewables Club" or "Club of Energiewende Countries" (Club der Energiewendestaaten). The members, chosen for their "pioneering ambitions", included IRENA, India, China, France, Morocco, Tonga, the United Kingdom, the United Arab Emirates, South Africa, Denmark and Germany. The founding document takes great care in underlining the Renewables Club is not intended as a competitor to IRENA, but rather as a "high-level political alliance" of "countries that have been at the forefront of the development and deployment of renewable energy technologies", "to advance the case of renewable energy in particular the objectives and programmes of IRENA" (BMU 2013: 2). The members of the Renewables Club expressed their aim to continue to set the agenda for the deployment of RES worldwide, to support the development of favorable regulatory and investment frameworks and to send a "strong political message of support for renewable energy's business case". In summary, "by forming this alliance", the members of the Club "aim to demonstrate conviction and excellence, and to lead by example" (BMU 2013: 2). Interviews suggest however that a main motivation for the club was discontent with IRENA, whose ambition seemed to have been diluted by its almost universal global membership (D4, D6).

Minister Altmaier announced his intention to bring together frontrunners in a forum outside IRENA as early as August 2012, underlining Germany's leadership in the effort and the direct link to its Energiewende: "It is decisive that countries willing to lead the way into a new energy era cooperate more closely. We need a sort of avant-garde, a 'Club of Energiewende countries' that lead the way and show what opportunities are linked to the Energiewende. […] The

Federal Republic of Germany wants to lead here with its Energiewende. [...] With every successful step the Energiewende takes, we demonstrate that there is an alternative to 'business as usual'[43]" (Altmaier 2012a). Altmaier was criticized by Green members of parliament for the exclusive character of the Renewables Club as well as for duplicating existing organizational structures, especially IRENA (Bundesregierung 2013a). According to the government, the Renewables Club could be followed in the future by similar high-level political initiatives on climate change and energy efficiency (Bundesregierung 2013a). However, with Minister Altmaier's departure from his position after the 2013 federal elections, any follow-up activities on the Renewables Club were effectively suspended. Despite calls for a revival of the club and its potential to be a cornerstone of German Energiewende leadership (Morgan et al. 2014, Tänzler & Wolters 2014, Steinbacher & Pahle 2015), the founding meeting in 2013 was to remain the only activity of the club to date.

5.2.7 *Other institutionalized transfer channels*

In line with the logics of "institutional leadership" targeted at the creation of "spheres of leadership" (Eckersley 2011), Germany has multiplied efforts to establish potential transfer channels with partner countries. In addition to international governance activities and energy partnerships described above, and the numerous initiatives in the framework of cooperation with developing countries, Germany has created a range of other channels for cooperation.

A lack of coordination with European neighbors has been singled out as one of the most important weaknesses of Germany's external Energiewende policy (Fischer & Geden 2011). Efforts to strengthen dialogue with European neighbors have noticeably intensified since Minister Altmaier's tenure in office. In February 2013, the "German-French Office for Renewable Energies" (Deutsch-Französisches Büro für Erneuerbare Energien, DFBEE) was established, on the basis of its predecessors, the "Coordinating Unit Wind Energy", a result of the

43 "Entscheidend ist, dass die Staaten, die auf dem Weg in ein neues Energiezeitalter vorangehen wollen, enger kooperieren. Wir brauchen eine Art Avantgarde, einen "Klub der Energiewendestaaten", die vorangehen und zeigen, welche Chancen in der Energiewende stecken [...] Die Bundesrepublik Deutschland will mit ihrer Energiewende dabei Vorreiter sein. Mit jedem Erfolg bei der Energiewende zeigen wir, dass es eine Alternative zum Weiter-so gibt."

renewables 2004 conference, and the "Coordinating Unit Renewable Energies" established in 2011 (Altmaier & Batho 2013a). The creation of office was presented as the symbol of cooperation on a "German-French Energiewende", an area that was hoped to become a "motor of German-French relationships" (Altmaier & Batho 2013b). The planned opening of the German EEG to participants from other European countries (Appunn 2016, BMWi 2015a, 2016f) and the signature, in June 2015, of a memorandum of understanding on electricity security of supply with "electrical neighbors" (BMWi 2015g) are additional signs of an increasing Europeanization of the Energiewende.

Of particular importance for California is the Transatlantic Climate Bridge, established in 2008 at the occasion of a preparatory conference for the creation of IRENA. Co-led by BMUB and the Federal Foreign Office, the Transatlantic Climate Bridge provides seed funding to German foreign missions abroad for activities related to sustainable energy and climate. Visits by German experts to California were funded through this program (D5, C7). Several delegation visits to Germany by groups of US policy-makers and journalists on different topics related to the Energiewende have been organized since 2014 (German Missions in the United States 2015).

The reasons for, as well as the success and legitimacy of Germany's leadership in the climate field have been the object of scientific and political debate (Eckersley 2016; Kübler 2014; Livingstone 2015; Schreurs & Tiberghien 2007; Steinbacher & Pahle 2015; Weidner & Mez 2008; Weimann 2012). Germany's leadership on the EU level (especially with regard to the adoption of the 2008 energy and climate package, see Cox & Dekanozishvili 2015) and within the G8, have earned Chancellor Merkel the (frequently disputed) title of "Climate Chancellor" (Klimakanzlerin, Dehmer 2007). Germany has initiated annual ministerial meetings in preparation of COP climate conferences. The so-called "Petersberg Dialogue" meetings started in 2010 in response to the failure of the Copenhagen COP15 and are hosted by Germany in partnership with the host of the subsequent COP to build momentum for the negotiations (BMU 2010).

5.2.8 Sustainable energy and German development cooperation

Bilateral development cooperation is at the core of Germany's active sustainable energy promotion efforts, both in terms of the variety of projects and partners and the financial volume of activities. The Energiewende's international dimension has always been linked to the idea of facilitating the uptake of renewable energy technologies in partner countries of German development aid. As shown in Figure 14, the amount of official development assistance (ODA) spent in the field of energy has increased[44] with the acceleration of the Energiewende in Germany. According to a 2015 OECD evaluation report of German bilateral development cooperation the share of energy in Germany's ODA doubled in the period between 2007-2013 (about 12%) compared to 6% in 2002-2006 (OECD 2015: 105). The report underlines the focus of Germany's international cooperation efforts in areas, such as renewable energies, where a particular value can be added, but concludes that support is tailored to the priorities of the partner country and that international tendering processes ensure aid is untied from procurement (OECD 2015: 40).

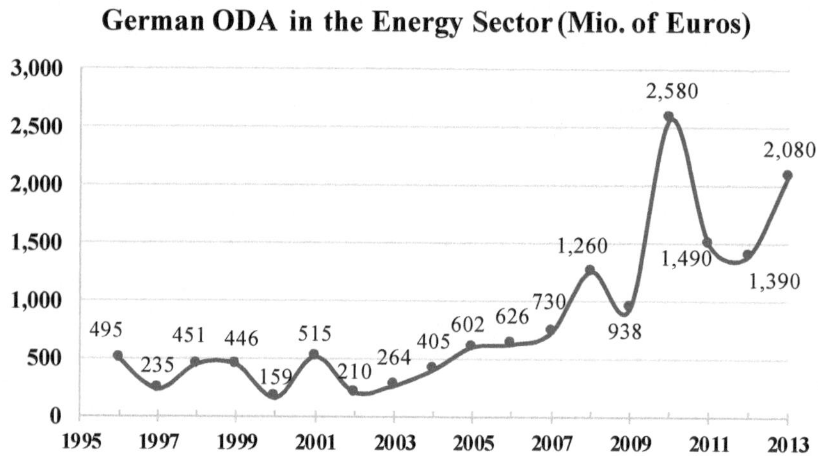

Figure 14: Evolution of German ODA expenditure in the energy sector in million euros. Graph by the author. Data Source: AidData (n.d.)

In 2014, Germany was the third-largest donor in the OECD (approximately €16.5 billion) behind the US and the UK, with German ODA increasing by 16% from 2013 to 2014 (BMZ 2016a). In recent years, energy has become the most important area of German bilateral cooperation (BMZ 2014). Further increases in

44 The peak in energy ODA in 2010 is due, i.a. to a 700 million commitment to the World Bank's Clean Technology Fund. Source: AidData.org

the budget for energy within development cooperation are foreseen: by 2030, Germany has pledged to double it to €3.6 billion *annually* with a goal of giving an additional 100 million people worldwide access to electricity and sustainable cooking facilities (BMZ 2014: 11). Germany's development aid in the energy sector is impartial in principle with regard to the origin of suppliers of technology (for interlinks between Germany's policy promotion and economic interests, see Morocco and South Africa case studies in this book), but it is partisan insofar as it is targeted at renewables and energy efficiency. Whereas there has been a policy since 2014 not to support nuclear energy projects abroad through export guarantees for private companies (BMWi 2014b), and even earlier on in the field of development cooperation, only more recently has a decision been reached to strictly limit involvement in coal-fired power plants in its partner countries. While the construction of new facilities such as the Medupi and Kusile plants in South Africa will no longer be supported, modernization measures can only be financed taking into account stringent criteria of efficiency and the presence of a climate strategy in the partner country (Bundesregierung 2015: 3).

Figure 15 shows the network of agencies and ministries involved in German development cooperation, including in the energy sector. For questions this book evolves around, the most relevant part of German development cooperation are individual projects implemented at country-level over several years by GIZ (technical cooperation) and KfW (financial cooperation). Within the 50 countries supported by German development cooperation, 24 have a focus area "energy" and 11 have energy cooperation projects outside a dedicated focus area. Three-quarters of GIZ projects and more than 80% of GIZ's project volume is financed by BMZ; in the energy sector, (co-)financing through BMUB or, for example in the case of energy partnerships, BMWi, is more common (GIZ 2016).

To channel additional development aid into the sustainable energy field, the federal government over the years has established several special initiatives. In addition to the one billion euros pledged for renewable energy and efficiency in developing countries at the WSSD in Johannesburg, an additional 500 million were added as a "4E facility" (*Sonderfazilität für Investitionen in Erneuerbare Energien und Energieeffizienz,* Special Facility for Investments in Renewables and Energy Efficiency). KfW committed funds within this facility within three years instead of the five-year payout phase initially planned (Meyer 2015: 3). The 4E facility was transformed into the Initiative for Climate and Environmental Protection (IKLU) in 2008 and by 2013, KfW had committed close to €2.5 billion on energy projects through this facility, mostly through preferential loans, triggering a total investment of well above €7 billion and the generation of sustainable electricity for the equivalent of 32 million people in the developing world (Meyer 2015: 3). The number of partner countries with an energy focus area increased from 7 in 2003 to 10 in 2006, 17 in 2007, 20 in 2009, and 24 in 2013; Meyer (2015: 6) sees this evolution as being directly correlated to the creation of 4E and IKLU and therefore as a result of the *renewables 2004* conference. Close to three quarters of funds from IKLU were used in the renewable energy domain, with the remainder earmarked for energy efficiency (Meyer 2015: 9).

Following a decision by the federal government in 2014, IKLU merged with the German Climate and Technology Initiative (DKTI) (Meyer 2015). DKTI, founded in 2011, is a common endeavor by BMZ and BMUB to finance climate technology-related projects, mostly in the energy sector, in partner countries. DKTI projects – such as a project supporting the Moroccan Solar Plan (MSP) (GIZ 2014b) – contain a technical and a financial cooperation component, and put an emphasis on "technologies that have been tried and tested in Germany" and which, "in close cooperation with German companies" are transferred in partner countries where their potential is greatest (GIZ 2012). According to information provided by BMUB (personal communication with BMUB information services, 24 March 2016), DKTI projects co-sponsored by the ministry were implemented in Morocco (Moroccan Solar Plan I and II), Russia (climate-friendly economy and best available technologies), Turkey (energy efficiency in buildings), Chile (solar energy), and Tunisia (Tunisian solar plan).

Finally, another large share of Germany's sustainable energy cooperation with developing countries is realized in the framework of BMUB's International Climate Initiative (*Internationale Klimaschutzinitiative,* IKI). Initially funded – as DKTI – by revenues from the auctioning of emission certificates in the European emissions trading scheme, its funds now come from BMUB's regular budget due to low CO_2 prices and the decrease in revenues from the European Emissions Trading System (ETS). Over 440 projects were financed through IKI according to a 2015 report, for a total volume of €1.6 billion since 2008 (BMUB 2015: 7). Slightly less than half of the overall budget (€719 million) was spent in

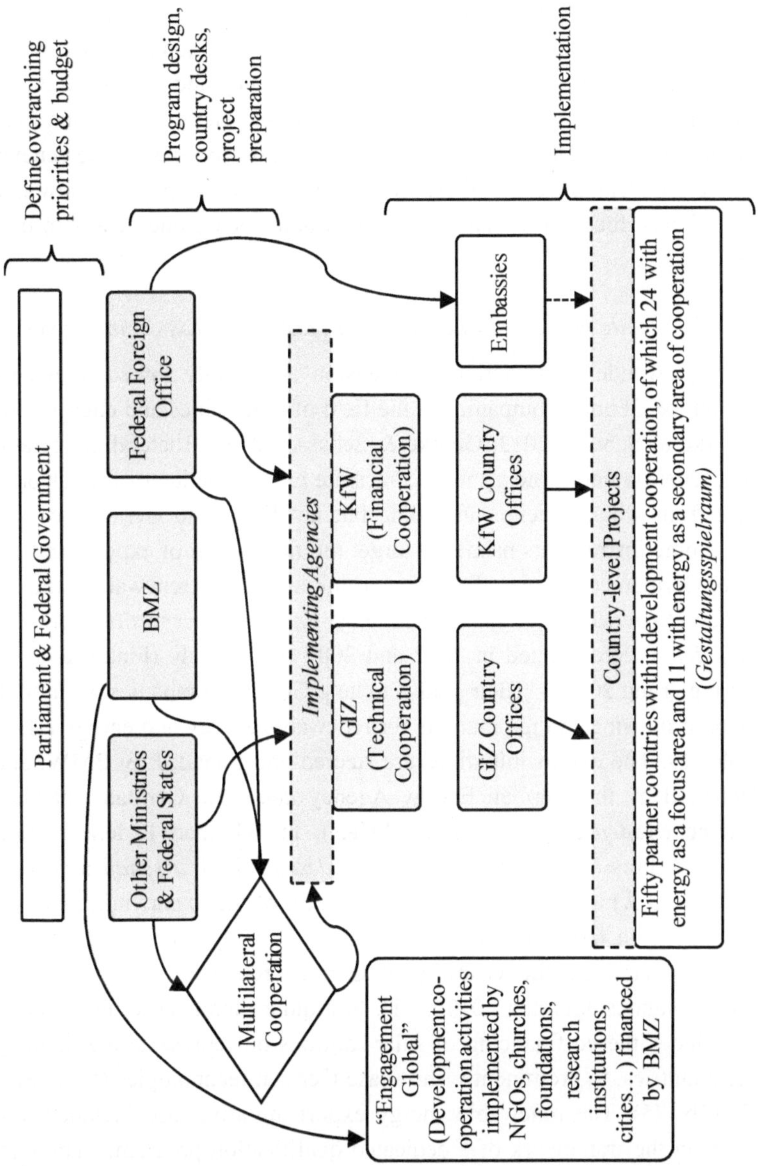

Figure 15: Structure of German development cooperation. Adapted from OECD (2015: 53) and BMZ (2014)

the field of GHG mitigation. Of the 455 projects in the IKI database as of spring 2016, 73 were in the field of renewable energy (International Climate Initiative 2016), but in other project areas such as climate policy and support to the development of national low-carbon strategies renewable energy also often plays an important role (D1, D6, D8) The change of responsibilities in the renewables field in 2013 from BMU to BMWi, also means that IKI projects now need to have a primary focus on climate, with renewables being one means to this end (D1, D2, M13).

5.2.9 Economic aspects of Energiewende leadership: Export initiatives

The Energiewende's international dimension is strongly linked to the export orientation of German companies in the field of renewables and energy efficiency (Gehrke & Schasse 2013, Gehrke & Schasse 2015). Bilateral cooperation in the field of sustainable energy may contribute to the creation of frameworks and markets from which foreign, including but not limited to German, companies benefit. Some instruments primarily target the facilitation of exports and market entry for German companies. Two "export initiatives" for renewables and energy efficiency (*Exportinitiative für Erneuerbare Energien / Exportinitiative für Energieeffizienz*) were created in 2002 and 2007 respectively (Finkel et al. 2013, Bundesregierung 2013b). Their goal is to help German companies position themselves on promising foreign markets for renewable energy and energy efficiency technologies. The export initiatives are steered and financed by BMWi and are implemented by the German Energy Agency *dena*, the German Chambers of Commerce (*Außenhandelskammern,* AHKs), the German Federal Office for Economic Affairs and Export Control (*Bundesamt für Wirtschaft und Ausfuhrkontrolle*, BAFA) and GIZ (BMWi 2016e). Part of the export initiatives is also the creation of a label "renewables made in Germany" with its own website, targeted at increasing the visibility of German renewable energy technology providers (Bundesregierung 2013b: 71). In a quite literal sense, increasing visibility includes the co-financing of solar rooftops on representative buildings in partner countries, to prominently showcase German technologies (Bundesregierung 2013b: 75). The renewable energy export initiative also included training measures in the framework of a dedicated qualification program, "Transfer Renewable Energy & Efficiency" (TREE). The program was discontinued in 2013, after having been transferred from BMUB's IKI initiative to the Federal Foreign

Office and eventually BMWi, where it was integrated into the export initiatives. In total, 1500 technicians and staff abroad where trained by a private company, RENAC, in all areas of renewable energy and energy efficiency operation and management since the program's inception in 2008 (D2).

The export initiative's activities, in addition to activities related to market analysis and marketing, include the development of projects in developing countries in the framework of a Project Development Program (*Projektentwicklungsprogramm,* PEP) (Stopper 2014, Rzepka 2013). Given that many growth markets for renewable energy are located in developing and emerging countries, the export initiatives tap Germany's longstanding relations with partner countries of development cooperation. The PEP component is implemented by GIZ and, in addition to target market analysis and study tours, also includes advice to foreign governments on the creation of attractive market conditions for renewable energy (Bundesregierung 2013b: 85). These activities are marginal compared to activities in the framework of traditional development cooperation, but synergies from longstanding partnerships are used. From a theoretical point of view, the PEP is a prototypical example of creating "export advantages", i.e., establishing favorable framework conditions abroad to leverage the strengths of a sector created as part of a lead market strategy at home (Beise 2003: 18, Jänicke & Jacob 2004).

5.3 Transfer Agents and Institutions

5.3.1 Active leadership: a network of institutions

A multitude of German transfer agents are involved in the promotion of renewable energies and energy efficiency in recipient countries, at different levels. Figure 16 presents the structure of Germany's "international Energiewende" landscape and puts agents in relation to the initiatives described above. The next sections address the main institutions and actors in more detail.

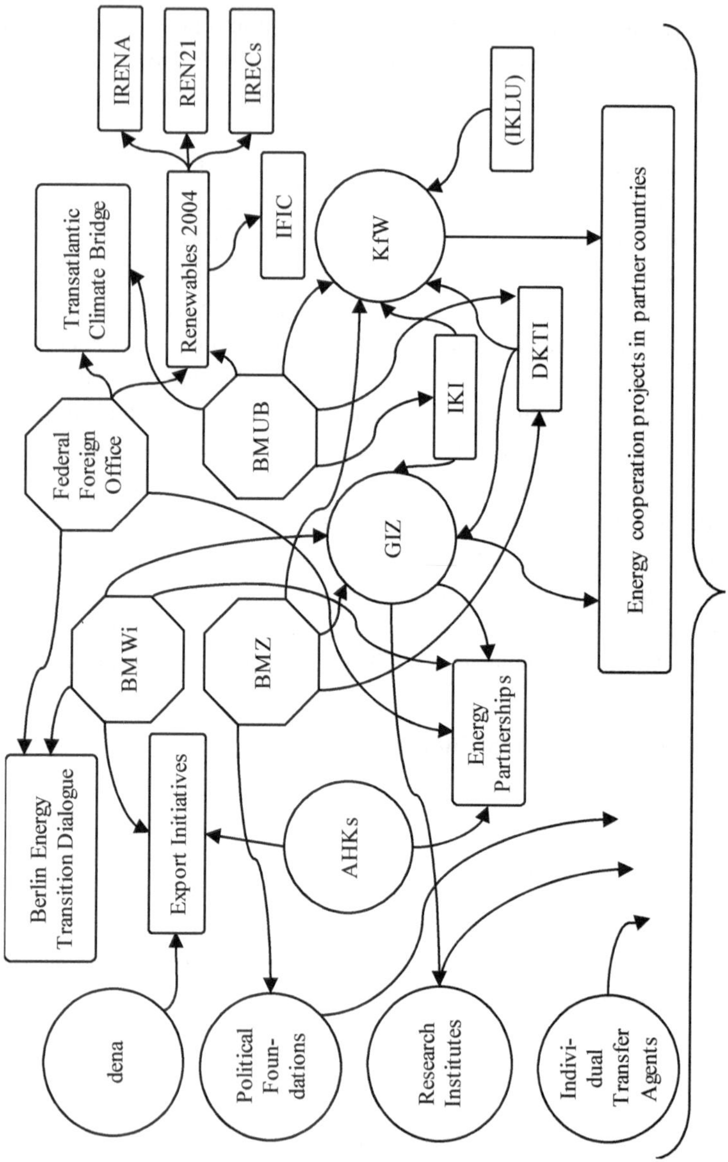

Figure 16: Network of institutions and programs involved in Energiewende outreach. octagons: ministries, circles: implementing agencies/actors, rectangles: projects, initiatives, initiated institutions

5.3.2 *Governmental departments*

Among the departments of the German federal government, the Environment Ministry BMUB (BMU until 2013) has traditionally played the driving role in international renewable energy policy (D6, D10). The organization of *renewables 2004*, its preparation and follow-up processes and the push for IRENA were decisively shaped by this ministry and its divisions such as the "International Affairs of Energy and Environment, Renewable Energy" unit. From its flagship initiative IKI to involvement in DKTI and frequent delegation visits carried out by ministerial staff (e.g., to California), BMUB has played a central role in international Energiewende leadership (D2, D4, D6, D10). The internationalization of BMUB's activities can be placed in a broader tendency of specialized ministries engaging in cross-border activities, especially through EU policy-making (D6, D10). BMUB continues to play an important role in international climate cooperation, but responsibility for renewable energy policy, including the unit in charge of international renewable energy policy has shifted to BMWi in 2013. In BMUB's 2016 budget, €338 million were allocated to projects with a focus on climate protection and biodiversity abroad (incl. loans), and another €400 million were already earmarked for the same issue for 2017-2021, 92% of which will need to qualify as ODA (Bundestag 2015b: 22). Another €5.5 million are to be spent on international cooperation, covering initiatives such as the Transatlantic Climate Bridge, whose annual budget of about €80,000 however appears moderate in comparison (personal communication with BMUB information services, 24 March 2016).

Within BMWi, there are now three units involved in international energy cooperation, one with a focus on industrial countries, one for developing and emerging countries, and one for export initiatives (BMWi 2016d)[45]. Eight staff members work exclusively on issues of bilateral energy policy cooperation and in 2016, a budget of approximately €7 million is available for "international cooperation for the deployment of renewable energies" (*Internationale Zusam-*

45 Units: II A 1 "General issues of international energy policy/external energy policy, multilateral cooperation on energy (excl. IEA), cooperation on energy policy with industrial countries"; II A 2 "Energy cooperation in the International Energy Agency, bilateral cooperation on energy policy with non-OECD countries"; II A 3 "Export Initiative Energy" (BMWi 2016d).

menarbeit Ausbau Erneuerbare Energien, personal communication with BMWi information services, 7 April 2016), with another €11 million already earmarked for this issue area for 2017-2019 (Bundestag 2015c: 52). This is in addition to €13.5 million for the renewables export initiative and €5 million for the efficiency export initiative in 2016 (Bundestag 2015c: 66).

The development ministry, BMZ, has been a driving force behind renewable energy promotion in developing countries, both in terms of policy advice and financing, since the Special Energy Program. Development Minister Heidemarie Wieczorek-Zeul stressed the link between domestic energy policy and development cooperation in a 2001 speech (Wieczorek-Zeul 2009 (2001): 36): "we need the same kind of goal-oriented promotion of renewable energy that we now have at home in the developing countries as well. I say this because no one should be under the illusion that the market will take care of things". In addition to major ODA contributions in the energy sector (see Figure 14), BMZ is also leading the management of Germany's global climate finance contributions. In May 2015, Chancellor Merkel pledged to double Germany's contribution to climate finance from the federal budget to €4 billion annually by 2020 and to mobilize an additional €10 billion by the same year (Kowalzig 2015: 4).

The Federal Foreign Office (*Auswärtiges Amt*, AA) has recently intensified activities related to international Energiewende leadership. In April 2013, a unit dedicated to the "foreign policy aspects of the German energy concept" (Unit 410-9) was created within the Foreign Office (D15, Auswärtiges Amt 2015c). Communication material on the Energiewende is prepared by the Foreign Office in Berlin for use by embassies and representations worldwide (Auswärtiges Amt 2015a, D15, S61), including a "Who is Who" of the Energiewende (Auswärtiges Amt 2015e). The Energiewende unit within the Federal Foreign Office also supports German representations abroad by providing presentation material, recommendations for speakers, and a helpdesk for embassy staff to answer technical questions on the Energiewende and assist staff with dealing with the growing number of requests regarding the Energiewende[46]. Some German embassies, including the one in Pretoria, now have their own energy/climate/environment units. In 2016, more than seven million euros are earmarked by the Federal Foreign Office for the issue area "Climate and Energy" (Bundestag 2015a: 16). This

46 Auswärtiges Amt (2015). Energiewende weltweit kommunizieren. Unpublished document.

does not include additional funds made for some delegation visits and workshops sponsored under the heading “Promotion of Germany’s image abroad”, an area to which more than €19 million were allocated in 2016 (Bundestag 2015a: 23, personal communication with Federal Foreign Office information services, 15 April 2016). The Federal Foreign Office also entertains a visitors’ program, in the framework of which several Energiewende related trips to Germany are organized annually for policy-makers, members of the civil society, and journalists from partner countries. Thirty-two official delegation visits to Germany with a focus on energy were sponsored or co-sponsored between 2010 and 2016, as shown in online material related to this book (personal communication with Federal Foreign Office information services, 15 April 2016; see also Auswärtiges Amt 2014, Ecologic n.d.). Although the majority of cooperation projects in the area of energy are sponsored by BMZ, the Federal Foreign Office has also carried out 59 projects with a direct connection to the Energiewende between 2011 and 2016, from capacity-building and training sessions, to the development of communication material. Project information, provided by AA, is presented in online material related to this book[47].

In 2016, the Federal Foreign Office launched an itinerant exhibition on the Energiewende, which will be shown in cities including Beijing, Cape Town, San Francisco, and Belgrade. Foreign Minister Frank-Walter Steinmeier (SPD, in office 2005-2009 and since 2013) officially inaugurated the exhibition in Beijing in April 2016: “When I talk about the German Energiewende, I have to worry that no one here in China will believe me… For that reason I have brought along […] a comprehensive exhibition on the German Energiewende[48]” (Auswärtiges Amt 2016b). A 20-page brochure made available to visitors stresses the growing number of requests regarding the Energiewende and its being “firmly embedded into an international framework” as motivations for the exhibition (Auswärtiges Amt 2016a). The most visible sign of the Foreign Office’s growing recognition of the Energiewende as a central issue of German foreign policy is the Berlin Energy Transition Dialogue (BETD). The BETD, co-sponsored by the Foreign

47 To access the book’s appendix, please follow the URL given in the imprint or visit www.springer.com and search for the author’s name.

48 “Wenn ich von der deutschen Energiewende erzähle, dann muss ich Angst haben, dass mir hier in China keiner glaubt… Deswegen habe ich zweierlei mitgebracht: Erstens, eine ausführliche Ausstellung über die deutsche Energiewende.”

Office and BMWi, took place as a two-day conference with over 700 attendees in 2015, and again brought hundreds of participants to Berlin in March 2016. Minister Steinmeier underlined the "dialogue" aspect of the format by stating that "each country, each region has to find a solution that suits its own requirements. The Energiewende doesn't come in 'one size fits all'" (Steinmeier 2016).

5.3.3 Implementing agencies

While cooperation programs are designed at the level of governmental departments, day-to-day interaction with stakeholders in partner countries takes place with staff from implementing agencies. GIZ is the most important interface of German technical cooperation, including in the energy field, in developing and emerging countries. GIZ was created after a merger between GTZ (German agency for technical cooperation), InWent (*Internationale Weiterbildung und Entwicklung gGmbh,* an agency focusing on capacity building) and DED (*Deutscher Entwicklungsdiesnt,* specialized in sending development workers) in 2011; it is – a private company with public-benefit status. With offices in 130 countries and total staff of over 16,000, of which 70% are local personnel from partner countries and 80% work abroad, GIZ is one of the largest development agencies in the world (GIZ n.d.-a). While more than 80% of GIZ's project volume comes from BMZ, other ministerial departments and institutions such as the European Commission also figure among its clients (GIZ n.d.-a). Whereas GIZ is responsible for the implementation of most of Germany's technical cooperation, including policy advice, training and capacity building, KfW Development Bank (abbreviated KfW) is responsible for the financial cooperation arm. It is the development branch of the *Kreditanstalt für Wiederaufbau*, Germany's public bank, and has offices in about 70 countries worldwide. The bank's other branches are also involved in the financing of exports and offer financing to private households and small businesses. This includes the implementation of government programs in the renewable energy realm, such as an electricity storage incentive program and preferential funding for renewable energy projects (KfW 2016a).

Other agencies, e.g., dena, are involved in the implementation of energy initiatives, like the export initiative for renewable energies as well as in the promotion of the "renewables made in Germany" label (BMWi 2016e). In addition to its activities in Germany, dena is active in international knowledge transfer with a focus on capacity building and training for energy efficiency, and has a geographical focus on Russia, China, Turkey and Central Asia (dena 2016).

5.3.4 Subnational actors

The focus of this book is on leadership activities stemming from the national level in Germany, but regional and local levels of government in the German federal system are involved in international cooperation for sustainable energy as well (on the role of municipal networks in climate policy diffusion, see Hakelberg 2014). At the level of *Länder*, energy agencies pursue internationalization strategies, to promote companies in the renewable energy and energy efficiency sectors from the region. The energy agency of North Rhine-Westphalia, for example, regularly organizes workshops on export markets and supports delegation visits (Energieagentur.NRW, n.d.). The Berlin Energy Agency signed a cooperation agreement with the Mongolian energy agency in 2012 (Berliner Energieagentur 2012). As shown in detail by Ralston (2013), German *Länder* have multiplied cooperation with other sub-national entities in the United States. The signature of a memorandum of understanding on cooperation in the energy and climate field between Baden-Württemberg and California in 2015 is one of the latest additions to sub-national cooperation (Under 2 MOU 2015, C2, C58).

5.3.5 Political foundations, NGOs, and research institutions

Non-governmental and research organizations as well as political foundations play important roles in international communication on the Energiewende from within Germany, as well as through their presence abroad. Each of the parties represented in the *Bundestag* (as well as FDP, who did not pass the 5% hurdle to enter parliament in 2013), collaborates with a legally separated, but affiliated, political foundation (*Stiftung)*. The foundations are highly involved in international cooperation, development cooperation and exchange with parties and civil society in other countries. Heinrich Böll Foundation, affiliated with the German Green party, has offices in 30 countries and projects in 60 countries (Heinrich Böll Stiftung n.d.-a). Konrad Adenauer Foundation, close to the CDU, has 80 offices abroad and projects in 100 countries (Konrad Adenauer Stiftung n.d.). Hanns-Seidel Foundation (CSU) entertains 4 international contact offices and supports projects in 60 countries (Hanns Seidel Stiftung n.d.). Friedrich Ebert Foundation (SPD) has 102 offices abroad (Friedrich Ebert Stiftung n.d.). Rosa-

Luxemburg Foundation (Left Party / *Die Linke*) operates 18 offices abroad (Rosa Luxemburg Stiftung n.d.). Friedrich Naumann Foundation (FDP) is present with offices in 60 countries (Friedrich Naumann Stiftung n.d.). Unsurprisingly, Heinrich-Böll Foundation is particularly active in addressing the international dimension of Germany's Energiewende and in communicating it to partners abroad (e.g., Nganga et al. 2013, Leidreiter 2012). The foundation's most emblematic activity in this regard is *www.energytransition.de,* a website that for some time was the only comprehensive source of information in English on the Energiewende. In addition to Heinrich Böll Foundation, Konrad Adenauer Foundation, and Friedrich Ebert Foundation are engaged in the international Energiewende field. They have conducted surveys on the perception of the Energiewende abroad and have pointed to the untapped potential of the Energiewende's international dimension (Hirsch 2015, Konrad Adenauer Stiftung 2013, Konrad Adenauer Stiftung 2014, Wettmann 2016).

The "international Energiewende" has been the subject of events and publications from research institutes and think tanks that are also involved in accompanying the implementation of the Energiewende in Germany. Agora Energiewende, a think tank dedicated to the Energiewende, has created a European team within its organization and an international team is to start work in 2016. Agora Energiewende has made available many of its publications on the Energiewende in English – a considerable improvement in the availability of information on the German model to stakeholders abroad (e.g., Agora Energiewende 2015a, Agora Energiewende 2015b, Agora Energiewende 2016b). Other research institutes and think tanks with activities and publications on the international or European dimension of Germany's Energiewende include the Potsdam-based Institute for Advanced Sustainability Studies (IASS, especially its Transdisciplinary Panel on Energy Change, e.g., Quitzow et al. 2016), Stiftung Wissenschaft und Politik-SWP (Röhrkasten & Westphal 2013; Westphal 2012), the German Institute for Development Policy-DIE (Messner & Morgan 2013; Messner et al. 2014), and the Environmental Policy Research Center at Freie Universität Berlin the author is affiliated with (e.g., Gullberg et al. 2014, Jänicke 2011, Jänicke 2012a, Marquardt 2014, Ohlhorst et al. 2012, Schreurs 2013, Solorio et al. 2014, Steinbacher 2015, Steinbacher & Pahle 2015, Tews 2015).

5.3.6 *Business organizations*

Since, as stressed above, much of the economic gain from Germany's lead market strategy comes from exports, the role of business organizations in international Energiewende efforts merits attention. The issue is most prominently represented by Germany's business-funded foreign Chambers of Commerce, the AHKs, in 130 locations worldwide (AHK n.d.). AHKs are involved in the implementation of energy partnerships and constitute a link between German businesses, including in the renewable energy and energy efficiency sector, and markets abroad. Specialized teams or even offices, as in the case of the AHK competence center in Cape Town deal exclusively with the renewable energy and clean technology sector. Although the promotion of policies was not a primary goal of AHKs in the case studies, there were regular contacts with decision-makers. German companies were brought in direct contact with partner country stakeholders in the framework of business tours, and AHKs got involved in policy debates, e.g., by submitting position papers in the framework of consultations (S23, S24, M29, M30). An example of international activities by the German renewable energy industry is the annual "Wind Energy and Development Dialogue" (Windenergie- und Entwicklungs Dialog, WEED), which is sponsored by the German Wind Industry Association (*Bundesverband Windenergie*, BWE), and carried out in cooperation with GIZ. Contacts, primarily from GIZ renewables projects worldwide, are invited to Berlin every year for a week of site visits and presentations about experiences with wind energy and renewable energy in partner countries and Germany (GIZ et al. 2015). Other business associations involved in international energy policy and business dialogue include the German-African Business Association (*Afrikaverein der Deutschen Wirtschaft*), with its annual German-African Energy Forum (Afrikaverein der Deutschen Wirtschaft 2016).

5.3.7 *Individual transfer agents*

In all cases analyzed in this book, individuals with a background in or experience related to the Energiewende, but unaffiliated with any of the implementation agencies played a – sometimes decisive role – in policy promotion and transfer. Hermann Scheer and Hans-Josef Fell stand out in this regard. Scheer's visits to California and interaction with California decision-makers at the local and state level as well as with renewable energy proponents had a lasting impact on the dynamics of how the German example was used in the debate (C23, C40, C45). His speeches and visits also brought the German feed-in tariff model to the attention of several interviews partners for the first time, even encouraging the creation of a pro-renewables advocacy organization, Renewables 100 (Renewables 100 Institute 2007). In South Africa, a German engineer with a prior Energiewende-related background in Germany strongly influenced Eskom's stance on renewables and promoted the idea of a "net FiT", directly transferred from the German model. The influence of these and other individual transfer agents will be discussed in the case studies.

5.4 Aims, Actions, and Followers: Discussing Energiewende Leadership

5.4.1 Motivations for Energiewende leadership and the German agenda

The question of what agenda Germany and different transfer agents pursue with international Energiewende activities is discussed on a case-by-case basis in the empirical part of this chapter. An analysis of *Bundestag* debates was carried out to identify overarching themes of Energiewende leadership[49]. In total, the protocols of 101 plenary debates were taken into account. These are all the debates that took place between the beginning of the 18th legislative term (fall 2013) and the end of field research for this book in spring 2015. The debates can only be an approximation of the importance and motives of Energiewende leadership, and a more longitudinal analysis of discourse surrounding the Energiewende's international dimension in the *Bundestag* and beyond is a highly relevant area for further research. As described in Chapter 4, the debate protocols were searched for terms including "Energiewende"; "Erneuerbar[e]" (renewable[s]); "Vorreiter" (frontrunner[50]); "Vorbild" (model); "Pionier" (pioneer); "global"; and "weltweit" (worldwide). Speeches in which the terms appeared in a relevant context were excerpted. The text segments are presented in online material related to this

49 See also Steinbacher and Pahle 2015: 8–10, where results for part of the debates used for this thesis are presented.

50 Due to a lack of direct, reasonable equivalent in German to the term "leadership", the term "Vorreiter" (frontrunner) is used.

book[51]. These passages were then coded with the set of codes presented in Table 8. As for the case studies, codes were developed from the data, after a review of all relevant text segments. One hundred and fifty-two passages were considered relevant for the topic of international Energiewende leadership (the number of coded segments is higher since overlaps of several codes for one segment are possible). Of the 150 passages, 63 were from speakers from CDU/CSU, 52 from SPD, 29 from the B90/Greens, and 6 from Die Linke (Left Party)[52].

A first relevant finding is that in more than one-third of all plenary debates in this time frame (35 out of 101 debates) at least one reference was made to the international dimension of the German Energiewende. Since none of the speeches was targeted *only* at the Energiewende's international dimension, this number is an indicator for how deeply intertwined the political debate on the Energiewende in Germany and international leadership are. Speakers overwhelmingly recognized and welcomed Germany's global leadership with regard to the Energiewende: "Of course we have to continue to play a leadership role, which Germany has always claimed and always fulfilled[53]" (Andreas Jung, CDU/CSU, Bundestag 2013b). Sometimes, this aim was even seen as a moral, natural obligation for Germany: "Everyone tells us: if the Energiewende does not work in your country, it will not work anywhere. So we have the [expletive] obligation and responsibility, to do something"[54] (MP [Member of Parliament] Eva Bulling-Schröter, Die Linke; Bundestag 2015d). Only 12 out of 512 codings included at least some criticism regarding the aim of finding followers: "We are exiting [nuclear], and of course we want to convince others globally to do that as well. But we cannot interfere with the sovereignty of other countries. We also don't want others to interfere with our Energiewende" [55] (MP Hiltrud Lotze, SPD;

51 To access the book's appendix, please follow the URL given in the imprint or visit www.springer.com and search for the author's name.

52 In the 18th legislative term, which began in October 2013, of the 630 seats in the Bundestag, CDU/CSU holds 310 seats, SPD holds 193 seats, Die Linke holds 64 seats, and B90/Greens hold 63 seats.

53 "Selbstverständlich müssen wir weiter die Vorreiterrolle einnehmen, die wir in Deutschland immer für uns in Anspruch genommen und immer ausgefüllt haben."

54 "Alle sagen uns: Wenn die Energiewende bei euch nicht geht, dann geht sie nirgends. Also haben wir die verdammte Pflicht und Schuldigkeit, etwas zu tun."

55 "Wir steigen also aus, und wir wollen weltweit natürlich auch andere gewinnen, das ebenfalls zu tun. Aber wir können uns in die Souveränität anderer Länder nicht einmischen. Wir wollen ja auch nicht, dass andere in unsere Energiewende hineinreden."

Bundestag 2014b). Statements that were somewhat critical of leadership aims were concentrated in debates on whether Germany should provide export guarantees to German companies involved in coal and nuclear projects abroad (BMWi 2014b).

A considerable number of speakers saw Germany's leadership as being in danger, especially among Green party speakers. The overlaps between party affiliation and codings are shown in Table 9 and illustrate that "threatened leadership" is a disproportionately frequent theme among Green members of parliament and is used as a critique of governmental climate policies: "You say Germany is a leader. I say: No, Germany used to be a leader, but in the meantime CO_2 emissions are rising again, and the development of renewable energies is unnecessarily slowed down"[56] (MP Katrin Göring-Eckhardt, B90/Greens; Bundestag 2014a).

As shown in Table 8, the most important theme overall was "leadership conditioned upon success". Speakers, especially from the Green party used international leadership as an argument to advocate for stronger efforts in the field of climate protection: "if it [the Energiewende] fails, the consequences would be disastrous, because other countries would of course then ask: 'If not even Germany succeeds in reaching its climate goals, why should we then do it?'[57]" (MP Bärbel Höhn, B90/Greens; Bundestag 2014d).

On the other hand, MPs from CDU/CSU, but also from SPD, used international leadership as an argument for reforms: "The next step will be a [reform of the electricity] market model. Once the market model functions, renewables will become a supporting pillar and will be competitive. Once we will have achieved this, then our Energiewende will also be an export hit"58 (MP Thomas Bareiß, CDU, Bundestag 2014d).

56 "Sie sagen, Deutschland sei Vorreiter. Ich sage: Nein, Deutschland war einmal Vorreiter, aber inzwischen steigen die Emissionen wieder, und der Ausbau der erneuerbaren Energien wird ohne Not ausgebremst."

57 "Wenn es scheitert, hätte das verheerende Folgen, weil die anderen Länder dann natürlich fragen würden: Wenn nicht einmal Deutschland es schafft, seine Klimaziele zu erreichen, warum sollten wir das dann tun?"

58 "Der nächste Schritt wird das Marktmodell sein. Wenn das Marktmodell funktioniert, werden wir es schaffen, dass erneuerbare Energien zu einer tragenden Säule und auch wettbewerbsgerecht werden. Wenn wir das geschafft haben, dann wird auch unsere Energiewende ein Exportschlager sein."

Table 8: Codes and memos for analysis of Bundestag debates on Energiewende leadership

Code	N° of coded segments	Memo (code description)
Implementation of leadership (LS)	2	Activities regarding how international Energiewende leadership could be implemented
LS = Economic advantages	8	Economic advantages stemming from intl. Energiewende leadership
GER is Energiewende leader	58	Germany as an intl. leader with regard to the Energiewende
Die Linke	6	Segment coded for the party "Die Linke"
External perception	34	Reference to perception from abroad
B90/Greens	29	Segment coded for the Green party speaker
CDU/CSU	63	Segment coded for CDU/CSU speaker
SPD	52	Segment coded for SPD speaker
LS aim	37	Leadership as explicit goal that should be achieved
LS threatened / in danger or maybe lost	33	Speaker sees Energiewende leadership as threatened
LS in Europe	12	German leadership in Europe related to Energiewende
LS aim refused / critical of LS aim	12	Germany should not aim for leadership, critique regarding leadership aim
LS - development nexus	5	Segment connects EW leadership and development cooperation
Leadership is / has been success if/because (general)	25	Energiewende LS has materialized
If climate protected -> followers	31	Leadership conditioned upon results in climate protection
If also eco success -> follower	28	Leadership conditioned upon maintaining economic success
Moral obligation / "natural" LS	23	Germany has a "natural" or "moral" obligation to be a leader in this domain
LS for climate reasons	54	Leadership is a German contribution to climate protection

The Energiewende's international leadership was described as depending on implementing it in a way that safeguards Germany's economic competitiveness: "We are showing the way: economic growth and climate protection can go hand in hand. The decisive thing is: Only if we show other countries of the world, who might not yet be doing as much in terms of climate protection, that both can go hand in hand, other states might import the 'German Energiewende', as it is called in English." (MP Anja Weisgerber, CSU, Bundestag 2014e). As this statement illustrates, climate protection was still the overarching goal of leadership in this type of statement, but simultaneous economic success was seen as a necessary condition to find followers.

Table 9: Overlaps between party affiliation and codes regarding Energiewende leadership in Bundestag debates. Bold and coloured fields highlight strong themes for a party

CODE	Die Linke	B90/Greens	CDU/CSU	SPD
Implementation of leadership (LS)				2
LS = Economic advantages			3	5
GER is Energiewende leader		1	**35**	**23**
Die Linke				
External perception (Außenwahrnehmung)	1	5	13	14
B90/Greens				1
CDU/CSU				
SPD		1		
LS aim	1	4	14	18
LS threatened / in danger / maybe lost	4	**22**	3	3
LS in Europe		2	6	4
LS aim refused / critical of LS aim		1	9	2
LS - development nexus	1	2		2
LS is/has been success if / because (general)		1	14	10
If climate protected-> Followers	2	**17**	5	6
If also eco success -> follower	1	2	**15**	10
Moral obligation / "natural" LS	2		10	11
LS for climate reasons	3	**16**	**24**	11

The themes that were present in parliamentary debates are also reflected in public statements by German decision-makers. During the 2011 press conference announcing the Energiewende package, Chancellor Merkel underlined that the Energiewende would make Germany "a frontrunner on the way to [...] an age of renewable energies" and that it can "show countries who decide either to phase-out nuclear energy or not to introduce it in the first place, that growth, jobs, economic prosperity, and an energy supply based on renewable energies can go together. That is the actual point we want to make"[59] (Bundesregierung 2011). Three years later, Merkel stressed the theme of leadership that is conditioned upon success, stating that the Energiewende would become an export hit if Germany succeeded in its implementation (Merkel 2014). Environment Minister Peter Altmaier (Altmaier 2013: 11–12) as well as Vice Chancellor and Energy and Economics Minister Sigmar Gabriel are other examples of top-level German decision-makers stressing that the "Energiewende should find followers" globally (BMWi 2014a).

One level deeper into the analysis, Germany has tightly linked its international leadership efforts in the sustainable energy domain to the way the Energiewende is designed and implemented. In particular, this has meant a focus on feed-in tariffs in policy advice and as a frequently used indicator of "successful" leadership (see section 5.4.3. below). The creation of IFIC or Germany's defense of feed-in tariffs at the European level (Solorio et al. 2014) illustrate this focus. Given the pioneering nature of Germany's energy transition and transfer agents' roots in the model, the important place of feed-in tariffs in outreach activities comes as no surprise. However, as the Morocco and South Africa case studies clearly show, Germany is becoming a "learning leader" and has adapted its advice offer to new models such as auctions. This broadened scope of leadership preceded the fundamental shift in Germany's own renewable energy policies, toward an auction-based mechanism to be used to determine tariffs for larger installations from 2017 onwards (Appunn 2016, BMWi 2015a). The case studies

59 "Wir glauben, dass wir als Land Vorreiter auf dem Weg zur Schaffung eines Zeitalters der erneuerbaren Energien werden können. [...] Wir glauben, dass wir den Ländern, die sich entscheiden, entweder aus der Kernenergie auszusteigen oder gar nicht erst einzusteigen, zeigen können, dass Wachstum, Arbeitsplätze, wirtschaftliche Prosperität und eine Energieversorgung in Richtung von erneuerbaren Energien zusammengehen. Das ist unser eigentlicher Punkt."

confirm the complexity of motives behind this leadership, but find a tendency toward perceived altruistic motives, in particular global climate protection, in line with the tenor of *Bundestag* debates.

5.4.2 Variants of German leadership

Germany's approach to Energiewende leadership can be compared to the "checklist" for leaders presented in Chapter 2, as shown in Table 10. While most criteria for leadership – a pioneering model, willingness to put it at the service of a collective goal, interest in finding followers, and actions to achieve this goal – can be clearly established, the effectiveness of leadership is to be assessed on a case-by-case basis. Germany's leadership covers a unilateral component linked to its domestic policy model and the aim to find followers. Active leadership spans the creation of "spheres of leadership" (Eckersley 2011) for example through its efforts in the creation of IRENA and the *renewables 2004* process, communication on the Energiewende and underlying policy problems, and structural leadership through technical and financial cooperation in the framework of development aid. Calls for a reinforcement of Germany's external Energiewende strategy have multiplied in recent years, in parallel to the extension of the scope of the project and an increase in its prominence in domestic debate (Li 2016, Messner and Morgan 2013, Morgan et al. 2014, Quitzow et al. 2016, Tänzler and Wolters 2014, Westphal 2012).

In summary, the development of Energiewende leadership over time can be illustrated as shown in Figure 17. An aim to have a global impact was present from the start of the Energiewende in Germany, followed by the creation of a strong domestic market for renewable energy (which contributed to driving down prices of renewable energy technologies, especially PV). On the basis of first visible success in Germany, the country took a more active role at the international scene. Bilateral cooperation and leadership efforts further increased with pressure to internationalize for German companies, the growing scope of the Energiewende, the evolution of the global climate change debate, and the resulting growing global attention for policy models that address this challenge.

The sequence shown in Figure 17 corresponds to a type of horizontal and vertical reinforcement (Jänicke 2013, Schreurs & Tiberghien 2007), where the domestic and the international arena for leadership are interlinked.

Table 10: Variants of German leadership through transfer. The symbol "⊙" denotes that a conclusion cannot be drawn without insights into recipient cases

<table>
<tr><td colspan="2">Pioneers</td><td colspan="2">...implement a policy innovation ahead of most other countries.</td><td>✓</td></tr>
<tr><td rowspan="5">Unilateral leaders</td><td rowspan="10">Active leaders</td><td rowspan="3">Provide a policy model that...</td><td>is visible and available.</td><td>✓/⊙</td></tr>
<tr><td>is perceived as legitimate.</td><td>✓/⊙</td></tr>
<tr><td>is framed to serve / serves a collective goal.</td><td>✓/⊙</td></tr>
<tr><td rowspan="2">Show willingness to...</td><td>contribute to address a collective goal.</td><td>✓</td></tr>
<tr><td>have others contribute to it as well.</td><td>✓</td></tr>
<tr><td rowspan="5"></td><td rowspan="5">Take measures to...</td><td>engage with recipients.</td><td>✓</td></tr>
<tr><td>create knowledge.</td><td>✓</td></tr>
<tr><td>communicate about policy and underlying issues.</td><td>✓</td></tr>
<tr><td>create channels through which transfer can take place.</td><td>✓</td></tr>
<tr><td>provide incentives to potential followers (material and non-material).</td><td>✓</td></tr>
<tr><td colspan="2">Effective Leaders</td><td colspan="2">... induce / facilitate action in follower countries that contributes to the achievement of the overarching goal / agenda for leadership.</td><td>⊙</td></tr>
</table>

5.4.3 *Effective Energiewende leadership?*

Given the importance of the international leadership theme in the Energiewende debate in Germany, surprisingly few attempts have been made to evaluate the effectiveness of its leadership activities. The most frequent indicator used to point to the international success of the Energiewende model is the global spread of feed-in tariffs. Although no causality can be inferred from this observation, the global adoption of feed-in tariffs is nevertheless the most frequent indicator used to point to Germany's actual Energiewende leadership.

The EEG is seen as "finding imitators in many countries" (Gabriel 2008) and accounts of the EEG frequently make reference to its being a "blueprint" and an "export hit". This narrative is shared across the renewable energy community in Germany – from NGOs, think tanks, and renewable energy advocates (Agentur für Erneuerbare Energien 2016, BUND 2014, Fell 2014, Greenpeace 2007: 4, Klima-Allianz Deutschland 2013, Leidreiter 2012, Rosenkranz & WWF Germany 2015: 7) to politics (Altmaier 2012c, BMU 2004, BMU 2007: 20–21, BMWi 2014, Bündnis 90/Die Grünen Bayern 2015, Krischer 2010, Land 2009), industry representatives (EUROSOLAR 2012) to research (Diekmann et al. 2012, DIW

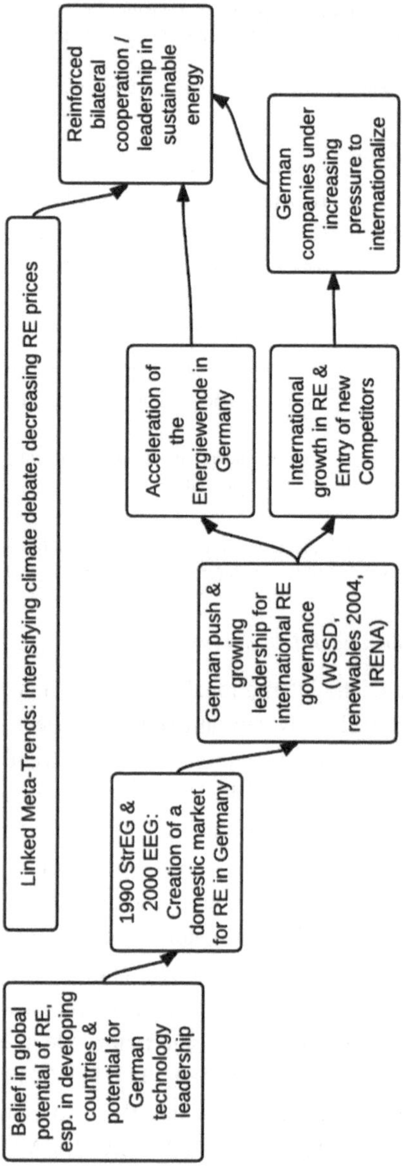

Figure 17: Evolution of German Energiewende leadership and meta-trends. Graph by the author

2010, Hirschl 2008: 562, Jacobs & Mez 2012: 264), development cooperation (Kückmann 2014, Wieczorek-Zeul 2009 (2001)) and media (Petersen 2010, Solarify 2014). As the analysis of Bundestag debates presented in this chapter shows as well, framing the EEG as a blueprint for policies around the globe and thereby appealing to Germany's responsibility has been a frequent strategy to underpin positions ahead of reforms of the law. The German Solar Industry Association (*Bundesverband Solarwirtschaft,* BSW), for example, invoked the EEG's global appeal in its position paper ahead of the 2008 amendment to the EEG (BSW 2008: 6). Government officials interviewed also pointed to the global diffusion of feed-in tariffs as a main indicator for Germany's leadership (D6, D7, D10, D15).

Jacobs and Mez (2012) see Germany's EEG as setting global trends, in particular with regard to policy design features such guaranteeing the tariff for a 20-year period and its calculation on a cost-recovery basis. The authors establish a link between the EEG and the global diffusion of feed-in tariffs through the fact that many feed-in tariff systems globally share crucial design elements with the German EEG and that the EEG is frequently referred to as best practice in review reports (Jacobs & Mez 2012: 267). While looking at Germany as a source is certainly justified, the diffusion of feed-in tariffs alone cannot serve as a sufficient metric for assessing the effectiveness of German Energiewende leadership. A causal link between the German EEG and the adoption of feed-in tariffs abroad is more frequently assumed than studied in depth (Bechberger et al. 2003: 18). Among the rare exceptions are Busch and Jörgens (2005a: 129–131) who provide additional evidence by pointing to the use of the German EEG example in policy-making in other European countries, including the Czech Republic, Austria, France, and the United Kingdom. Jacobs (2012b) addresses the co-evolution and convergence of feed-in tariff systems in Germany, Spain and France through in-depth qualitative case studies.

Konrad Adenauer Foundation, Friedrich Ebert Foundation and the German Agency for Renewable Energy have carried out surveys of the international audience's look at the Energiewende (Agentur für Erneuerbare Energien 2014a; Hirsch 2015; Konrad Adenauer Stiftung 2013; Konrad Adenauer Stiftung 2014). The surveys all show high interest in the Energiewende, pointing to the prominence of the model, which is a factor for policy transfer. The studies of emerging countries' views on the Energiewende conducted by Konrad Adenauer Foundation however also show considerable differences between countries, with stakeholders in India and China expressing far more admiration for the model than those in Russia and Brazil (Konrad Adenauer Stiftung 2014). A common thread across this survey was the perception of the Energiewende as a multi-objective project, with goals ranging from industrial policy to climate protection. Climate protection and environmental policy are nevertheless considered the most important objectives the Energiewende pursues (Konrad Adenauer Stiftung 2014: 6). This is an important result given the requirement of leadership being based on a model that is perceived to contribute to a collective objective.

The Energiewende also plays an important role in more general surveys of Germany's image abroad. Stakeholders interviewed in a publication on Germany's image in the eyes of stakeholders worldwide (GIZ 2015a: 49) stated great interest in the Energiewende, with more positive views among stakeholders in countries outside Europe. The strong interest in the Energiewende observed in these studies is in line with the experience of German officials interviewed (D2, D4, D6, D15). Frequent requests for information and a focus on the Energiewende in personal interaction with stakeholders from abroad are part of the day-to-day business of these officials. The information services of BMUB, BMWi, and the Federal Foreign Office all stated requests for information on the Energiewende were often informal and too numerous to be monitored systematically for statistical purposes (personal communication, Federal Foreign Office, BMWi, BMUB information services, March and April 2016). Although the 2011 Energiewende package and Chancellor Merkel's change in position further reinforced interest in the German energy policy model (D6), interviewees did not perceive 2011 as a turning point in their interaction with foreign officials (D2, D6, D7, D10). This is in line with case study results.

To the best of the author's knowledge, this book is the first study that focuses on an in-depth analysis of Energiewende leadership through policy transfer to particular recipient cases. The case studies presented in Chapters 6 (Morocco), 7 (South Africa), and 8 (California) explore Energiewende transfer as a rich, complex, and political process. The effectiveness of Germany's Energiewende leadership takes more subtle forms than a simple copy pasting of the German model, but is visible when considering different policy layers and the value of selective lesson drawing (see discussion, Chapter 9).

6 Case Study: Morocco

6.1 Case Introduction

6.1.1 Case overview

Morocco is a lower middle-income country with a per capita GDP of $3,190 in 2014 and a population of about 34 million (World Bank 2016b). The country's geographical location, stretching from the strait of Gibraltar and the Mediterranean in the north to the contested territory of the Western Sahara and the Mauritanian border in the south give it a strategic position. Morocco's claim to the territory of the Western Sahara (its "Southern Provinces") led it to leave the African Union, which recognizes the status of the Sahrawi Arab Democratic Republic as a state (Zisenwine 2011: 72). Morocco is a constitutional monarchy. The Moroccan king, Mohammed VI, ascended to the throne in July 1999, succeeding his father, King Hassan II. During the events of the Arab spring in 2011, protests in Morocco were mainly targeted at constitutional, social and economic reform rather than regime change and were contained by a referendum on constitutional reform reinforcing the powers of parliament and government in 2012 (Auswärtiges Amt 2015b, Ben-Meir 2015). Approximately 40% of Moroccans are employed in the agricultural sector and another 40% work in the services sector; general unemployment reaches close to 10% and youth unemployment approximately 20% (UNDP 2015b). The country ranks 126th out of 188 countries in the Human Development Index (World Bank 2016b).

The case study is structured into four sub-chapters. Subchapter 6.1 provides background information about the specificities of the Moroccan energy landscape. Subchapter 6.2 presents the framework for cooperation between Germany and Morocco in the energy space as well as German leadership activities. Section 6.3 traces policy transfer and leadership from the drafting of the first Moroccan renewable energy law in 2007 to recent debates regarding decentralization. Subchapter 6.4 draws preliminary conclusions and discusses findings with regard to this book's main research question.

K. Steinbacher, *Exporting the Energiewende*, Energiepolitik und Klimaschutz.
Energy Policy and Climate Protection, https://doi.org/10.1007/978-3-658-22496-7_6

6.1.2 Moroccan energy challenges

Morocco's energy system is under pressure from several angles. Since the country has only minimal domestic fossil fuel resources, its energy dependence on imports exceeds 90%. Morocco's economic and demographic growth caused a steep increase in total energy demand, which is also growing on a per capita basis because of rising living standards (Direction de l'Observation et de la Programmation n.d.). According to a 2014 IEA report, primary energy supply is dominated by oil (67.6%), followed by coal (16.1%), biofuels/waste (7.4%), natural gas (5.7%), electricity net imports (2.2%), hydro electricity (0.7%) and wind (0.3%); energy supply increased by almost 60% between 2002 and 2012 (IEA 2014: 15). Growth was even more pronounced with regard to electricity demand, with a rise of almost 80% between 2002 and 2012 (IEA 2014: 15). Morocco's energy consumption per $1,000 of GDP is at about two thirds of the IEA average (0.09 toe vs. 0.13 toe) while electricity generation per capita is more than 10 times lower than the IEA average (0.8MWh/capita per year vs. 9.3 MWh) (IEA 2014: 15); electricity consumption per capita was more than three times lower than the World average in 2013 (World Bank 2016). Domestic electricity generation in 2012 was mainly based on coal (43.4%), followed by oil (25.3%), natural gas (22.7%), hydro (6%) and wind (2.7%) (IEA 2014: 15). Recent data shows that the share of wind energy in the electricity mix increased to 6.18% and that of hydro to 7.19 % at the beginning of 2014 (Direction de l'Observation et de la Programmation 2014: 12).

The share of solar energy will only start to play a role in the electricity mix from 2016 onwards, with the official opening of the first phase of the Ouarzazate solar complex in February 2016 (KfW 2016c). Renewable energy targets in Morocco are expressed as a share of installed electricity generation capacity: by 2020, 42% of capacity is to be based on renewable sources. This goal translates into 2,000 MW each of solar, wind, and hydro to be installed by 2020 (MEMEE 2013b: 33). In 2009, the country also fixed a goal of 22-25% of electricity generation to come from renewables by 2020 and 27 to 32% by 2030 (MEMEE 2009: 15). By the beginning of 2016, the total capacity of hydro-electricity plants amounted to 1,700 MW, while wind capacity reached 750 MW and the first 160 MW of CSP were online. Total electricity capacity installed was about 8,000 MW (ONEE 2014a). In early 2016, Energy Minister Amara announced an expansion of renewable energy goals for 2030, including a goal of 10,000 MW renewable capacity to be installed between 2016 and 2030 (52% of total installed capacity by 2030), of which 4,500 MW would be solar, 4,200 MW wind and 1,300 MW hydro (MEMEE 2016), including the capacity covered by the 2020 targets.

Despite low per capita CO_2 emissions of just under 2t per year in 2012 (IEA 2014: 30; about five times lower than the global average), Morocco is an active voice in international climate governance. It ratified the Kyoto protocol in 2002 (Fritsche & Schmidt 2008: 45), hosted the COP7 in Marrakech in 2001 and will host the COP22 in the same city in 2016. A national climate change control plan was developed in 2009 and attention is increasingly shifting to climate adaptation since the country is heavily affected by changing climatic conditions, in particular decreasing precipitation (IEA 2014: 31).

6.1.3 Structure of the Moroccan energy landscape

The Moroccan energy sector is dominated by a vertically integrated, state-owned utility, the National Office of Water and Electricity (*Office Nationale de L'Electricité et de l'Eau potable,* ONEE; formerly: Office national de l'électricité, ONE). ONEE acts as a single buyer for electricity and produces about 41% of Morocco's electricity in its own plants, with the remainder being imported from Spain (18%) or purchased from independent power producers (IPPs), who operate a very large coal-fired power plant (Jorf Lasfar, 1,320 MW), a combined-cycle gas-fired plant (Taharddat, 380MW) and a 50 MW wind park (IEA 2014: 54).

Morocco's renewable energy act, Law 13-09, entered into force in 2010 and allows IPPs to feed electricity to the high-voltage grid and sell it directly to consumers – in principle including to consumers abroad (MEMEE 2010). The mid- and low-voltage level, where smaller installations could inject electricity to the grid, remained effectively closed at the time of field research. The situation is only slowly starting to change with the adoption in January 2016 of Law 58-15, which allows for the opening of the lower-voltage grid under circumstances to be defined by ordinances (MEMEE 2015). Due to a lack of a regulatory framework for small-scale installations, renewable energy development in Morocco has so far been limited to large-scale plants that sell electricity to the Moroccan Solar Agency (MASEN) or to ONEE (wind) through negotiated PPAs as part of the Moroccan solar and wind plans (Andriani et al. 2013: 6–7). Like in the case of

South Africa, electricity distribution in Morocco is only partly carried out by the state-owned utility ONEE (55% of distribution) while all electricity transmission is ONEE's responsibility. Private utilities at the municipal level, the so-called "régies" are in charge of electricity distribution to the remaining 45% of the market (ONEE 2014b). One of the specificities of the Moroccan electricity system is the existence of two intercontinental electricity transmission lines connecting the Moroccan electricity system to Spain underneath the strait of Gibraltar, with a capacity of 1,400 MW. Following a successful rural electrification program that partly relied on solar home systems and was supported by international donors including Germany, over 98% of Moroccan households had access to electricity in 2012, compared to just under 20% in 1995 (IEA 2014: 21).

Morocco's key institution for the development and implementation of energy policies is the Ministry for Energy, Mining, Water and Environment (*Ministère de l'Energie, des Mines, de l'Eau et de l'Environnement*, MEMEE). MEMEE is responsible for renewable and conventional energy, energy statistics and planning and has oversight over ONEE. It is a key partner for German development cooperation in the energy sector, for example for the energy advice project "Promotion of Renewables and Energy Efficiency for Sustainable Development in Morocco" (Promotion des Énergies. Renouvelables et de l'Efficacité Énergétique pour un Développement durable au Maroc, PEREN). Several state agencies were created to accompany the implementation of the Moroccan solar and wind plans and the 2009 Moroccan energy strategy. The Agency for the Development of Renewable Energies and Energy Efficiency (*Agence Nationale pour le Développement des Energies Renouvelables et de l'Efficacité Energétique,* ADEREE), replacing the former Center for the Development of Renewable Energy (*Centre de Développement des Energies Renouvelables*, CDER), was created by Law 16-09 in 2010. The agency has a practical focus, especially in the area of energy efficiency, including specialized training and education, labeling as well as cooperation with municipalities. Also in 2010, MASEN was established by Law 57-09 to oversee the implementation of the Moroccan Solar Plan, including the organization and oversight of tenders for solar energy projects. MASEN participates as a shareholder in solar energy projects, negotiates power purchase agreements (PPAs) and acts as an intermediary between ONEE and project developers (Andriani et al. 2013: 7). In addition to MEMEE and ADEREE, MASEN is a main project partner for programs implemented by GIZ such as the

"Accompanying the Moroccan Solar Plan" (*Accompagnement du Plan Solaire Marocain,* APSM) or the "German Climate Technology Initiative" (DKTI 1-3) projects. The state-owned Moroccan Society for Energy Investments (*Société d'Investissements Energétiques,* SIE) is in charge of identifying investment opportunities for private investors in the field of renewable energy and acts as a financial facilitator. With an initial capital of one billion Moroccan dirhams (approximately 100 million euros in 2015), SIE participates in wind energy, energy efficiency, and PV projects at the mid-voltage level. In 2014, SIE signed a cooperation agreement with Dii, the follow-up organization to the Desertec industrial initiative, to cooperate on identifying projects and investment opportunities in Morocco (M16, SIE 2014). Another institution created in 2010 to support the implementation of renewable energy in Morocco is the Institute for Solar and New Energies Research (*Institut de Recherche en Energie Solaire et Energies Nouvelles*, IRESEN). It focuses on bridging the gap between basic and research and commercial application by granting financial support to research projects on a competitive basis.

Private sector and civil society organizations play a limited role in Moroccan renewable energy policy making and implementation. The Association of Moroccan Solar and Wind Industries (*Association Marocaine des Industries Solaires et Eoliennes,* AMISOLE) represents the interests of the renewable energy industry and has lobbied for an opening of the low voltage level to small-scale solar installations, together with German transfer agents. Morocco's monarchical political system leaves little room for civil society actors and open political debate (Dalmasso 2012). As an interview partner working with Moroccan municipalities put it, "energy policy is not something everyone is talking about, it is not a societal project. Everything happens far away from the individual, something gets [agreed] on at a high political level, from the perspective of local actors, and they then implement it" (M25). Only recently have small grassroots organizations emerged that address current energy policy issues, such as the citizen initiative "Morocco without nuclear" (Maroc sans nucléair).

Table 11: Organizations covered through interviews in Morocco

ADEME: French Environment and Energy Management Agency	IDE-E: Institute for Development, Environment and Energy in Morocco
ADEREE: Moroccan Agency for the Development of Renewable Energy and Energy Efficiency	IRESEN: Moroccan Institute for Research on Solar and New Technologies
AFD: French Development Agency	KfW, Rabat office
AfDB: African Development Bank	Maroc Sans Nucléair
AHK: German Chamber of Commerce in Morocco	MASEN: Moroccan Agency for Solar Energy
Amisole: Moroccan Solar and Wind Industry Association	MEMEE: Moroccan Ministry for Energy, Mining, Water and the Environment
Cegelec: Morocco branch of Cegelec (French company)	Embassy of the Kingdom of Morocco in Berlin
CFCIM: French Chamber of Commerce in Morocco	Saharawind
CNRST: Moroccan National Centre for Scientific and Technical Research	SIE: Moroccan Society for Energy Investments
Consultants	Siemens Morocco & Tunisia
EIB: European Investment Bank	Economy and Trade section, Spanish Embassy in Morocco

6.1.4 Overview of field research

Field research was carried out in Morocco in February and March 2014. In total, 48 officials, experts and stakeholders from Morocco, Germany and third parties (France, Spain, international financial institutions), were interviewed in Rabat and Casablanca. Interviews were carried out in French and German primarily and direct interview quotes in this chapter were translated by the author.

Forty interviews included a ranking exercise on the objectives of Moroccan renewable energy policy. While conducting field research, the author was hosted by GIZ's Secretariat of the Moroccan-German Energy Partnership in Rabat. Interviews were carried out independently and questionnaire design, the choice of interview partners and analysis were the decision of the author alone. Interviewees were primarily chosen for their involvement in renewable energy policy

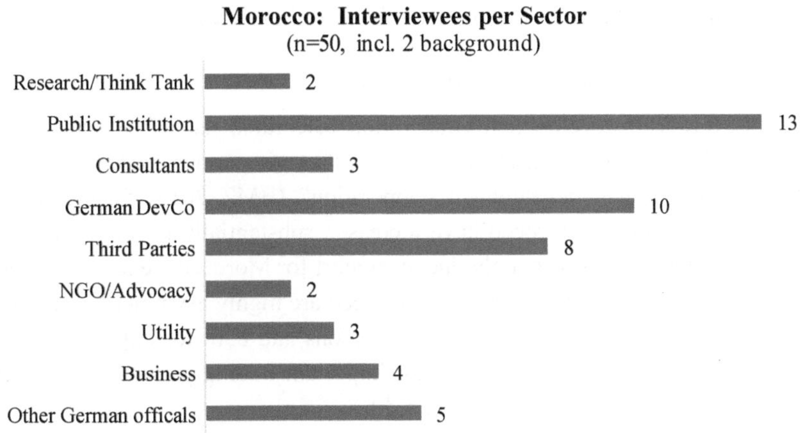

Figure 18: Number of interviewees per sector (Morocco). DevCo: Development cooperation

debates and implementation, which for almost all interview partners also meant involvement in programs and projects of Moroccan-German bilateral cooperation. To triangulate findings, especially given the lack of secondary literature in the Moroccan case, interviews with third parties (France, Spain, multilateral donors) and independent consultants were carried out.

Table 12: Interviewee affiliation by sector (Morocco). DevCo: development cooperation, MA: Moroccan, Inst.: Institutions

Identifier	Sector	Id.	Sector	Id.	Sector
M1	German DevCo	M18	MA Public Inst.	M34	MA Public Inst.
M2	MA Public Inst.	M19	Third Parties	M35	German DevCo
M3	German Others	M20	Third Parties	M36	German Others
M4	Third Parties	M21	MA Public Inst.	M37	German DevCo
M5	MA Public Inst.	M22	Third Parties	M38	MA Public Inst.
M6	MA Public Inst.	M23	German DevCo	M39	Research
M7	MA Public Inst.	M24	German DevCo	M40	German DevCo
M8	Consultant	M25	Consultant	M41	MA Public Inst.
M10	Utility	M26	Business	M42	Utility
M11	Third Parties	M27	Research	M43	Utility
M12	Business	M28	MA Public Inst.	M44	NGO/Advocacy
M13	German DevCo	M29	German Others	M45	Consultant
M14	Third Parties	M30	German Others	M46	MA Public Inst.
M15	MA Public Inst.	M31	Third Parties	M47	German DevCo
M16	MA Public Inst.	M32	Business	M48	German DevCo
M17	Third Parties	M33	NGO/Advocacy	M49	German DevCo

6.2 Moroccan-German Cooperation in Sustainable Energy

6.2.1 General framework for cooperation

German-Moroccan development cooperation celebrated its 50th anniversary in December 2013. Morocco has been seen as a strategic partner country "from the first hour of German development cooperation" (BMZ 2013: 5). Germany's ODA commitments to Morocco have increased substantially in recent years, as shown in Figure 19 below, mainly due to support for Morocco's energy program. Trade relations between Germany and Morocco are highly asymmetric. Morocco ranks 60th among Germany's export destinations and 65th for imports in 2014, while Germany was Morocco's 8th most important trading partner, with a clear upward trend (German Embassy Rabat n.d.).

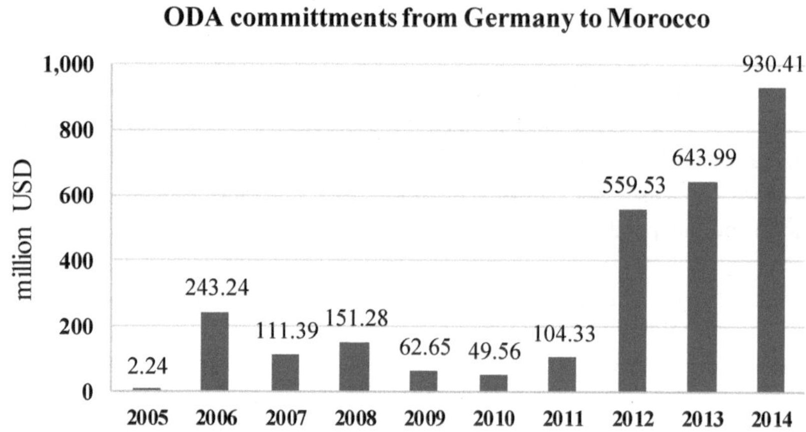

Figure 19: Annual ODA flows from Germany to Morocco, 2005-2014. Data Source: OECD, DAC (2016)

Due to Morocco's colonial past, close ties still link Morocco to France and Spain, with French being the commonly used language in media and administration (Zisenwine 2011: 77). Several international financial institutions, including the World Bank, the European Investment Bank, the European Bank for Reconstruction and Development, the Islamic Development Bank and the African Development Bank operate offices in Morocco. A major factor influencing Morocco's external relations is its geographic proximity to the European Union, with only 14 kilometers separating Morocco from the European continent. Morocco was granted an "advanced status" (*statut avancé*) by the EU in October 2008 and a liberalization agreement is currently being negotiated (Zisenwine 2011: 77).

Morocco entertains a particularly strong relationship with Germany in the energy field. It was the first of the Middle East – North Africa (MENA) countries to sign an agreement on the establishment of an energy partnership with Germany, in July 2012 (BMWi 2012a). In June 2013, Morocco's particular link to Germany and its commitment to developing sustainable sources of energy on its territory led to its membership in the Renewables Club initiated by German Environment Minister Peter Altmaier (BMU 2013). Morocco is one of two African countries to have been invited to join the now inactive Club, the other one being South Africa. Sustainable energy has been an element of German development cooperation with Morocco since the 1980s (Osianowski 1989), but efforts have considerably intensified since 2007-2008, first in the field of technical cooperation and policy advice and massively in the field of financial cooperation since 2012 and the launch of the Noor solar complex close to the city of Ouarzazate. Germany, through KfW and BMZ, is the biggest financier of the project, contributing €864 million out of the €2.2 billion total cost (KfW 2016b). The total portfolio of German financial cooperation commitments in the energy sector in Morocco currently amounts to over 1 billion euros.

6.2.2 Renewable energy: Germany's active leadership efforts in Morocco

In the area of technical cooperation and policy advice, numerous institutionalized channels of cooperation with Germany (see list of projects in Table 13) make Morocco a likely case for policy-transfer. On the other hand, its status as a developing country in a different linguistic and cultural area might hinder transfer processes. Morocco therefore constitutes a test case for the possibility of lesson drawing between industrialized and developing countries in the area of sustainable energy policies and the effectiveness of leadership in different contexts.

Germany's strong focus on Morocco as a strategic energy cooperation partner in the MENA region is no coincidence. Morocco's political will to implement a sustainable energy strategy, the political turmoil in other countries of the region during the Arab spring, and the vision of green electricity exports under the umbrella of the Desertec idea made Morocco a particularly attractive ally in the sustainable energy realm. The importance of energy as a field for Moroccan-German cooperation has increased tremendously. To reflect the dominance of energy as a field of cooperation, "renewable energy" has replaced "environment" in 2014 as one of the three focus areas of development cooperation between the two countries, the other ones being sustainable economic development and water (BMZ n.d.-a). Germany suggested this switch prior to the 2014 bilateral governmental negotiations, since (renewable) energy had already held a dominant position within the former "environment" focus area for years and should become more visible as a focus area (M23, M36). The focus of German efforts in its environment portfolio had shifted to renewable energy prior to the adoption of the Moroccan energy strategy in 2009. The shift coincides with additional funds made available to the energy department within BMZ, as part of an overall strengthening of energy cooperation and the Energiewende, from where money was then awarded to countries with a strong energy portfolio (M36). The reinforcement of sustainable energy cooperation was seen as a function of demand for specific support from Germany with its particular expertise (M35).

The main German actors in this bilateral cooperation in the energy sector with Morocco are GIZ in the domain of technical assistance and KfW for financial cooperation. GIZ is present with about 130 employees in Morocco; about 30 of these are local staff. Most GIZ projects as well as KfW loans to Morocco are funded or leveraged through BMZ, with the important exceptions of the German-Moroccan energy partnership (funded by BMWi) and 50% of the €8 million German Climate Technology Initiative project DKTI Moroccan Solar Plan, funded by BMUB. Table 13 provides an overview of technical cooperation projects in the field of energy that were active or being prepared at the time of field research.

The German Chamber of Commerce (AHK), based in Casablanca, co-leads the secretariat of the Moroccan-German energy partnership and assists German companies interested in entering the Moroccan market. AHK identifies market opportunities, organizes visits to Germany and Morocco for companies from both countries and provides information about the legislative framework and business conditions for German companies in Morocco.

Table 13: List of technical cooperation projects carried out by GIZ (bilateral and regional involving Morocco). Data Source: GIZ (2014c) and GIZ (n.d.-b)

Project Name	Spon-sor	Duration	Budget in EUR	Objective / Indicator for success
Promotion of renewable energies and energy efficiency (PEREN)	BMZ	01/2008-12/2014	7 Mio.	Strengthening institutional and professional capacities for the intensified use of renewable energies and energy efficiency.
Accompanying the Moroccan Solar Plan (APSM)	BMZ	01/2012-12/2014	3 Mio	Companies, applied research and training are better integrated in a "promotion policy" for the Moroccan solar sector.
German Initiative for Climate Technology – Moroccan Solar Plan (DKTI)	50% BMZ 50% BMU	10/2013-12/2017	8 Mio	The skills base and competences of companies, research institutes, training centers and employment agencies in the domain of solar technologies are reinforced.
Twinning MEMEE-BMWi	EU	03/2010-12/2013	850,000	Building capacities for energy planning and forecasting as well as monitoring within MEMEE. Strengthening institutional and legal framework conditions for integration with the EU framework.
Support to the Secretariat of the Energy Partnership (PAREMA)	BMWi	07/2013-12/2015 (next phase until 2018)	625,000	Cooperation projects which aim at increasing the use of renewable energies, energy efficiency and the reduction of greenhouse gases are identified and launched in the framework of the energy partnership with support by the Secretariat. The objective is a long-term partnership. The private sector should be actively involved in the dialogue.
DESERTEC and solar plan, consulting and support to the Union for the Mediterranean	BMU	08/2010-08/2013	1.5 Mio	Support of BMU in the framework of accompanying the process of developing and introducing international structures and framework conditions for the promotion of renewable energies leading to investments in renewable energy installations in the MENA region.
Development of a regional center for renewable energies and energy efficiency (regional project, RCREE)	BMZ	07/2008-08/2013	6 Mio	RCREE, the Regional Centre for Renewable Energies and Energy Efficiency is established as a regional center based in Cairo. With the help of RCREE, the diffusion of renewable energies and the use of energy efficiency are encouraged in the whole MENA region.

Project Name	Spon-sor	Duration	Budget in EUR	Objective / Indicator for success
Energy efficiency in the building sector in the Mediterranean region	EU	12/2009-12/2013	4.9 Mio	Contributing to the development of energy efficiency and renewable energies in the building sector and associated sectors in order to have an impact on the reduction of greenhouse gases.
Employment promotion through energy efficiency and renewable energies in mosques	BMZ	2015-2019	5 Mio	Increased business and employment opportunities in the field of energy efficiency and renewable energies on the basis of the Moroccan programme "Energy efficiency in mosques"
DKTI 3: Renewable Energy and Energy Efficiency in the Provinces of Tata and Midelt	BMZ	2015-2020	6 Mio	The capacities to make use of the development potential of Renewable Energy and Energy Efficiency (RE-EnEff) in the Provinces of Tata and Midelt are increased.
Employment through development of renewable energy (regional project) - REACTIVATE	BMZ	2013-2016	5 Mio	Improved framework for the development of markets for employment-intensive renewable energy technologies.
Support of the Moroccan Energy Policy	BMZ	2015-2017	2 Mio	The legislative and institutional framework for a market development in the energy sector, particularly for the deployment of renewable energies and the increase in energy efficiency, are improved.

On a subnational level, the German *Land* of Schleswig-Holstein initiated a cooperation project with CDER in the wind sector (funded through the International Climate Initiative IKI and the EU) in 2009. In 2014, a cooperation agreement between ADEREE and the Land of Hessen was signed, with a view to implement Clean Development Mechanism (CDM) projects in Morocco. Several German political foundations have offices in Morocco. GIZ also cooperated with Moroccan municipalities through ADEREE's "Jiha Tinou" ("My Region") program. The Konrad Adenauer Foundation (CDU), Friedrich Ebert Foundation (SPD), and Hanns-Seidel Foundation (CSU) operate offices in Rabat and Heinrich Böll Foundation (Green party) opened an office in mid-2014. Energy-related events have been organized by those foundations, such as a conference on a Moroccan-European energy partnership organized by Konrad Adenauer Foundation in June 2013 (El Aidi 2013).

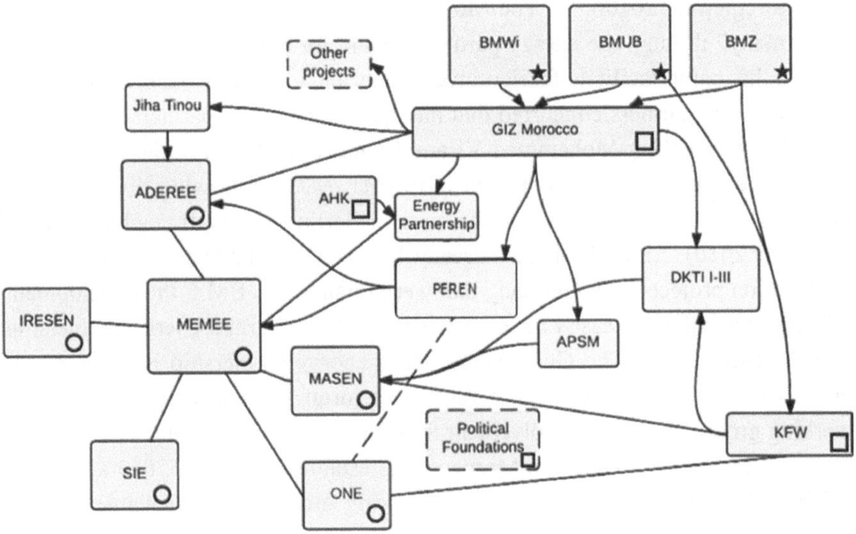

Figure 20: Moroccan renewable energy policy network and main transfer channels. Indicators: circles: Moroccan institutions, squares: German implementation agencies in Morocco, stars: German ministries, no indicators: projects. Graph by the author

6.2.3 The German-Moroccan energy partnership

Germany's and Morocco's interest in creating framework conditions for electricity exports and collaborate more closely on sustainable energy led to the creation of a bilateral energy partnership in July 2012 (BMWi 2012a). As can be seen from Figure 21, the creation of the partnership falls into a period of intensification of cooperation between Germany and Morocco in the field of energy. The creation of the partnership was influenced by lobbying efforts from the Desertec Industrial Initiative (M1, M13), but was also strongly desired by Moroccan counterparts. Dii had repeatedly contacted BMWi to deplore slow progress on the Mediterranean Solar Plan and stressed the necessity of a memorandum of understanding allowing exports via the Spanish-Moroccan interconnector and further into the European grid (M1, D4). The German government "support[ed] the Desertec initiative in establishing necessary contacts to politics and business in the North African countries" and accompanied the initiative on the political side

(Bundesregierung 2010b: 7). The Moroccan side hoped for a "direct connection to Germany" through the energy partnership and for a more direct political partnership that could facilitate foreign investments and electricity exports (M21).

Interview partners concurred that the idea of an energy partnership was discussed between King Mohammed VI and Chancellor Merkel at the UN Millennium Development Goals summit in 2010 (M1, M21), following discussions at the ministerial level in January of that year in Berlin (Botschaft des Königreichs Marokko 2010). Since Morocco already had a well-established set of energy cooperation projects with Germany that were financed by BMZ, the development ministry was all but pleased to see BMWi establish a formal energy partnership with Morocco (M3). The German-Moroccan energy partnership is not a development cooperation project, but a high-level forum for exchange organized in working groups, and with involvement from the public and private sector (M30). The partnership's secretariat in Morocco was established in July 2013 and is co-led by GIZ and AHK. Initially, three working groups on "renewables", "Desertec", and "sustainable development" were created and led by Moroccan and German co-chairs (M30). A fourth working group on electricity grids was added in 2013 with a focus on the integration of renewables to the grid. Reflecting the different priorities (and some competition) between German ministerial departments, the working groups are managed by BMU/BMUB, BMWi and BMZ respectively. The October 2013 energy partnership meeting included an "energy transition dialogue", where experiences from the German Energiewende were translated to the Moroccan context[60]. In the framework of the partnership, a study tour to Germany took place in December 2013 to transfer German experiences regarding the management of electricity grids with higher shares of renewables (D4, M13). A workshop organized that same month discussed opportunities for an opening of the low-voltage level and the deployment of PV in Morocco[61]. In 2015, a dialogue between German municipalities who are pioneers in the energy transition and Moroccan municipalities was organized (M48).

60 Atelier Maroco-Allemand sur la transition énergétique. [Moroccan- German Energy Transition Workshop]. 29.10.2013, Hotel La Tour Hassan, Rabat. Agenda. Unpublished Docum.

61 Atelier d'information et de préparation pour un "Programme intégré du PV" dans le cadre du Partenariat énergétique maroco-allemand, 5-6 décembre 2013 – Hôtel Ibis, Gare Rabat-Agdal, Rabat. [Information and preparation workshop for an "integrated PV program" in the framework of the German-Moroccan energy partnership] Agenda. Unpublished Document.

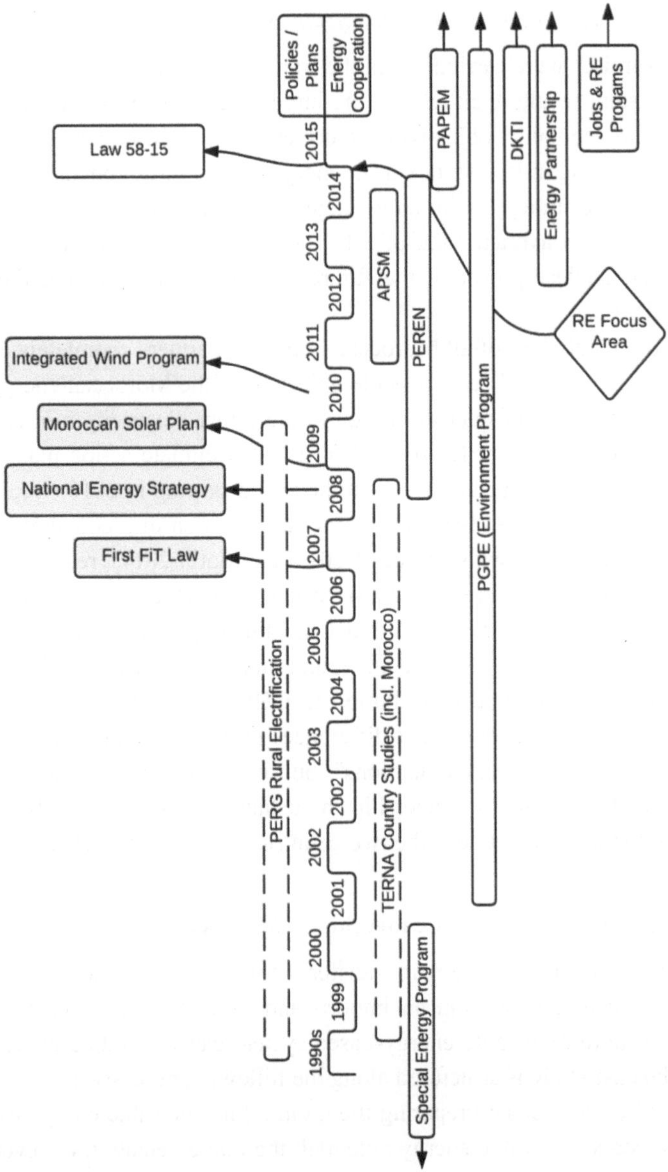

Figure 21: Timeline of main renewable energy policy events and cooperation projects in Morocco. Graph by the author

6.2.4 Moroccan energy transition objectives

Ranking exercises were carried out in the framework of 40 of the interviews and led to consistent results highlighting the importance of energy independence as well as socio-economic objectives as drivers behind Morocco's new energy strategy. Increasing independence from energy imports was considered the main driver behind the strategy, with an average rank of 1.9, followed by the goal of creating jobs, at an average rank of 3.1. The least important objectives were an aim to avoid nuclear power in the future (8.0) and decentralizing the energy system (7.4).

Approximately half of all Moroccan interview partners completely excluded "avoiding nuclear power" as a possible objective of the Moroccan energy strategy. The potential role of nuclear energy in the future electricity mix is still unclear in Morocco (Steinbacher 2015: 41). Its 2009 climate action plan included plans for 1,300MW of nuclear energy to be added starting in 2013 (Département de l'Environnement 2009: 14). But despite the adoption of a cooperation agreement on nuclear energy with France in 2014 (Le Matin 2014), recent announcements point to a reinforcement of the Moroccan renewable energy strategy rather than a shift toward nuclear (MEMEE 2016). Climate protection as a goal for the Moroccan energy strategy received a medium rank (4.9), which is still significant given the country's limited contribution to GHG emissions on a global scale. Interviewees were conscious about differences in the motivation behind Morocco's and Germany's "energy transitions", but these did not preclude transfer. In particular, a shared aim for leadership in sustainable energy provided a fertile ground for cooperation between the two countries (M16, M19, M46).

6.3 Tracing Transfer and Leadership in Morocco

The role and influence of German transfer actors and experience in Morocco's early efforts to transition from an import- and fossil fuel-based energy supply system to a more renewable energy-based system can be studied along several phases. The case study is structured along the following processes:

Phase 1 (1986-2004): Preparing the ground for renewable energy by identifying Morocco's renewable energy potential; the implementation of several pilot projects and Moroccan-German cooperation in the Special Energy Programme (section 6.3.1).

Phase 2 (2006-2008): Development of the first renewable energy law and of decisive energy scenario studies sponsored by Germany (sections 6.3.2 and 6.3.3).

Phase 3 (2008-2010): The legal framework for renewables and efficiency is defined, based on a new Moroccan energy strategy. The Moroccan solar and wind plans are announced; Germany's role in technology choice is visible and controversially discussed. The Desertec idea gains momentum (sections 6.3.4 to 6.3.9).

Phase 4 (2014-present): German efforts concentrate on consolidating the implementation of the national solar and wind plans, but also on bringing back decentralization and PV to the agenda (section 6.3.10).

6.3.1 Beginnings of German-Moroccan energy cooperation

As BMZ officials expressed it at the occasion of the 50th anniversary of cooperation between the two countries, Germany's experience in the field of renewable made it "the natural partner" and a "pioneer" in this domain in Morocco (BMZ 2013: 5). GTZ was involved in Morocco's first steps toward identifying its renewables potential and hence in in preparing the ground for future developments (Enzili et al. 1998, Marquardt et al. 2015). Other countries also became engaged early on in energy cooperation with Morocco, in particular the United States via its development agency USAID. A USAID-Morocco Renewable Energy Project was put in place as early as 1981 to "assist the Ministry of Energy and Mines to create a Center for Renewable Energy Development" (Kohler & Kreith 1985: 1), following recommendations from several "feasibility studies and analyses" from the late 1970s (Kohler & Kreith 1985: 4, USAID 1981: 2). Germany reacted to USAID's and the World Bank's activities by signing an agreement on the establishment of a Special Energy Program with Morocco in 1986 (*Sonderenergieprogramm,* SEP, see Osianowski 1989, Osianowski 1997), to "cooperate on the adaptation and deployment of systems for the use of renewable energy sources" (Osianowski 1989: 90). GTZ's focus was on supporting CDER, which had been established with the help of USAID in 1982 (Osianowski 1989).

Among the major contributions of the Special Energy Program were wind measurements taken between 1991 and 1994 for sites along the Atlantic coast and in the North of Morocco (Enzili et al. 1998: 44). They clearly showed that Morocco has some of the most promising sites worldwide for the deployment of wind farms (Enzili et al. 1998: 43). German support was decisive for translating the excellent potential for wind energy into a first small pilot wind energy project on Moroccan territory in Al Koudia Al Baida (Malgas et al. 2008: 22). ONEE, which received a €4.35 million loan from KfW for the project, became familiar with renewable energy technologies through this and other pilot projects and considered it a success (Malgas et al. 2008: 22).

In parallel to the activities of the Special Energy Program, lesson drawing based on the German example took place on a more micro-level in the late 1980s and 1990s. Aziz Bennouna, a pioneering Moroccan engineer in renewable energy trained in Germany, pushed for German-Moroccan cooperation in the energy sector in the late 1980s. Bennouna submitted a policy paper on the opportunities for mutual cooperation to the Moroccan Ambassador in Bonn in 1989, from where it was transferred to German president Richard von Weizsäcker (M39) and the Moroccan King Hassan II (Bennouna et al. 2007: 8). Positive reactions in both countries subsequently led the Bavarian Ministry for Environment to finance a study on the "large-scale use of solar energy" and to organize a visit by the Moroccan energy minister to Bavaria and companies such as Flagsol in 1991 (M39), starting of a history of cooperation in the field of solar thermal electricity.

Following encouraging results from wind measurements, CDER's director general from 1994, Ali Fassi Fihri, who was to become the director general of ONE in 2008 "politicized the issue" of renewables with the support of GTZ's Special Energy Program (M18, M23). The first use of renewable energy outside pilot projects then took place within Morocco's highly successful rural electrification program (*Programme d'Electrification Rurale Global*, PERG, where the installation of 16,000 solar home systems was co-financed by a preferential loan from KfW (Komoto et al. 2009: 137, ONE 2014). Despite these achievements, the overall performance of CDER during the Special Energy Program was judged weak and no dedicated energy project was in place between the end of the program in 2000 and the start of a new program in 2008 (M23). During this period of time, Morocco was however part of GTZ's global Technical Expertise for Renewable Energy Application (TERNA) project assessing wind power potential and framework conditions in several countries including Morocco and South Africa (GTZ 2009).

In summary, Germany contributed to laying the groundwork for future renewable energy policy developments in Morocco through the identification of renewable energy potential and pilot projects (IfaS & GTZ 2010, Jäger 2011, Liptow & Remler 2012: 93). These were essential for mitigating reluctance to adopting policy frameworks for the introduction of these new technologies: "in the framework of this rural electrification program with decentralized solar energy, we organized exhibitions at the markets for citizens, trained installers and launched pilot projects at mosques. All this was the fruit of Moroccan-German cooperation. (…) And all this contributed a lot to launching the big programs later on" (M34).

6.3.2 A game changing study on energy scenarios

Cooperation between Germany and Morocco picked up speed again in 2005, with a change in leadership within CDER. CDER's new director-general reactivated cooperation between the center and GTZ that had come to an end five years earlier. Together with GTZ's environment program staff in Morocco, the idea was developed to carry out a large-scale study on energy scenarios and perspectives for renewable energy in Morocco (M2, M23, M47). GTZ was approached specifically because "Germany was a leader in renewable energies" (M2). The main motivation was to "be able to provide arguments to decision-makers in support of a new dedicated strategy for renewable energies […] to deliver the arguments to show Morocco had to build a strategy based on [its] resources" (M2). Earlier assessments of Morocco's renewable energy potential supported by Germany, albeit useful, could not "serve as a global argument, as a specified, strategy, a scenario. We needed to launch a serious study" (M2). The study was awarded to a consortium of researchers, in particular from Fraunhofer Institute for Systems and Innovations research (Fraunhofer-ISI), a leading German research institution in the field.

A first version of the 240-page energy scenario study[62] was published in June 2007 and was finalized at the end of 2007 (Roller et al. 2007). Ministerial staff and Morocco's new Energy Minister Amina Benkhadra, who entered office in October 2007, were involved in the genesis of the study, with "[German] experts being present at all meetings and discussing directly with the minister", according to a ministry official (M2). Study coordinators saw the attention from

62 The study was commonly referred to in interviews as the "Fraunhofer" or "Fraunhofer ISI" study.

the minister as a decisive factor for the study's impact, but other channels of influence were used as well. The Royal Institute of Strategic Studies (*Institut Royal des Etudes Stratégiques*), which provides advice directly to King Mohammed VI, examined the study and "was convinced" (M2). Interview partners stated the study was even "read with a lot of interest in the royal cabinet" (M1). A high-ranking official from MEMEE confirmed that the study played a "very important role, be it for the development of renewables or for putting in place a legislative framework [...] they showed the existing capacity and the potential in Morocco, the necessity of putting in place a legal and regulatory framework - it was very, very important" (M5).

The study includes four detailed energy scenarios, with different shares of renewables and technologies (wind, solar PV, CSP, biomass) and outlines the necessary regulatory steps to implement a new energy strategy based on more renewables. It also provides estimates of expected job creation and industrial policy potentials – crucial elements given the policy priorities revealed through the ranking exercises. The study thereby introduced clarity about expected effects and options and decisively strengthened the knowledge base for decisions when pressure for energy policy action mounted rapidly in 2007 and 2008, due to rising fuel prices and demand. The study's specific influence on technology choice, in particular CSP, and on the controversies around this choice are discussed later in this chapter.

The impact of the GTZ-sponsored study can best be appreciated in comparison to an energy study by McKinsey, a global consulting firm, which was commissioned by MEMEE (M5, M18). The results of the study are not publicly available, but several officials independently stated it recommended putting coal at the core of the Moroccan energy strategy and that it emphasized stronger liberalization of the energy sector (M2, M5, M18). The impact of the McKinsey study was put into question by a number of observers. In contrast, the 2007 GTZ study was seen as key since it was "the first concrete study to be available before the launch of the new energy strategy, the first one that intervened before that" (M5). An official from MEMEE also confirmed that coal would have continued to play an important role in the Moroccan energy mix anyway to respond to the exceptional situation of rapidly rising demand in 2008 and that – contrary to the

GTZ study and its longer term impact – the McKinsey study did not trigger any major change (M5)[63]. Another official confirmed: "I would say they put the McKinsey study in a drawer and we never heard of it again" (M2). Interview partners recalled World Bank staff as seeing the GTZ efforts very critically in the beginning, especially since World Bank officials in Morocco at that time had a strong preference for wind and the energy scenarios developed in Roller et al. (2007) put a stronger focus on solar (M47). A change in staff and overall strategy at the bank was to reduce this opposition in the following years.

6.3.3 Morocco's first renewable energy law

In 2006, the World Bank formulated the adoption of a law on renewable energy and energy efficiency as an indicator for its 100-million-dollar energy policy development loan to Morocco (World Bank 2009a). In the framework of the World Bank's energy loan, consultants for the bank developed a draft renewable energy law for Morocco "behind closed doors", to provide a regulatory framework for investments, especially in wind energy, which the World Bank's loan could finance (M47). Becoming aware of the, in their eyes "totally insufficient", draft law in 2007 (M47), GTZ added discussions on the legislative framework for renewables to work that was already ongoing for the abovementioned energy scenarios study. The aim, for German and Moroccan counterparts, was to counter the unilateral move by the World Bank with a more inclusive and participatory process in the framework of a working group that was in place anyway for the study; it included ONEE, MEMEE, CDER and other officials (M2, M47). Since no specific energy-focused bilateral project was active at that time, a "special studies fund" and the environment program were used to provide advice for the development of a first law on the promotion of renewable energy and energy efficiency in Morocco (Fritsche & Schmidt 2008: 59). While the World Bank "put the topic [of a renewable energy law] on the agenda" its draft did not translate into a legal framework (M47), contrary to the draft law prepared in the process headed by GTZ. To accompany the legislative drafting process, GTZ orga-

63 It did however serve as the basis for the urgency measures taken in the framework of the National Priority Actions Plan (PNAP), such as differentiated tariffs and demand-side measures to alleviate pressure on Morocco's electricity generation capacities in the short term (World Bank 2008b: 13).

nized meetings with relevant Moroccan stakeholders, ensured and sponsored the participation of national and international short-term experts to those meetings, which were interwoven with the development of the energy scenario study (M1, M47). GTZ's internal success indicator for this project was to present a full legislative draft for renewables and efficiency to the Moroccan government, through MEMEE. The Moroccan government then adopted the first renewable energy law in Northern Africa in May 2007. The law foresaw a simplified permitting process for renewable energy auto-production installations below 2 MW (instead of 50kW before), the possibility for independent producers to export electricity abroad and some form of compensation for the feed-in of renewable energy, to be defined later (Fritsche & Schmidt 2008: 56).

Not only was GTZ's important role in the draft recognized (Fritsche & Schmidt 2008: 56, Kalman 2007). News reports at the time even stated that Morocco had "imported" the German EEG (Klimaretter 2007, Kalman 2007), calling it the "Moroccan EEG" (Mena-Institut 2007). It was hoped the law would provide attractive conditions also to German investors, as underlined by the GTZ program leader in charge of supporting the drafting process (Kalman 2007). However, the legislative process was eventually halted after the adoption of the draft law by the government. Due to national elections in the fall of 2007, the parliamentary adoption process of the law could not be completed. Given the 2008 energy crisis and the urgency of adopting a new National Energy Strategy first, a modified renewable energy law, Law 13-09, was only adopted in 2010.

6.3.4 The 2008 National Energy Strategy

A combination of shocks required urgent political action and led to a new Moroccan National Energy Strategy in 2008. The share of electricity imports in the country's electricity supply had more than tripled between 2005 and 2008 (MEMEE 2009: 65). As a head of unit from MEMEE put it, there was also a "special dynamic in 2008, characterized by erratic variations of oil prices, which reached up to 150 dollars per barrel" (M34). Interview partners agreed that by 2008, a new energy policy "wasn't a choice anymore, but [there was] an obligation to do something else" (M2). Foreign observers such as the US ambassador to Morocco perceived the urgency as well, as becomes clear from a leaked document: "the 2008-2025 [energy] strategy has consumed the Ministry of Energy for the last five months, with key officials working eighteen hour days without respite [...]

key decisions could not be delayed […] action to invest in new production facilities and to select the fuels on which they would be based was essential" (Riley 2008: 1). The influence of the energy scenarios study and of German transfer actors was to show concrete, detailed policy options (M45) and to demonstrate that renewables could make an important contribution to Morocco's energy needs while bringing socio-economic benefits at this time of crisis. The presence of ambitious renewable goals in the 2008 National Energy Strategy can therefore be regarded as an achievement of the energy scenarios study GTZ had sponsored and of German transfer agents working closely with Moroccan partners.

The first comprehensive energy strategy was presented to the public in March 2009, but already a year earlier, on 15 April 2008; a first draft strategy had been presented to King Mohammed VI. It received his confirmation in July 2008 (MEMEE 2008: 7). In his annual speech from the throne, the king underlined the need for short-term urgency measures and pledged to "vigorously pursue efforts to make alternative and renewable energies the cornerstone of national energy policy" (MEMEE 2009: 4), setting the tone for a new era in Moroccan energy policy. The king summoned his government to "accelerate the implementation of necessary legislative and regulatory measures to institutionalize relevant provisions for energy efficiency and renewable energies in all economic and social sectors" (MEMEE 2009: 5).

Of particular relevance for the question of policy transfer is the king's recognition of the role of foreign knowledge for "the choice of strategic options […] in a spirit of concertation and consensus" in the draft energy strategy (MEMEE 2008: 6). The accelerated transfer of technologies between developed and developing countries is also mentioned as a necessary condition to respond to energy challenges of the future (MEMEE 2008: 5). The draft energy strategy still had coal as central to the energy mix in the short and medium term (MEMEE 2009: 22), but fixed several objectives for renewables: their share was to represent 10% in the total primary energy mix by 2012, and 20% of electricity consumed (MEMEE 2008: 7).

German advisors also had a very positive view of the study's impact on Moroccan policy choices: "as far as the conception of the energy strategy is concerned […] they built upon our 2007 study on what was feasible and implementable in Morocco […] our study […] reached the royal cabinet. A counselor to the King in the environmental field was highly interested and the King himself received it [the study] with 'open ears'. And then the royal speech of 2008 was all focused on renewables and energy efficiency. [...] It is impressive to see how the puzzle pieces go together, sometimes it is crazy stories and coincidences" (M1).

6.3.5 An unexpected turn toward solar in Moroccan energy policy

While the energy strategy is one turning point in Moroccan energy policy, even more dramatic change was to come from decisions later in 2009 on how the plan would be implemented. The March 2009 energy strategy saw solar energy – "including for the European Union" – as a "long-term option without immediate investment" (MEMEE 2009: 22). The strategy therefore set a long-term target of only 740 MW for CSP and 400 MW of PV by *2030*[64] (MEMEE 2009: 25). For CSP, the overall *realizable* potential "excluding projects for [electricity] exports" is indicated in the energy strategy with 470 MW by 2020 (MEMEE 2009: 107).

A strong role for solar energy, let alone for CSP in the Moroccan short-to mid-term energy mix could therefore not have been expected based on the March 2009 National Energy Strategy. But only six months later, King Mohammed VI announced the Moroccan Solar Plan. Not only does the Moroccan Solar Plan put 2,000 MW of solar energy by *2020* at the very heart of Morocco's energy policy, but also decides its implementation be started with a 580 MW CSP complex (KfW 2016b) – more than what was seen as *realizable* half a year earlier. Wind energy and hydro capacities were also to be increased to 2,000 MW each, but from higher levels (ONEE 2010).

The important place for solar energy in the mix (14% of installed electricity capacity by 2020, up from virtually zero in 2009) was as surprising to observers as was the choice of CSP for the first Moroccan Solar Plan project, "Noor[65]". An interview partner heading a research unit met with the minister five days before the decision "but absolutely nothing made [me] think this announcement could

64 Wind energy targets in the plan (2,280 MW by 2020) are in line with later decisions on the Moroccan integrated wind program.

65 With a capacity of 580 MW developed in three phases close to the central Moroccan city of Ouarzazate, Noor is the world's largest concentrated solar complex. The first part of the park was officially inaugurated in February 2016 and the project is heavily supported by Germany financially: KfW development bank, upon demand from BMZ and BMUB, funds 40% of the loans provided for it (KfW 2016b).

be made only five days later [...]. Nothing leaked. If the minister herself had known, I would have noticed something. [...]. Before, nobody wanted to listen and all of a sudden – the solar plan. [...] Until today, I have no clue who insinuated this idea to the King. [...] It remains a puzzle to me. I think it came from outside, from abroad. [...] nothing presaged this would happen" (M27).

6.3.6 Germany's role in the Moroccan Solar Plan

Germany's role in the Moroccan energy policy decision for solar energy and CSP was crucial. It unfolds along the following dimensions: the creation of the epistemic basis for CSP in Morocco through studies, awareness-raising among key officials, structural incentives through preferential loans, in addition to incentives provided by the multilateral Clean Technology Fund, and very importantly the vision of exporting "green" electricity to Europe and Germany in the framework of the Desertec project. Although the choice of CSP could be seen as a question of technology transfer rather than policy transfer, it needs to be kept in mind that Morocco's energy policy *is* the sum of top-down energy strategies and projects, and that the choice of this technology equals to a particular approach to an energy transition based on large-scale generation.

Moroccan interviewees also drew a direct link between the Fraunhofer-ISI study (Roller et al. 2007) and the design of the Moroccan Solar Plan: "it is this work [the study] that allowed to launch the new strategy published in 2009 and the solar program. You always need to put things into their context [...] even with the result of coal being the cheapest option, there was this decision to choose solar" (M2). Among the four energy scenarios for Morocco presented in the study, the aptly named "balanced portfolio" scenario is presented in a particularly positive light. The scenario is based on the addition of 768 MW of CSP, over 3,133 MW of wind and about 559 MW of biogas and biomass to the Moroccan system, but excludes PV (Roller et al. 2007: 197). The potential for job creation – a crucial priority for Morocco – was estimated at 10 jobs per MW of CSP installed, more than for any other technology (Roller et al. 2007: 122). Since wind and biogas were considered mature technologies with a well-developed world market, the study concluded it "would be difficult for Morocco to play an ambitious regional role on the renewable energy market" with these technologies and that "a certain industrial and technological delay" would arise from choosing a strategy without (concentrated) solar power (Roller et al. 2007:

217). The predicted creation of about 13,000 jobs by 2020 in the "balanced portfolio" scenario very likely was a strong argument in the eyes of Moroccan officials. In comparison, the most ambitious scenario in the study, the "solar strategy", with 2000MW of PV (which was still very expensive at the time), 3133 MW of wind and 775MW of CSP was immediately refused by the Moroccan energy minister (M1) who was shocked by the cost of up to 900 million EUR per year for this strategy (Roller et al. 2007: 202). With a maximum cost of 250 million EUR per year in 2020 (Roller et al. 2007: 198), the "balanced portfolio" strategy therefore appeared as a much more attractive option.

The energy scenario study's implicit preference for CSP over PV was seen critically by a GIZ official who arrived shortly after the study was finalized: "I flew to Morocco […] to meet the minister and because this […] study was being presented […] In the end it turned out the study was too focused on CSP. Maybe I should have financed another study, but I missed on that occasion. I reacted too late" (M1). The 2007 study was not the first one to highlight the economic and energy potential of CSP in Morocco. The three "MED-CSP", "TRANS-CSP", and "AQUA-CSP" studies financed by BMU and carried out by the German Aerospace Center (*Deutsches Zentrum für Luft- und Raumfahrt,* DLR from 2004-2007 (Bundesregierung 2009a: 166) were not only decisive in pushing the vision of electricity exports from Morocco to Europe, as highlighted in section 6.3.7, but also firmly placed CSP on the agenda of Moroccan-German energy cooperation from the mid-2000s.

Besides the creation of knowledge about the potential of CSP in Morocco, German support played out through other channels in the run-up to the Moroccan Solar Plan. In September 2009, only weeks before the announcement of the Moroccan Solar Plan, GIZ invited a head of department at MEMEE to participate in the CSP conference "SolarPACES" in Berlin. The head of department reportedly was "completely overwhelmed and enthused, said it was incredible he had not been aware of all those innovations, all these large [CSP] installations" (M1). Immediately after his return to Morocco, the official reportedly drafted a report on his experience at the SolarPACES for the minister. The report reached the minister at a time when "everyone […] was thinking about how the National Energy Strategy could be implemented with a lighthouse project" and it therefore "fell on fertile ground, exactly at the right moment, in a vacuum" when an idea was needed (M1). Other interview partners alluded to the influence of this visit to SolarPACES: "I know they [GIZ] organized common trips with relevant actors to Germany and visited fairs shortly before the launch of the plans. Yes, this might have played a role in the decision" (M35). Germany actively canvassed its partners from the MENA region through SolarPACES. Environment Minister Sigmar Gabriel in a letter to the conference participants stated: "in North Africa [...] unusable land becomes a valuable source of energy and income" with CSP plants (SolarPACES 2009: 5). He also underlined Germany's leadership role in "driving the expansion of solar thermal power plants" with "no other country [...] steering this course [...] as resolutely and successfully as Germany" (SolarPACES 2009: 5). Enthusiasm for CSP in Northern Africa also has to be seen in the light of the announcement of the Desertec Industrial Initiative (Dii) two months earlier, in July 2009 – a "sign of crucial progress" according to the minister (SolarPACES 2009: 5).

German research institutes actively lobbied for CSP prior to the Moroccan Solar Plan, both in Morocco and among German decision-makers (M13, M47). At a conference organized by the World Bank's MENA "CSP scale-up" initiative in Rabat in June 2009, a DLR researcher underlined CSP had the "advantage of producing a generation profile that largely matches that of most demand profiles", is "substantially more predictable than [...] many other renewables" and therefore facilitates utilities' work (World Bank 2009b: 20). It is likely that these arguments were particularly appealing to ONEE representatives who were also present at the meeting: "among the conditions from ONEE there was storage because the peak is at night time [...] that is why CSP was chosen" (M5).

Most importantly, in addition to the positive light shed on CSP by the Fraunhofer-ISI study and the official's visit to Germany, it is the perspective of being able to export electricity from CSP plants in Morocco to Europe/Germany that has prompted design choices in Moroccan energy policy.

6.3.7 *The perspective of electricity exports as a policy driver*

Germany's efforts to promote CSP in North Africa are strongly linked to the then widespread vision of importing "green" electricity from Northern Africa to further Germany's and Europe's own energy transition goals. Such exports were to support a twofold aim. On the one hand, they would contribute to financing renewable energy projects in Northern Africa through higher prices paid by European buyers. On the other hand, they should allow European countries, and above all Germany, to cover part of their green electricity needs through a dispatchable technology and possibly at cheaper prices than their domestic renewable energy production. Electricity imports from Northern Africa were seen as a contribution to Germany's own long-term energy transition (Bundesregierung 2009a: 72, Nitsch & Wenzel 2009: 33).

The idea of electricity exports from Morocco had been discussed among Moroccan and German energy experts at least since the mid-1990s (M39). Efforts to promote electricity exports from Morocco, in particular through CSP, intensified at the beginning of the 2000s. In 2003, the Trans-Mediterranean Energy Cooperation (TREC) was initiated by the German Association for the Club of Rome and the Hamburg Climate Protection Foundation, with scientific support from DLR (The TREC Development Group 2003). TREC involved Moroccan experts such as Abdelaziz Bennouna, who had been lobbying for the idea of energy cooperation and trade between Morocco and Germany since the 1980s. In April 2004, Germany sponsored the first MENA Renewables Conference (MENAREC), as part of the preparations to the Bonn renewables2004 conference. Three fundamental studies on CSP sponsored by BMU (MED-CSP, TRANS-CSP and AQUA-CSP) were discussed at this conference (Trieb 2004) and at subsequent editions of the conference in 2005, 2006 and 2007 (Trieb & Müller-Steinhagen 2007). The studies assessed the potential of CSP for electricity production and desalination in Northern Africa and Europe and focused on the feasibility of trans-Mediterranean transmission lines to integrate the European and Northern African electricity grids. They were crucial in laying the ground for the idea of importing renewable electricity from Northern Africa, but this vision is probably best known through the activities of the Desertec foundation, which emerged from TREC in January 2009.

The creation of a dedicated industrial branch of Desertec, the Desertec industrial initiative (Dii), was announced in July 2009. German companies dominated the consortium until its decline, and included Siemens, Schott Solar, Solar Millennium, E.ON, MunichRe and Deutsche Bank. The announcement of Dii was perceived as the concretization of the Desertec vision and marked a turning point in Moroccan-German energy cooperation. Interview partners from GIZ recalled that the announcement in July 2009 was the "push over" or tipping point for CSP and the export idea in Morocco (M13, M47) and delegations from Germany lined up to meet with Moroccan decision-makers. While Desertec and Dii relied on German initiatives, the idea of electricity exports by expanding renewable energy in Northern Africa was also introduced at the European level, through the Mediterranean Solar Plan within the Union for the Mediterranean. Germany was particularly attached to (and eventually succeeded in) placing a German official to lead the energy portfolio at the Union for the Mediterranean's secretariat: "the German federal government has expressed its interest in the placement of a German expert for the area of renewable energy in various conversations with partners of the Union for the Mediterranean" Bundesregierung (2009b: 6).

Dii had lobbied the German government to intensify cooperation efforts with Morocco in particular and promote the uptake of CSP there (M1, M37). CSP-focused research institutes like DLR, together with the Desertec foundation, Dii and the Deutsche CSP, an industry association created in 2012, were perceived as a "huge lobby" (M13). The enthusiasm for CSP was directly linked by interviewees to the domestic situation of solar energy companies in Germany, whose market share sharply declined at the time. Germany's involvement in CSP around 2009, but also in the following years, was therefore seen by several interview partners as a means to open new markets to its ailing solar industry (M10, M13, M32, M37).

Germany's interest in electricity exports from Northern Africa, and Morocco in particular, becomes visible in a number of responses to parliamentary enquiries (Bundesregierung 2010b: 7, see also Bundestag 2010, Bundesregierung 2009a: 72): "The federal government empathically supports the targets of this [the Desertec] business initiative [...]. Germany is committed to an intensified use of renewable energies, nationally and worldwide. The federal government also has an interest in leading German technologies in the field of renewable energies – in particular wind and solar energy – being used in the MENA region." The strategic relevance of CSP for meeting rising energy demand and climate change challenges – "CSP is the most promising renewable energy source for developing countries in the world's sunbelt" (World Bank 2006: 51) – led the World Bank to reinforce investment in this technology as well, within a "CSP scale-up" program. World Bank staff in Morocco was first perceived to be skeptical about CSP until 2007/2008, but changes in staff and the creation of the

Climate Technology Fund[66] (CTF) with a strong focus on CSP in 2008 (Pariente-David & Zakou 2012) changed the position of the World Bank in Morocco (M47). It is plausible that the outstanding conditions offered to Morocco by a CTF loan – “40 year maturity, a 10 year grace period, a 0.25 percent per annum service charge and a one-time management fee of 0.25 percent” (World Bank 2011: 45) – and funding from other donors including Germany did facilitate the choice of CSP in Morocco. But interview partners agreed the availability of finance was not the trigger for this decision and financing started to pour in only after the decision for the Moroccan Solar Plan (M5, M19, M21, M45). King Mohamed VI’s aim for his country to become “a key actor in the Euro-Mediterranean energy cooperation” and to “insert itself in the Mediterranean Solar Plan to present an offer [based on the] availability of renewable sources” (MEMEE 2009: 5) is more closely linked to the perspective of electricity exports, hopes for economic gains from investing in a new technology and the electricity storage available in CSP plants.

6.3.8 *Germany’s controversial role in technology choice*

Germany’s role in the choice of CSP was the most controversial topic addressed in the interviews, among an otherwise very positive assessment of Germany’s role in Morocco’s energy policy. Interview partners voiced criticism concerning this technology choice, in particular because the electricity export vision, crucially promoted by Germany, had not materialized[67] by the time of field research and

66 The history of actual CSP projects in Morocco started well before the creation of Desertec, with the idea for an Integrated Thermo Solar Combined Cycle power plant in Ain Beni Mathar, developed with the World Bank in 1999 (World Bank 2006, Fritsche & Schmidt 2008: 44). The project received more than 40 million dollars of funding from the World Bank’s Global Environment Facility (GEF) and was one of GEF’s four CSP flagship projects worldwide.

67 The Moroccan state bears a burden stemming from the financial gap between the cost of electricity generation by CSP and the resale price paid by ONEE, despite preferential loans. This gap was estimated at 685 million MAD (approximately $80 million) or 0.3% of Morocco’s total budget per year (AFD 2012: 10). The tariff granted in the first PPA for CSP was 1.62 MAD per KWh, while the resale price paid by ONE only amounted to 0.82 (base time) to 1.14 MAD (peak times). The financial gap is partly closed by contributions from International Financial Institutions, but 70% of the differential – or approximately one billion dollars over the 25 years of the PPA – are to be covered by the Moroccan government (World Bank 2011: 20). This is a heavy burden, given that wind and PV at resource-rich sites have reached grid parity in 2014 and could thus be operated without subsidies – a de-

expected price declines in CSP technology had not materialized. Before addressing the German role in the choice of CSP, it should be noted that Germany also supported other renewable energy technologies in Morocco, including photovoltaics, where it also contributed to an assessment of the technology's potential (e.g., Jäger 2011). KfW is also a major contributor to the three phases of the photovoltaics program of ONEE, which started after the initial Moroccan Solar Plan decisions discussed in this section (Ben Hayoun 2016).

Despite critical voices from within GTZ in Morocco, the general trend in German energy cooperation around the years 2008 and 2009 was in favor of CSP. A study advising BMZ on the energy strategy to adopt in the MENA region defines the "development, demonstration and diffusion of CSP systems" as an area Germany should focus on, "taking into account the comparative advantages of German development cooperation" (Fritsche & Schmidt 2008: viii). These comparative advantages originate in the fact that "German companies are currently very well positioned in the area of CSP technologies" (Fritsche & Schmidt 2008: 108) and in earlier activities of BMU and KfW, regarded as "key in the region" since they "made an important contribution for demonstration and promoted interest in realizing the potential [for CSP] in [...] Morocco" (Fritsche & Schmidt 2008: vi). Even the earliest project of German-Moroccan energy cooperation, the Special Energy Program, saw solar thermal electricity production as the "most important program for Morocco" in the long term and as an opportunity to "export solar hydrogen" to Europe (Osianowski 1997: 324).

The view on CSP in the renewable energy community had changed considerably at the time of field research in 2014. German transfer actors were largely critical of the choice of CSP for the first implementation phase of the Moroccan Solar Plan, also because it had become clear that prices for PV – the main alternative to CSP – had decreased dramatically compared to the time of the Moroccan Solar Plan decision. Several interview partners (M13, M24, M37) raised strong concerns over the financial burden of the Ouarzazate project for Morocco and its socio-economic benefits. German agents said "CSP was cheaper [than PV] in 2008, everyone was convinced prices would go further down and remain

velopment that was not expected at the time of the CSP decision. In 2016, a 850 MW wind energy tender in Morocco yielded a globally record-low average price of 3 cents USD (0.29MAD) per kWh (Parkinson 2016).

cheaper. There was this 'ghost' going round, that CSP would create more jobs [...]. Large power plants like CSP do not create more jobs, these are big industries with an old mindset" (M13) and "the projects are politically motivated, the job arguments are only façade" (M37). CSP was also seen as incurring a risk of Morocco moving into a dependence on foreign technology instead of energy imports (M1, M24).

Observers from third countries also had a critical view on the choice of CSP for Morocco and on the role Germany played in this choice, through different channels. Since "Germany's policy was to say they already did PV in Germany, so now they invest in CSP: 'You cannot do CSP in Germany, so we do it in Morocco'. That's the reason Morocco chose the most expensive technology option, because the donors required solar thermal" (M22). Part of this argument was shared by German advisors, who saw some German reluctance to financing PV given the country had already contributed to bringing down the cost of PV through domestic policies: "Of course you need to see that at the time massive financial means were available for CSP, through the CTF and KfW as well. The [German] Federal Government had decided, with DKTI, that they would not finance PV any more, where they had already contributed to price reductions" (M24). For some Moroccan observers, the most critical aspect was the hope of electricity exports rather than the choice of CSP as such for the Moroccan Solar Plan: "Since the 1990s they [Germany] launch studies on the possibility to import renewable energies. They are the first ones to go for that strategy. They have a strategy; they have a vision. And they try to impose their vision [...]. One must say they are correct in their vision to 80 or 90%, but they need to discuss it" (M27).

Other stakeholders' views on the role of Germany in CSP selection differed, and in particular the role of GTZ advisors was seen as positive. A MEMEE official saw GTZ as in fact being in favor of PV and explained the choice of CSP was above all driven by internal Moroccan factors, especially ONEE's condition that the solar capacities added had to be dispatchable (M5). Another MEMEE official recognized Germany had played an important role in the choice of CSP, but did not necessarily see it negatively: "To me it seems natural, logical, that a big country like Germany seeks to develop itself outside its borders. One could say in quotation marks that Morocco is a laboratory for Germany to test new technologies" (M6). Germany's interest in promoting CSP as a technology that could not be implemented in Germany was also seen as relevant in order to strengthen the relationship with the MENA region and play a positive role in an instable world region (M18). The Desertec idea, "an expression of German expectations" (M3), and the question of electricity exports were thus clearly seen more critically than CSP in general by Moroccan stakeholders. German industry and government were perceived as the driving forces behind the idea of electricity exports, as a direct consequence of Germany's Energiewende (M2, M6, M32, M36). As the head of IRESEN noted in a 2012 interview: "Germany in its 2050 energy strategy has a tremendous energy need especially if they stop their nuclear energy. And from 2030 onwards they will import green electricity. […] Germany's support [to Morocco] is not disinterested. Beyond the fact that they are very competent and advanced in this field, there is also their own energy strategy that will require energy imports and there are several studies and negotiations ongoing to prepare the ground, be it from a legal or institutional or technical perspective, it's something that is going to happen" (Hafidi 2012).

Some Moroccan interview partners still believed in the potential of electricity exports to Europe at the time of field research (M15, M16), while others recognized the Desertec concept was "a myth" (M7) since Spain had made it clear in 2013 that it would not allow the transmission of imports from Morocco in the framework of the Mediterranean solar plan (Carafa 2015: 1700). The infeasibility of this idea was not seen as putting into question the Moroccan energy strategy (M10, M15) nor the role Germany, through its implementing agencies, played in the renewable energy field in Morocco. A distinction thus needs to be drawn between a general "German agenda" and the role of German transfer agents on the ground, who had become well-accepted members of the local policy network and where overwhelmingly seen as a positive factor.

6.3.9 *Law 13-09 and the feed-in tariff debate*

While the preceding sections have shown an important role for Germany in design choices and the general direction of Morocco's energy strategy, non-transfer occurred with regard to policy instrument selection. As outlined above, Law 13-09 on renewable energies (MEMEE 2010) followed up on the draft renewable energy law from 2007, in which German transfer agents had already played an important role. Work on the new renewable energy law intensified in 2008 and a draft was adopted in April 2009 by the Moroccan government, before passing parliamentary approval and being promulgated in February 2010. Law 13-09 allows private parties to produce electricity for consumers, including for export outside Morocco, or for auto consumption and to use the mid-, high-, and ultra-high voltage grid for transmission (MEMEE 2010). The introduction of a specific article on the right to use the national grid with a view to *export* electricity came as a surprise to parties involved in the working group charged with drafting the law (M1), and can be seen as linked to the export perspective Germany decisively created, although the article was not proposed by a German transfer agent.

As Moroccan and German interview partners critically noted, Law 13-09 does not foresee any incentive mechanism for renewable energies and does not allow feed-in into the low-voltage level grid, thereby precluding the possibility of small-scale decentralized installations (M26, M37). Renewable energy projects that are needed to reach Morocco's national renewables goals in the Moroccan solar and wind plans compete for PPAs, but no specific subsidies beyond the guarantees of each PPA are foreseen. To encourage a broad uptake of renewables in Morocco, the 2007 Fraunhofer study discussed above had advocated for feed-in tariffs in a very clear manner: "To ensure the development of renewable energies through sustainable and incentivizing financing, we recommend feed-in tariffs" (Roller et al. 2007: 221), arguing that "virtually all countries who successfully developed renewable energies […] used feed-in tariff systems" (Roller et al. 2007: 211). The example of the German feed-in tariff experience was mentioned in the study as well (Roller et al. 2007: 183). As stated in Chapter 5, Germany actively promoted the use of feed-in tariffs worldwide, and initially preferred this option also for the MENA region, where "reliable promotion schemes as demonstrated in an exemplary manner by Germany with its EEG [could have] considerable impact and improve the framework conditions for RE" (Fritsche & Schmidt 2008: vii). The idea was therefore brought into the process in Morocco by Germany in the framework of the development of the first renewable energy law in 2007.

As a high-ranking Moroccan official directly involved in the making of the subsequent 13-09 renewables law confirmed, "they [Germany] recommended the use of a feed-in tariff as implemented in Germany. They recommended a state aid for photovoltaics" (M5). Another interview partner from the Moroccan side felt that, at the time, "the Germans tried to sell the idea of a feed-in tariff. There was an enormous amount of missions [to Morocco] with German experts, B2B days, bilateral conferences […] it's normal that everyone tries to defend their approach" (M28).

Although GTZ officials had a general preference for feed-in tariffs based on their own positive experience in Germany (M1, M24, M47), this policy option was rapidly excluded given the firm opposition by the Moroccan energy minister to any increase in end-user prices for electricity (see also Liptow & Remler 2012). GIZ was sensitive to the minister's opposition to feed-in tariffs: "Energy Minister Benkhadra made it clear that she could not propose a law that, through a redistribution of costs, would lead to higher electricity prices. […] And she is right in saying so: with a majority of households earning 200-600€ per month, the existing electricity tariffs are already perceived as very high" (Jäger 2011: preface). Very strong opposition to feed-in tariffs also came from the World Bank, with officials reportedly stating in meetings that the bank would take all necessary measures to avoid the choice of feed-in tariffs in Morocco (M1, M13, M47). The World Bank's opposition was however not a game-changing factor, since feed-in tariffs had always been excluded by decision-makers in Morocco.

In the light of rapidly falling prices for PV modules, the focus shifted to net-metering as an instrument to facilitate the uptake of decentralized PV generation, leading GIZ to announce one had to "leave the entrenched patterns of thought associated with the 'feed-in tariff' logic and to theoretically examine a new approach, the 'net-metering'-mode, more closely" (Jäger 2011: preface). The identification of commercially realizable PV potential required a "system change from the logic of a guaranteed feed-in tariff to a logic of 'net metering'" (Jäger 2011: 11). Those statements can be regarded as a proof of adaptation among German transfer actors, and their focus on general orientations of the Energiewende rather than on exporting instruments, which is comparable to the South Africa case.

In addition to general opposition for domestic, internal reasons, feed-in tariffs were also the only clearly negative lesson drawn from the German energy transition experience: "the example of feed-in tariffs that would rather be a negative lesson. They had a massive exit of cash flow" (M15). As presented in Steinbacher (2015: 44), a negative view on the German experience was however not the root cause for opposition to feed-in tariffs. Rather, general structural differences between Morocco and Germany led to an assessment by Moroccan stakeholders that feed-in tariffs were not an advantageous or feasible policy for the country. On the one hand, feed-in tariffs were seen as particularly favoring a domestic solar industry and hence interesting for a technology importing country: “As we don’t have any renewable energy technology industry in Morocco [...] they are not interesting for us at all. Although their effect in terms of massive renewables deployment, GDP, job creation, added valued and wealth creation is really important” (M28). On the other hand, different degrees of citizen involvement and financial capacities made feed-in tariffs an ill-adapted policy instrument for Moroccan interview partners: “In Germany, it started with individual citizens. Here it is rather the contrary [...] We should encourage this ‘culture of cooperatives’, but this culture is not yet well established in Morocco. [...] In Germany it worked. Why? Because people were aware, there was a history, a culture, it didn’t happen from one day to the other. [...] Once the big projects [in Morocco] are implemented, maybe the population will be sensitized and can organize itself to develop projects. [...] So it’s mandatory to develop the big projects first, make the technology viable commercially and then the population can get involved. But you cannot do it exactly the way Germany did” (M27).

6.3.10 Agenda-setting for decentralization and small-scale renewables

The Moroccan energy strategy so far has a clear focus on large-scale renewable electricity plants. The controlled addition of predefined amounts of capacity to the grid through auctions is linked to the importance of energy independence and energy security as objectives behind the transition. The question of how Morocco’s renewable energy potential could be tapped fully by also including small-scale installations and opening the low-voltage level grid to a larger group of producers has been present in debates within the renewable energy network in Morocco for years. Solar home systems were used as part of the Moroccan rural electrification program PERG (ONE 2014), but other small-scale installations are still widely absent from the energy landscape. Access to the low-voltage grid remained excluded because of concerns expressed by the Ministry for the Interior and out of consideration for the municipal utilities and private electricity distributors, the “régies” (M1, M13). Not only these municipal distributors, but also ONEE could lose considerable parts of its revenue if auto-production of electricity grew (M10). The loss of control on how much electricity capacity is added, and perceived challenges of operating the grid with higher shares of decentralized generation are additional reasons for the reliance on large-scale plants in the Moroccan renewable energy strategy (M42).

German transfer agents, joining forces with Moroccan industry representatives and some MEMEE officials, have effectively kept the topic on the agenda despite the overwhelming focus on the Moroccan solar and wind plans and their large-scale projects (M1, M26). German-Moroccan cooperation in the area of decentralized generation has been reinforced since late 2013. Recent developments like the adoption of a PV roadmap in 2015 and of Law 58-15 on opening the low-voltage level at least hint at the preliminary effectiveness of these efforts (MEMEE 2015). Already during the development of the first renewables law in 2007, and then Law 13-09, GIZ advisors pushed for an inclusion of the low-voltage level and hence small installations to the law: "From the beginning, they [GIZ] suggested developing residential PV. They showed numerous times that we could do 2000 MW through residential projects only, which is equivalent to what MASEN does [in the Moroccan Solar Plan]. The Germans said that over and over again, I confirm" (M5). The opening of the low voltage grid to private production was repeatedly recommended and advocated by GIZ: "the creation of conducive legislative framework conditions such as the inclusion of the low voltage level to Law 13-09 through by-laws and simple administrative procedures are indispensable preconditions for tapping the existing [PV] potential" (Jäger 2011: 23).

Given the resistance from ONEE and the private electricity distribution companies, small installations were not considered in Law 13-09. In the light of falling PV prices at that time, GIZ decided to keep pushing the issue of decentralized PV (M1) by commissioning studies on the market potential for PV (Jäger 2011) and the idea of a million solar rooftops program for Morocco (Sidki 2011). Despite the interest the studies created within MEMEE, frequent announcements of an intention to open the low-voltage level to small projects were not followed by regulatory action. The issue was thus introduced into the newly created German-Moroccan energy partnership in 2013, to generate momentum and political support for the idea[68]. GIZ also hired three consultants – a Moroc-

68 Atelier d'information et de préparation pour un "Programme intégré du PV" dans le cadre du Partenariat énergétique maroco-allemand, 5-6 décembre 2013 – Hôtel Ibis, Gare Rabat-

can lawyer, a German engineer (who also played an important role in transfer activities to South Africa), and a Tunisian economist – to accompany the work of a task force bringing together public and private sector energy stakeholders and ministry officials on the question of decentralized generation (M1, M8). A task force participant explained GIZ's central role in driving its work: "These are restricted meetings, with GIZ and a few key people from MEMEE. GIZ goes there with concrete suggestions. They call on the ministry; they sketch out how the program could look like. The people from MEMEE take this information and communicate the proposals to a higher hierarchical level. […] GIZ constantly reminds MEMEE. MEMEE simply lacks human capacity, that explains the 'stop and go' [but] […] from a categorical 'njet' we shifted to a possibility of at least being able to discuss" (M8). Reducing resistance was also possible because of information provided by GIZ and its consultants on how the program could be designed to avoid losers: "we want to collect arguments, show the relevance of opening the low-voltage level, show that the country is benefitting […] This study allows us to facilitate dialogue and inform actors. […] The role of GIZ, as I see it, is to get MEMEE to think and they did a good job at that" (M8).

An increasing number of demands "from well-informed circles in Morocco to open [the mid-and low-voltage level]" added to the pressure for policy action: "There is pressure from Moroccan companies on government because they see business opportunities [but] at the same time also from German and international companies" (M25). A Moroccan private sector representative recognized Germany's important role in the opening of this potentially large business segment: "To be honest with you, there is a confluence of pressures. There is the local private sector, here. And then there are the countries with whom we exchange, be it Germany or Spain. Germany has a large contribution […] I know someone [from GIZ] who lobbied for the low-voltage level. Who commissioned studies, who knocked on doors, who insisted. In the end it is certain that this had an effect. I can't say if it's rather him or the two of us together, but it is clear that cooperation between governments leads them to listen eventually, to be influenced by similar institutions in other countries […] Germany can help us open

---

Agdal, Rabat. [Information and preparation workshop for an "integrated PV program" in the framework of the German-Moroccan energy partnership] Agenda. Unpublished Document.

the eyes of public authorities. As a leading country, it can inspire our authorities, it can lobby for us in order to have the regulatory framework evolve" (M26).

German advisors saw the job creation potential of decentralized solutions as a major reason for their support to this segment, where experience from Germany with its almost 50% share of installations owned by citizens, farmers, and small businesses, would be particularly useful (M35).

6.4 Case Discussion: Morocco

6.4.1 Germany's agenda for leadership in Morocco

Understanding what agenda drives leadership is important to assess its effectiveness. The agenda behind Germany's engagement in Morocco remained subject to questions for interview partners (e.g., M10, M11, M14, M42, M21). A Moroccan official in almost daily contact with German counterparts was one of many interview partners expressing uncertainty about the overarching motives of Germany's outstanding involvement in Morocco: "I thought about it, but do Germans tell me what they want? No. […] Do they tell themselves 'we'd like other countries to do the same thing we do and we start with Morocco'? I don't know if that is the reason. I am asking you the question!" (M21). As in the case of South Africa, a clear distinction needs to be drawn between the perception of the motivation of *individual German agents* on the ground and the broader picture of drivers for Germany's leadership as a country. Given their longstanding presence in Morocco, German advisors seemed to have blended into the energy policy network and their agenda was predominantly perceived as promoting renewables together with actors from the Moroccan side, as longstanding and trusted advisors, whose support was highly valued not only because of the lack of capacities within Moroccan institutions.

On a more general level, Germany's activities were mainly seen as motivated by its global climate protection efforts, foreign policy strategy, and development priorities, rather than as a way to give German companies a particular advantage. It was also important that cooperation efforts were financed by development-oriented BMZ, and by BMUB with its climate-focused mission. An exception was the inception of the BMWi-led Moroccan-German energy partnership, which was primarily driven by an aim to accompany the efforts of the Dii consortium in Morocco politically. The budget and relevance of this partnership are however far smaller than the multi-million euro cooperation projects sponsored by BMZ and BMU. Moroccan interview partners on the one hand considered economic motivations as a *potential* logical explanation for Germany's (or any country's) involvement in Morocco, but on the other hand did not believe this had actually been a main driver, given the limited success of German companies in the Moroccan renewable energy tenders. Moroccans interviewed noted that, looking at the numbers, "there is a very limited economic interest" (M21) and that "Germany is very, really very dynamic. There are agreements, financial institutions financing projects" even though this "hasn't really benefitted German companies" (M26). Industry representatives were unsurprisingly skeptical about the supplier-neutral German approach: "at least France's interests are clear. [...] The French get involved when the part for French companies in a project is sufficient. [...] GIZ's approach does not at all go in that direction. They help the Moroccan administration implement their projects. Really, I don't know what that is good for. For me, it doesn't make any sense" (M12).

Even though German companies do not seem to have benefitted proportionally to the degree of German engagement, there is no doubt that a stronger renewable energy market is of interest to German companies in the longer term (M23, M42). That longer-term perspective was seen as a potential reason for German involvement, but not as a main driver: "For Germany, it's probably also about searching for markets for its companies, but also about taking away terrain from the French. There is certainly also a part of believing in North-South energy cooperation" (M32). The part of German involvement that was linked to economic interests was seen as related to the downturn of Germany's own solar industry due to competition from China (M10).

The day-to-day interface of German leadership in Morocco is primarily technical cooperation through GIZ, and, to a lesser degree, financial cooperation through KfW, whose experts do not work from within Moroccan institutions. GIZ advisors in particular were not seen as driven by an aim of exporting German technology. A Moroccan official noted: "GIZ does not export German technology. When I compare that to what French and Spanish agencies do... no, it really does not, it is really something else. [...] GIZ rather plays its role as a tranquil NGO; it exports some know-how when they make experts come here, but promoting German industry? No." (M7). The weak link between German support in the sustainable energy field and immediate benefits for its companies did not mean German efforts were seen as completely neutral. As a German advisor put it, "German development cooperation *is* partisan. The target is to minimize negative consequences for the environment and the climate" (M36). Germany's activities in the energy field are limited exclusively to renewables, efficiency, and fundamental questions of grid management and regulation. That specialization was perceived by interview partners as part of a broader German foreign policy strategy: "Primarily, I think they have developed a diplomacy that encourages renewables and supports the presence of renewables elsewhere in the world" (M11) and "every country in the world searches for its niche in cooperation. The US does a lot on democratic transition, on women, the Japanese do education. Every country positions itself and GIZ positions itself on environment. That's totally natural since the Germans are known for being leaders in environmental affairs and the fight against climate change. They make their know-how available to Morocco. At the same time, I believe countries like Morocco can have a demand; they are naturally looking for the partner with the most experience. So Germany's activities respond to a need. It's win-win, it's a response to a real need" (M8). In that regard, Morocco was seen as a test ground for Germany's approach to leadership to achieve global objectives: "With PV, Germany thought certainly that one should make sure that new markets emerge. It was clear from the start that these were not going to be in Europe. That was one aim. At the moment it is more about getting involved in Morocco to support a new energy model that can be duplicated elsewhere" (M18).

In terms of preferences for particular policies and technologies, Germany's controversial role in the decision for CSP has already been described. By the time of field research, the perception of German preferences for this technology had evolved given the limited success of German companies in renewable energy tenders in Morocco – a striking parallel to the South African case. Also, although a personal preference of German GIZ advisors and consultants for feed-in tariffs and decentralized generation, based on the German experience, was perceived (M2, M5), the main motivation behind individual German transfer agents' activities was clearly seen as supporting a Moroccan path to an energy system transformation and working with Moroccan partners within the policy network.

German and Moroccan interview partners described a situation in which German activities reinforced "demand" for sustainable energy advice, which in turn provided a motivation for German transfer activities: "First of all, we act in a partner-oriented way. In Morocco, there simply was and is a very clear window of opportunity. Germany has the know-how and the instruments to serve this demand" (M3). The unprecedented availability of energy and climate funds through IKI and a lack of viable alternatives in the region due to the Arab Spring also directed Germany's focus to Morocco (M35). Taken together, "the mix Morocco was offering, the stability, the policies, the existing instruments and planned projects, that was unique" (M3). As another German interview partner put it, "I have been working in the area of financial cooperation in energy [...] in America, in Asia, also in Europe and South-Eastern Europe and now in Northern Africa. And the quality of projects that are being proposed here and the structures behind them, the partners who implement the projects, are outstanding. So I believe it is easy for the Federal Government to commit funds because you know they are falling on fertile ground" (M35). Germany's role in triggering that very demand for sustainable energy policy advice and financing was discussed as well. A third-party consultant said Moroccans believed Germany "wants to sell their expertise, use their experience here and that is valued very highly. In the eyes of Moroccans, Germans help define problems and then also have a solution ready right away. It is clear that no Moroccan could buy German technology with the current purchasing power. So demand is generated and in parallel financing is offered. You generate a demand you then serve as well" (M25). A Moroccan interview partner confirmed that view almost word-by-word saying, "I sincerely believe that the impulse for many projects comes from the international level. Funds are available, the assistance that is here is offered to us and – "hop" – a need is created" (M7), but did see this impulse in a positive way as speeding up necessary developments.

Interview partners formulated concrete expectations for future cooperation with Germany. Among the wishes were continued assistance, going beyond the implementation of the 2020 energy strategy goals: "we expect [...] a long-term support in our strategy. We don't have a strategy for 2030, 2050. [...] We know Germany has precise and scientific techniques to determine a clear long-term plan with scenarios. I would like to have the support of Germany in that sense" (M5). That wish corresponds to what German advisors viewed as necessary next steps in their support: "it is important [...] to promote long-term scenarios and strengthen simulations so Morocco doesn't need to be afraid. It's clear that ONEE is afraid of losing costumers" (M24). Stakeholders with a background in research pointed out that Germany's help should not focus on creating networks among Moroccan researchers or on training, which was deemed unnecessary (M18, M27), but rather invest directly in research equipment and infrastructure.

6.4.2 Hypothesis I: Mediated consideration of the German model

The assessment of consideration as a variant of transfer from Germany to Morocco needs to be addressed at two levels. First, it concerns the potential of German "leadership by example" triggering interest and consideration of the Energiewende model in Morocco. Secondly, and somewhat more challenging conceptually, the possibility of "mediated consideration", i.e., German transfer agents translating the model into the Moroccan context, needs to be explored. The role of the Energiewende example as a direct source of lessons for Moroccan stakeholders was limited. Unilateral leadership did however impact German-Moroccan energy relations in several ways and – as in the case of South Africa – is an indispensable basis for active leadership efforts and transfer. Unilateral leadership played out as a factor in policy transfer to Morocco along the following lines, which are discussed in more detail below:

Shared aim of leadership between Morocco and Germany facilitating transfer;

Legitimation of active leadership activities through domestic experience in Germany and recognition of leadership by Moroccan interviewees;

Vision of electricity exports from Morocco to Germany directly linked to plans for the implementation of the German energy transition;

Energiewende as the reference model for German advisors in Morocco and availability of funds for strong active leadership efforts linked to importance of sustainable energy in Germany.

Moroccan interview partners overwhelmingly recognized Germany's leadership in the sustainable energy field based on its longstanding energy transition efforts: "I, personally, I know that there is no one better than Germany, at the level of technologies but also apart from that" (M7). Germany and Morocco were seen as belonging to the same group of countries worldwide that are going forward with ambitious sustainable energy policies (see also Steinbacher 2015, p. 49-50) "I would say today Morocco follows Germany's path. Germany is a leader, a model, with its deliberate, ambitious and courageous policy" (M28). Both sides were seen as winning from close ties and cooperation in the sustainable energy field: “Morocco presents opportunities for Germany, it can involve itself here, that’s legitimate. Germany has the technology, the leadership in this domain, and Morocco has the appropriate framework conditions. So the two have to seize opportunities; the German side has a lot to win from it as well” (M6).

The German experience in renewable energy served as a legitimacy base to transfer agents on the ground. Germany’s involvement in Morocco was seen as a direct consequence of its unilateral leadership, up to the point where interviewees said Germany had possibly reached a maximum level of renewables development at home and now saw it as its mission to develop renewable energy in “developing countries […] the best example being Morocco” (M5). Also, direct links were drawn between Germany’s energy transition, a resulting need for electricity imports from Northern Africa, and the relationship between Germany and Morocco: “The German choices correspond to Moroccan choices. Germany’s idea of having renewable energy exported from Northern Africa is of great interest to us. Morocco is ready to export” (M46).

While some Moroccans “look very closely at what is happening in Germany and want to do the same thing” (M32), and interviewees were aware of the general lines of debate in Germany (some also in great detail), consideration of the Energiewende occurred mainly through the intermediary of German transfer agents in Morocco. Interview partners expressed a desire to learn from the German experience and mistakes: “all we want is to know the errors [Germany made], to save time and do things well on our side” (M7). But only a few interviewees reported they actively and regularly searched for information and actively informed themselves outside of contacts with German counterparts in Morocco (M18, M45). Moroccan energy sector stakeholders who had spent time in Germany during their studies were an exception and said they informed themselves of developments in Germany (e.g., M15, M18, M39). One explanation for the general lack of purely demand-driven learning without intermediaries is the sheer amount of information from and interaction with German transfer agents that reduces the need to use resources, time and capacities for this type of research. Capacities in Morocco were seen as stretched to their limits, with ministerial officials covering vast topic areas, and means for studies, scenario-development or energy statistics being very limited (M7, M28).

It is important to underline that for German transfer agents in Morocco, in particular from GIZ, the Energiewende played a tremendously important role as a reference model and shaped the understanding of their role and work in Morocco. This mechanism of "mediated unilateral leadership", with the Energiewende as the background from which German transfer agents operate was even more pronounced in Morocco than in South Africa. In the latter case, despite the Energiewende still being a crucial source of lessons for German transfer agents, views on the German experience were slightly more critical and experiences from other countries than Germany were more frequently brought into the process compared to Morocco. This is also linked to the fact that activities intensified in Morocco around 2007-2009 and that numerous other country examples only started to become relevant later, when renewable energy policy debates emerged in South Africa.

Although the Energiewende example rather played out in the background, some direct communication on the experience in Germany took place, namely in the framework of an energy transition workshop in 2013, a study tour to Germany in the same year and day-to-day communication between German and Moroccan staff (M21). Communicating on the Energiewende experience is becoming increasingly important with questions of renewable energy integration to the grid and the promotion of decentralized electricity generation[69]. An interview partner also underlined that Germany "communicated and informed about its model at all levels, not only in bilateral cooperation [but also] in all international institutions, like the UN […] I think it's clear for all countries which are at least a little bit interested in sustainable development that Germany is exemplary. One can think of the Bonn conference in 2004 or the one in Beijing in 2005. Germany really has done a lot. They played a crucial role for the creation of IRENA, who is also an important actor" (M28). More important than mere communication on the Energiewende were cognitive leadership efforts of producing information and evidence adapted to specific Moroccan needs, to showcase the potential and feasibility of high-renewable energy scenarios. The creation of knowledge with

69 Atelier d'information et de préparation pour un "Programme intégré du PV" dans le cadre du Partenariat énergétique maroco-allemand, 5-6 décembre 2013 – Hôtel Ibis, Gare Rabat-Agdal, Rabat. [Information and preparation workshop for an "integrated PV program" in the framework of the German-Moroccan energy partnership]. Report. Unpublished Document.

the help of Germany was seen as crucial to “accelerate and facilitate things” (M6).

With regard to nuclear energy as an element of the Energiewende, interview partners were well aware of the phase-out and clear position against nuclear in Germany, but did not see German efforts in trying to shape Moroccan decisions in this field. Nor did Moroccans draw any lessons from Germany on this point. The possibility of nuclear energy looms in Morocco, but Germany’s strategy has been to focus fully on promoting renewables and facilitate their uptake to make nuclear appear as a less interesting option. GIZ advisors saw French attempts to promote nuclear (Le Matin 2014) as a motivation to reinforce their own efforts regarding renewables and avoid that the “seven-headed Hydra called ‘nuclear’” could get hold of Morocco (M1). The phase-out decision in Germany was seen by a third-party observer as a driver for cooperation with Morocco: “it is clear that Germany with its nuclear phase-out objectives has every reason to maintain very good relationships with countries who are trying as well. […] Germany needs partners of choice […] and why not to import electricity from Morocco. Germany has to gain as well from supporting a country like Morocco to achieve this energy transition” (M20).

A final direct link between German domestic energy policies and Germany’s engagement in Morocco is the availability of additional funds for such cooperation. The DKTI projects, co-sponsored by BMU/B and BMZ to accompany the implementation of the Moroccan Solar Plan, support MASEN and ensure socio-economic benefits arise from the Ouarzazate solar project, are partly funded through the International Climate Initiative (IKI) and “are to be understood as part of the Energiewende. There is this connection” (M35).

6.4.3 Hypothesis II: Instrumental use of transferred knowledge

Before assessing the utilization of transferred knowledge in Morocco, the role of German transfer agents in the policy network merits consideration. As Howlett and Ramesh (2002: 34) point out, internationalization can bring “new ideas” and “new actors” to policy networks, and both are relevant in Morocco. GIZ energy advisors are an integral part of the small Moroccan energy policy network. They operate from within the energy ministry, where only one other foreign advisor, a staff member from the French Environment and Energy Agency (ADEME), was based at the time of research. Germany’s continuing cooperation with Moroccan counterparts, starting in the 1980s, was mentioned as a significant advantage when a window of opportunity opened in the face of an energy crisis around 2008. For example, a member of the GIZ energy advisory team had been involved already in the Special Energy Program in the 1980s, working alongside Moroccan officials, many of whom also look back to long careers in the energy field (M5, M23). Germany's early support to the Moroccan rural electrification program was also mentioned as a basis for an important role of German agents later on (M5, M23, M34). The strong reliance on personal exchange and close, longstanding working relationships – "you cannot get any closer" (M20) – also grant individual GIZ advisors a particularly important role as members of the policy network: "technical cooperation very much depends on the person. With our long-term deployments and positioning people within the institutional structures of our partners, within ministries, the German approach to technical cooperation is singular" (M3). This means that advisors have room to bring their understanding of desirable policy frameworks and Energiewende background into the process and use it, within the limits of their project's mandate. A second particularity of active German leadership is its sheer volume. The diversity and number of projects in GIZ's energy portfolio by far exceeds that of any other multilateral or bilateral cooperation partner in Morocco, leading a French expert to the conclusion that GIZ "is a battle force" (M17). German leadership in Morocco also shaped the very structure of the policy network, by supporting the renewable energy agency ADEREE (formerly CDER) since its beginnings, helping to create MASEN, and building capacity within MEMEE (M23, M47). These structural changes within the Moroccan energy landscape reinforced the use of renewable energy policy knowledge in a sort of virtuous circle.

This excurse on the role of German transfer agents in the policy network is important to assess "utilization" as a variant of transfer in the Moroccan case. Evidence produced by studies that were sponsored by Germany was used in the development of the Moroccan energy strategy, as well as the Moroccan renewables Law 13-09, the 2007 renewables law, and the creation of an initial framework for smaller-scale installations in Law 58-15. Rather than transferring evidence from Germany directly, the studies targeted the Moroccan context and translated the general principle of "more renewables" to scenarios and recommendations for Morocco. The energy scenarios study described above is the most striking case for this (Roller et al. 2007), but other studies on the potential of smaller installations (Jäger 2011, Sidki 2011) also played a role. These publications, according to Moroccan interview partners, introduced new knowledge and evidence, as well as concrete policy proposals and scenarios, into the debate and could not be dismissed because of their perceived scientific quality and the lack of any comparable source of knowledge: "Obstacles always arise when you cannot prove something, when you don't have the arguments. That is why the feasibility studies, with German and international support are so important: they are a light that is guiding" (M6).

The transferred knowledge clearly closed epistemic gaps in Morocco. Moroccan pro-renewables members of the policy network in particular, also used the studies to strengthen their position, and German transfer agents relied on them in their arguments to make the case for the feasibility of a high-renewables future for Morocco. Interviewees did not report a strategic negative use of the Energiewende example or of knowledge transferred by German agents, although ONEE representatives expressed a skeptical view about the energy future it sketched. The state of the Moroccan grid, ONEE's capacities of managing high shares of renewables, and the risk of losing business, were mentioned as concerns, but – for example contrary to the case of California – the experience of German utilities or the Energiewende were not used negatively to point to these concerns.

In addition to helping close epistemic gaps, and empowering pro-renewables actors within the institutions, some Moroccan interviewees reported transfer and cooperation provided a "label" for the quality of their work. A Moroccan public official stated that initiatives supported by Germany "gain momentum, they are more important. It gives us a kind of label, a guarantee, it sends a signal when we succeed in attracting international partners to a project […] having international partners confirms we are doing a good job" (M7). A German advisor noticed that effect as well, with Germany "granting Morocco national and international legitimacy [in the energy field]. Germany is a kind of certification body for Moroccan energy policy" (M37).

The effect of German activities facilitating and speeding up implementation of sustainable energy policies in Morocco and of "labeling" was also observed for financial cooperation. Morocco's renewable energy agency mobilized two to three times more funds from international than from Moroccan sources. This financial support "got things moving in Morocco and continues to do so" (M28), by allowing agencies to "advance more rapidly on concrete projects and – and that is important – to depoliticize the projects" by adding a "neutral" partner (M7).

The question of whether financial incentives prompted the reflection of transferred knowledge in policy decisions cannot be answered with certainty (see below). As far as the drivers for the use of that knowledge are concerned, the need to act due to the 2008 energy crisis and to devise a new energy future for Morocco can be seen as decisive. It is nevertheless clear that active leadership, in particular the presence of German transfer agents, but also training, support to institutions, and the certainty that Morocco would find partners for the implementation of a high-renewables energy policy, made this option more feasible and increased the likelihood of utilization in the Moroccan case.

6.4.4 Hypothesis III: Reflection in policy orientations rather than instruments

Knowledge sponsored and promoted by Germany was intensively used in Morocco, but reflection in policy instrument decisions was selective. Before discussing reflection for the Moroccan case, it needs to be recalled that strategic policy decisions are made by the Moroccan king and decision-making processes at the very top can hardly be traced from the outside.

While the differences between Germany and Morocco in terms of economic development and policy objectives played a role in the choice of policy instruments, namely the opposition to feed-in tariffs, these differences did not preclude the transfer of general policy orientations, such as "more renewables" (Steinbacher 2015: 45). An industry representative expressed this general feeling: "what you hear here, is 'we want to do it like Germany' and there is no reason that we couldn't do it as well, especially if, in addition, we have the sun" (M32). Based on longstanding cooperation, Germany was "at the right place at the right time" with its pro-renewables message in 2008, when increasing demand, rising fossil fuel prices and import dependence pushed for policy change. There was an element of inevitability in the Moroccan decision to "do something", but the general orientation of policy choices toward renewables in the Moroccan energy strategy and the Moroccan Solar and Wind plans was greatly facilitated by knowledge and support from Germany: "The energy transition is not a choice, it's a necessity. We don't have choice. Like any other country we want to appropriate new technologies. We have a strongly increasing demand so we want to adopt these technologies – but also the processes, the German 'way of doing things' – into the Moroccan context" (M6). In particular, the degree of ambition of Moroccan objectives (42% of capacity installed is to be renewable by 2020, and 52% by 2030) and the importance of solar and especially CSP in the Moroccan plans can be traced back to the involvement of German advisors, working along Moroccan officials. In particular, the 2007 GTZ study on energy scenarios and different channels of communication and financial support for CSP were perceived as having shaped these decisions. Much earlier on, potential analyses brought the issue of renewable energy to the realm of possibilities to tackle the country's energy challenges and thereby prepared the ground for future developments. Following the identification of Morocco's renewable energy potential, the financing and support of thousands of solar home systems and of a first wind power plant at the Al Koudia al Baida site demonstrated the feasibility, at least at niche level, of renewable energy in Morocco (Marquardt et al. 2015). Al Koudia al Baida also was an important learning opportunity for ONEE (Mouline 2007), reducing (although not eradicating) resistance to renewables.

Policy transfer was selective when it comes to policy instruments and the "how" of the Moroccan energy transition. German advisors on the ground had a clear preference for German-style feed-in tariffs, but soon had to recognize that this approach was not feasible in Morocco. Related to the issue of feed-in tariffs, which were hoped to enlarge the base of the Moroccan energy transition to a bigger group of actors, is the issue of decentralized generation. Here again, GIZ advisors, but also Moroccan industry representatives had a preference for adding small-scale installations to the Moroccan mix. Resistance to an effective opening of the low-voltage level from 2007 onwards lasted at least until 2015, when Law 58-15 was adopted, but its implementation is still unclear. The push by German and Moroccan actors to decentralize the Moroccan energy transition did not result in immediate transfer mainly because of opposition from ONEE and municipal electricity distributors who were worried about losing revenues and encountering difficulties with managing the grid. But willingness to follow Germany's example of high shares of decentralized generation was also limited within the administration. On the one hand, the model was deemed less adapted to a developing country (although this has changed due to greatly reduced PV prices). On the other hand, an aim to maintain top-down control of the energy system, from the highest levels of government, was also seen as an explanatory factor.

German advisors' view on the effectiveness of their work in terms of policy outcomes was in line with the assessment by Moroccan counterparts – strong with regard to the general direction of renewables and more nuanced with regard to decentralized generation and policy instruments. Effectiveness was linked to Germany's particular proximity to ministry staff and the legitimacy the Energiewende at home provided to outreach activities: "Whether there is an influence on the framework [for renewables] here? In all humbleness: clearly yes. The cooperation has existed for a very long time, we helped identify the renewables potential in Morocco, there was the Special Energy Program. At the same time, we don't drag our partners by the hand, there simply is a big demand here in Morocco" (M3). A stakeholder from the industry side noted, "it is a great advantage that GIZ is located within MEMEE, that ensures smooth communication. That way MEMEE can be sensitized for German ideas and technologies [...] even though the developments so far do not advantage German industry" (M30).

Interview partners were rather reluctant to speak of open competition between German cooperation initiatives and those of other countries or international organizations. However, examples of competition were given, in particular in the field of financial cooperation, where a real "rush" to Morocco took place during the Arab spring. Interview partners explained that international financing institutions and bilateral development banks were under pressure to spend their money on valuable projects and Morocco offered some of the few opportunities in this sector in the region. Projects such as the Ouarzazate solar park are co-sponsored by several financiers and collaboration worked smoothly once financing agreements were signed, after donors had competed for being chosen to finance the project (M35).

In summary, nevertheless, conditionality as a transfer mechanism appeared to play a secondary role at most in Morocco. The decision to adopt an energy strategy that gives an important place to renewables occurred in the context of an energy crisis. It was decisively facilitated by years of "preparing the ground", longstanding and trustful working relationships between Moroccan pro-renewables forces and German transfer agents, and the provision of information on energy scenarios at the right time. The perspective of exporting renewable electricity to Europe, in particular to Germany, and the availability of preferential financing played a role in the choice of CSP for the first phase of the Moroccan Solar Plan. In response to priority objectives of Moroccan decision-makers, in particular with regard to socio-economic development, in particular job creation, Germany has adjusted its offer, e.g., by the creation dedicated projects linking renewable energy and energy efficiency to employment opportunities (BMZ 2016b). This adjustment is likely to grow in importance in the future in order to sustain the nascent Moroccan energy transition.

7 Case Study: South Africa

7.1 Introduction to the Case Study

7.1.1 Case overview

South Africa's first free and democratic elections after decades of apartheid in 1994 and the entry into force of a progressive constitution in 1997 marked the beginning of a new era in South Africa (Deegan 2011). While the country's international contacts were very limited during the apartheid years due to international boycotts, bilateral and multilateral cooperation resumed quickly after the country's transition toward democracy. Germany was among the first donors to launch bilateral cooperation after the end of apartheid. The German-South African binational commission (Binationale Kommission, BNK) was established as a bilateral, government-level forum in 1996 (BMBF 2015). Bi-annual, high-level meetings have since then framed cooperation between the two countries in several working groups, ranging from environment and energy to economy, science and research, to cultural affairs and vocational training (BMZ n.d.-b). Due to its exceptional role on the African continent – South Africa is Germany's most important trading partner in Africa and the continent's second largest economy after Nigeria – South Africa, since 2012, has the status of a "Global Partner for Development" for Germany (Bundesregierung 2012). The term "Global Partners for Development" (*Globale Entwicklungspartner*) replaced the prior "anchor country" concept in German development cooperation for partners who not only receive development aid from Germany, but also join forces in implementing common projects in third countries and in addressing global challenges. The new terminology reflects South Africa's status as a regional leader and a relevant partner in global challenges of sustainable development (BMZ 2015: 5). As stressed in a BMZ strategy paper on Global Partners for Development, cooperation with this type of country can only reasonably occur in areas where "German policy has a comparative advantage […] Experience can be offered, for example, in the area of Energiewende, the sustainable use of resources […]"(BMZ 2015: 10).

K. Steinbacher, *Exporting the Energiewende*, Energiepolitik und Klimaschutz.
Energy Policy and Climate Protection, https://doi.org/10.1007/978-3-658-22496-7_7

Bilateral trade between Germany and South Africa is important, but unbalanced, with exports from Germany to South Africa amounting to €8.3 billion while imports from South Africa were worth €4.9 billion in 2014 (Auswärtiges Amt 2015d). Germany is South Africa's second largest trading partner behind China, and an estimated 100,000 jobs have been created by the 600 German companies present in South Africa (Auswärtiges Amt 2015d). With a per capita GDP of $6,483 in 2014, South Africa is an upper middle-income country (World Bank 2016). Due to factors such as high inequality and short life expectancy due to the HIV/AIDS crisis, South Africa nevertheless only ranks 116th out of 188 countries in the UNDP's Human Development Index (UNDP 2015a: 2). Despite encouraging progress in some areas of development, South Africa experiences major social and economic challenges. The distribution of income and wealth in South Africa is highly unequal, leading to the highest Gini coefficient (63.4/100) of all countries for which data was available in the World Bank Development Indicators (World Bank 2016b). The lack of jobs, in particular for South Africa's youth, is a persistent and urgent societal problem. More than a quarter of the population were officially unemployed in 2014, with youth unemployment reaching 52.6% (World Bank 2016b). Economic growth has been slowing down, from 2.2% GDP growth in 2013 to 1.5% in 2014 and 1.3% in 2015 (Statistics South Africa 2016). Recent signs of social unrest, expressed in student protests and petitions asking president Jacob Zuma to step down, cast shadows over the country's development process (Govender 2016).

ODA to South Africa in 2013 amounted to $1.3 billion USD (all donors combined), roughly half a percent of South Africa's GDP that year (World Bank 2016b). The United Kingdom, one of South Africa's most important donors, has recently announced a phasing-out of its bilateral aid, while the United States and the World Bank are also planning to decrease their commitments given South Africa's status as an upper middle-income country (Piccio 2013). Germany's total ODA to South Africa until 2014 amounted to €1.14 billion (Auswärtiges Amt 2015d) and €72.5 million were promised for the 2014-2015 period (South Africa-Germany Binational Commission 2014: 6), roughly 60% of which was energy related. A total of 2,109 cooperation projects between Germany and South Africa (including regional projects) were registered in the AidData database as of 2013 excluding Germany's major contributions to multilateral cooperation through the EU or the World Bank's Global Environment Fund (AidData n.d.).

7.1.2 South African energy challenges

South Africa's electricity supply system almost entirely depends on the use of locally sourced coal, making it one of the most polluting worldwide. Average CO_2 emissions per capita amounted to 9.3 metric tons per year in 2011, a value close to Germany's, but at much lower levels of economic development and electrification, and well above the average for upper middle-income countries (World Bank 2016a).

Electricity generated in the mostly amortized coal-fired power plants resulted in some of the least expensive electricity prices in the world for many years, amounting to virtually only the long-run marginal cost of electricity production (Edkins et al. 2010: ii). Average electricity prices have however risen substantially since the 2008 energy crisis, increasing by two-digit rates annually (Edkins et al. 2010: ii), and reached above 8 dollar cents in 2015.

Pressure for change in the South African electricity sector, toward diversification and hence the introduction of renewables, comes from several sides. Given South Africa's rank as 12th largest emitter of greenhouse gases (GHG) globally (Sustainable Energy Africa 2015: 12), the country has come under international pressure to follow up on pledges to reduce GHG emissions in the run-up to the 2011 Conference of the Parties (COP17) hosted in Durban. The commitments made by President Zuma at the 2009 COP15 in Copenhagen (minus 34% of CO_2 emissions compared to a business-as-usual scenario by 2020 and minus 42% by 2025) led to the inclusion of emissions caps to the national electricity planning tool, the Integrated Resources Plan[70] (IRP) for 2010-2030 (Eberhard et al. 2014: 7). The development of South Africa's climate change mitigation policies has been supported by Germany (GIZ 2015b: 10), in addition to cooperation in the field of renewable energy and efficiency.

70 As discussed below, the IRP is a crucial strategic document in that its projections and models lay out necessary capacity additions to the electricity sector by technology and inform how much capacity is auctioned in the REIPPPP in each technology.

An even more important driver for change in the South African electricity system is its dire security and quality of supply record, which has even worsened since the 2008 electricity supply crisis (Eberhard 2013). The absence of sufficient reserve margins in the electricity system since the early 2000s put heavy constraints on security and quality of supply, increasing the frequency and duration of “load shedding”, a “euphemism for planned blackouts imposed by Eskom”, South Africa’s state-owned energy utility (Eberhard 2013: 1). Since 15% of households in South Africa still do not have access to electricity (World Bank 2016c) – 12 percentage points more than Morocco, whose GDP per capita is roughly half – demand is set to increase further and will put additional pressure on the electricity system. South Africa’s economy is only growing at low rates at the moment, but since this might be partly due to a lack of quality of electricity supply, unserved demand needs to be taken into account when evaluating the actual state of security of supply in the country (Calitz et al. 2015: 29). The South African electricity sector is dominated by Eskom, a quasi-monopoly, which operates all parts of the electricity value chain, apart from electricity distribution in about half of all cities[71]. Eskom is also a core player of South Africa’s particularly closely-knit network of (conventional) energy, the mining industry, and politics known as the Minerals-Energy-Complex (MEC). The MEC has dominated the South African energy sector and, in fact, large parts of the economy for decades (Baker 2016, Fine & Rustomjee 1996). The introduction of independent power producers outside the MEX need to be seen as a major challenge to this fundamental characteristic of the South African economy (Baker 2016:3). In addition to pressure for climate action and security of supply concerns, other factors such as local air pollution, have created urgency for change in the South African electricity system. A more detailed analysis of the targets pursued through renewables is provided in section 7.1.4, based on ranking exercises carried out in the framework of interviews.

7.1.3 Renewable energy policy in South Africa

The need to diversify South Africa’s electricity mix and to tap the country’s renewable energy potential was recognized already in the first White Paper on

71 This electricity market structure is similar to Morocco’s.

Energy Policy developed after the end of apartheid in South Africa and presented in 1998 (Department of Minerals and Energy 1998). In 2001, given Eskom's poor financial performance and investment backlog, the South African government approved proposals to introduce a set of liberalization measures including a goal of at least 30% of electricity to be produced by IPPs (Vagliasindi & Besant-Jones 2013: 222). A first concrete objective for renewable electricity, 10,000 GWh by 2013, was set in the subsequent 2003 White Paper on Renewable Energy (Department of Minerals and Energy 2003: i). However, the target was not further specified with regard to its baseline, scope and timing, and did not lead to any significant development of renewable energy in South Africa (Sebitosi & Pillay 2008: 2514). This is despite the country's outstanding solar and wind resources (Department of Energy 2015).

From a temporal perspective, analysis presented in this chapter focuses on the years between 2008 and 2014, during which the two main initiatives for renewable energy deployment in South Africa were developed, namely the Renewable Energy Feed-In Tariffs (REFIT) and the Renewable Energy Independent Power Producers Procurement Program (REIPPPP). The launch of the REIPPPP in 2011 marks the most effective development in the renewable energy field in South Africa. In the program, IPPs bid for a pre-defined amount of new renewable electricity capacity in different technologies. Tenders take place on a yearly basis and bids are not only evaluated on the basis of the price at which each bidder proposes to produce one kWh of renewable electricity, but 30% of the evaluation points are based on socio-economic criteria such as jobs created or the share of black management (Eberhard et al. 2014, WWF South Africa 2015). The REIPPPP is the first successful attempt to introduce IPPs to the South African energy system, thereby threatening Eskom's electricity generation monopoly (see section 7.3.2).

Despite the REIPPPP's importance, an analysis of South African renewable energy policy would be incomplete without thorough consideration of the prior REFIT program. These feed-in tariffs for renewable electricity were announced and published by the National Energy Regulator of South Africa (NERSA) in 2009, but were never effectively paid out to concrete projects. This was due to uncertainty surrounding the level of the tariff, its constitutionality, and opposition from the Department of Energy (DoE) as well as the National Treasury, South Africa's ministry of finance (see section 7.3.2). The REFIT however played a crucial role in preparing the ground for the country's current renewable energy program and is therefore selected as the second major point structuring the process tracing exercise presented in this case study. Prior to the announcement of the REFIT, German transfer agents also played an important role in preparing the ground for these two policy developments, in particular by providing support to pro-renewables advocates in the provincial government since the mid-2000s, as well as to individual researchers since the 1990s. These activities, as well as cooperation in the fields of energy efficiency and climate policy, are important in order to understand the conditions for German energy leadership in South Africa and are treated as part of the process tracing.

7.1.4 Objectives pursued through renewable energy policy in South Africa

Since policy objectives pursued are likely to influence instrument selection and policy transfer, stakeholders' views on these objectives were gathered through ranking exercises as part of 41 of the interviews conducted in South Africa in 2014.

Job creation ranked highest as a perceived objective of renewable energy development in South Africa, with an average rank of 2.2, followed by security of supply (2.4) and creating an industry (2.5). Climate protection was seen as an objective the government pursues with a medium priority (4.1). This goal was often conflated in the interviews with the objective of "responding to expectations from the international community" and was seen as intrinsically linked to the COP17 summits hosted in Durban in 2011: "I think really the main reason for the REIPPPP programme was the COP17 in Durban, being able to host that. Without that, we would still be struggling and there would be no renewables. I think South Africa cared about its role and image in the COPs, to regain reputation from the international community. It was really this one event that was the main driver (S19)." The sincerity with which the government pursued climate protection efforts and in particular the actual degree of ambition of the country's climate goals were frequently put into question by interviewees.

The summit was indeed mentioned by a large number of South African interview partners as the immediate driver for the speedy implementation of the REIPPPP. This aim led to the announcement of the first bidding round results at the sidelines of the summit (Eberhard et al. 2014: 3). As outlined above, the South African electricity system has come under tremendous pressure from a security of supply point of view, while Eskom is financially constrained. Introducing renewables by bringing in IPPs was a means to respond to the primary objective of securing supply and/or increasing energy independency and thereby secure supplies in the future. The possibility offered by tenders to control the amount of capacity added and, to a certain extent, the siting of projects, preserves the top-down, central-planner approach predominant in the South African energy system. The design of the REIPPPP also clearly reflects the policy objective of job creation and industrial policy. Local content requirements and local economic development criteria account for 30 percent in the evaluation of submitted offers for renewables procurement in the REIPPPP (Eberhard et al. 2014; Stands 2015; WWF South Africa 2015).

It is noteworthy that the potentially important contribution of decentralized renewables to job creation and local economic development has not constituted an important enough argument to remove obstacles to small-scale renewable energy installations yet, partly due to the potentially detrimental consequences on municipal budgets (Sustainable Energy Africa 2013: 2). The REIPPPP itself has been extended to "small" installations of more than 1 MW (more than 140 times the size of an average residential solar rooftop installation in Germany, see Seel et al.: 13) and below 5MW, a category for which a total of 200 MW have been set aside in the 6,925 MW REIPPPP (Fourie et al. 2015: 12). The Small IPP program is particularly supported by KfW (Global Environment Facility 2015).

7.1.5 Framework for South African – German cooperation in sustainable energy

Cooperation between Germany and South Africa in the field of sustainable energy started in 1996, with first pilot projects within the BMZ-sponsored "Program for Biomass Energy Conservation" (ProBEC) on efficient cooking stoves (Wentzel & Pouris 2007). In 1999, South Africa was among the first set of developing countries for which GTZ developed market analyses and country studies regarding wind energy potential and the respective regulatory frameworks in the framework of the BMZ-sponsored TERNA project. In the framework of this global project, cooperation on "new renewables" between Germany and South Africa was pushed on the provincial level in the Western Cape province. The TERNA project continued to form the framework for energy cooperation between Germany and South Africa until the end of the 2000s. Against the background of several years of rather loose cooperation in the energy field, and motivated by the upcoming COP hosted by South Africa as well as recurrent blackouts, the governments of South Africa and Germany agreed in 2008 to define "energy and climate" as one of the three focus areas for cooperation. The South African-German Energy Program (SAGEN) was launched in 2011 and now provides the framework for most energy-related technical cooperation.

Separate programs are operational for climate policy advice and skills development for the "green economy". During binational negotiations in autumn 2014, it was decided to rename the "energy and climate" focus area to "green economy" to reflect South Africa's developmental priorities (South Africa-Germany Binational Commission 2014: 7). Table 14 provides an overview of funding agreed during this round of negotiations for the "green economy" focus area.

Table 14: Funds committed in 2014 by Germany to South Africa for the "Green Economy" focus area. Data: South Africa-Germany Binational Commission (2014: 8)

Programs for the "Green Economy"	**Million EUR**
Small IPP Support Program – loan component. Financial Cooperation (FC)	20
Small IPP Support Program – accompanying measure. Technical Cooperation (TC)	2.5
Preparation of the Inga 3 Low Head Project (FC)	4.0
South African – German Energy Program (SAGEN). (TC)	10
Support of the South African International Renewable Energy Conference (SAIREC) through SAGEN. (TC)	2.2
Skills for Green Jobs (TC)	5.0

Germany's strong entry into the field of renewable energy advice in South Africa with the start of SAGEN in 2011 coincides with the Danish decision to pull its development aid out of the sector given South Africa's status as an emerging country. Denmark has since strengthened trade cooperation and cooperation with its energy agency, and is recently trying to reconnect to its prior strength in the sector through projects in energy efficiency and wind energy in particular (S37). Denmark's central role in preparing the early ground for renewables in South Africa during the 2000s, in particular through wind measurements and support to the first White Paper on renewables was recognized (S2, S22, S25, S28, S44).

However, windows of opportunity for change were missing during the time and Denmark's influence was therefore judged disproportionately small compared to its efforts: "The Danes were here very early, they spent lots of money and pushed wind a lot. I think they had little influence" (S44). The Danish decision to phase out development cooperation in the energy field was already known when SAGEN started (S17, S37), hinting at a coincidence in timing rather than the crowding out of the Danish program by Germany: "I think it was really a confluence of different factors. The British have a beautiful word, 'serendipitously' things happened. With other donors leaving, Germany could take the space. The Danes had pushed for years and nothing happened. It is sad for them that they left just before things started to move and Germany arrived" (S2).

In February 2013, South Africa and Germany signed an agreement for the establishment of a bilateral energy partnership (BMWi 2013). Activities are coordinated through a high-level energy group involving ministry and industry representatives in semi-annual meetings. A yearly steering committee group sets the work program (Nold 2014). The energy partnership is sponsored by BMWi and focuses on renewable energy and energy efficiency (BMWi 2013). Its aim is to facilitate and frame private sector activities through political exchange – an aim that has been clarified in response to criticism regarding the presence of German companies at the launch of the partnership and as part of its working group (Bundestag 2013a: 44). Facilitating the creation of sustainable market conditions, enhancing communication on private sector initiatives, promoting climate protection and "communicating the Energiewende" were also quoted as aims (Nold 2014: 3). Since the implementation plan for the energy partnership was still under development at the time of field research in South Africa, its impacts are not reflected in analysis presented in this chapter. Activities foreseen (Nold 2014: 9–11) included study tours, ad-hoc workshops and training programs with a focus on companies, which are complementary to the much larger SAGEN program. Figure 22 provides an overview of policy decisions and of milestones in German-South African cooperation.

Since the formal establishment of "energy and climate" as one of the three focus areas of cooperation between Germany and South Africa in 2008 (the other ones being HIV/AIDS and "good governance"), energy has been by far the most important sector of German ODA. Germany has committed over €600 million in grants and preferential loans to cooperation in the energy field with South Africa

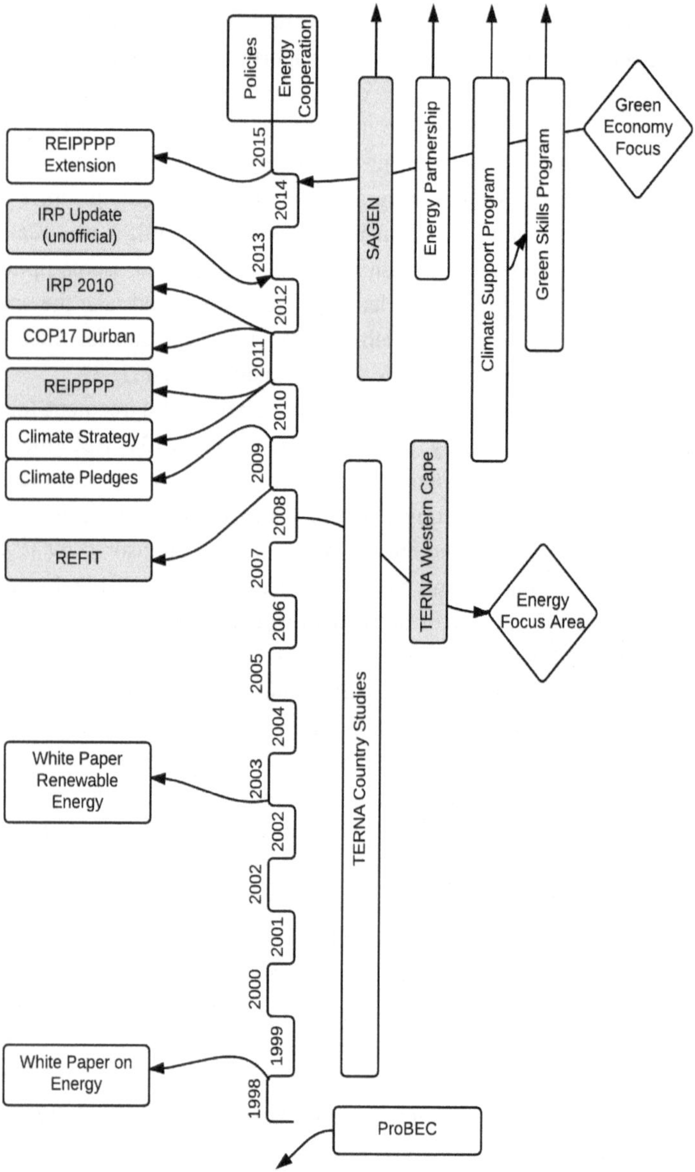

Figure 22: Timeline of renewable energy policy developments and South African-German cooperation. Graph by the author.

until 2014, which means that about half of all contributions from the EU and member states in this sector come from Germany[72]. The large amounts of ODA spent in the energy sector compared to other areas of cooperation are also, but not exclusively, due to the high capital intensity of investments in this area. Examples for capital-intensive investments are KfW's contribution of €100 million to Eskom's CSP plant in Upington or a new commitment agreed in 2015 on a €300 million loan for the modernization of South Africa's power grid (KfW 2015). Table 15 provides an overview of organizations covered through interviews in South Africa.

Table 15: Organizations covered through interviews in South Africa

AFD: French Development Agency	Industrial Development Cooperation
AHK: German Chamber of Commerce	IPP office
Camco	Juwi
CSIR: Council for Scientific and Industrial Research	KfW
Danish Embassy	Konrad Adenauer Foundation
DoE: Department of Energy	LEREKO
Department of Environmental Affairs	NERSA: National Energy Regulator of South Africa
German Embassy in Pretoria	RECORD: Renewable Energy Centre of Research and Development
Earthlife	Rosa Luxemburg Foundation
Embassy of the Republic of South Africa in Berlin	SAIIA: South African Institute of International Affairs
EMVELO	SANEDI: South African National Energy Development Institute
Energy Research Centre, University of Cape Town	SASTELA: South African Solar Thermal Energy Association
ENSafrica	SAWEA: South African Wind Energy Associations
EoN	Siemens
Eskom	Spanish Commercial Office
EU Delegation to South Africa	Sustainable Energy Africa
Genesis Eco-Energy	University of Cape Town
GIZ	University of Stellenbosch
Globeleq	Unlimited Energy
Heinrich Böll Foundation	

72 KfW (2014). Überblick Energie und FZ in Südafrika. Unpublished Document.

7.1.6 Transfer channels and South African energy landscape

As shown in Figure 23, Germany's main cooperation program in the energy field, SAGEN, is nested in a network of different actors in the South African (renewable) energy policy landscape. The central organization within the network for the definition of the South African energy strategy is DoE, which, in cooperation with the National Treasury, sponsors the implementing unit for the REIPPPP, the so-called IPP Office. SAGEN also cooperates with the association of local governments, different municipalities, provincial governments, and with dedicated institutions in the sustainable energy field, in particular the South African National Energy Development Institute (SANEDI). KfW's focus of cooperation is with South African financial institutions and Eskom, but many projects with municipalities have been sponsored and implemented over the years as well, in particular for energy efficiency and solar water heaters (Gerding 2011). Elements in Figure 23 that are identified with a star hosted German transfer agents within their organization at the time of field research.

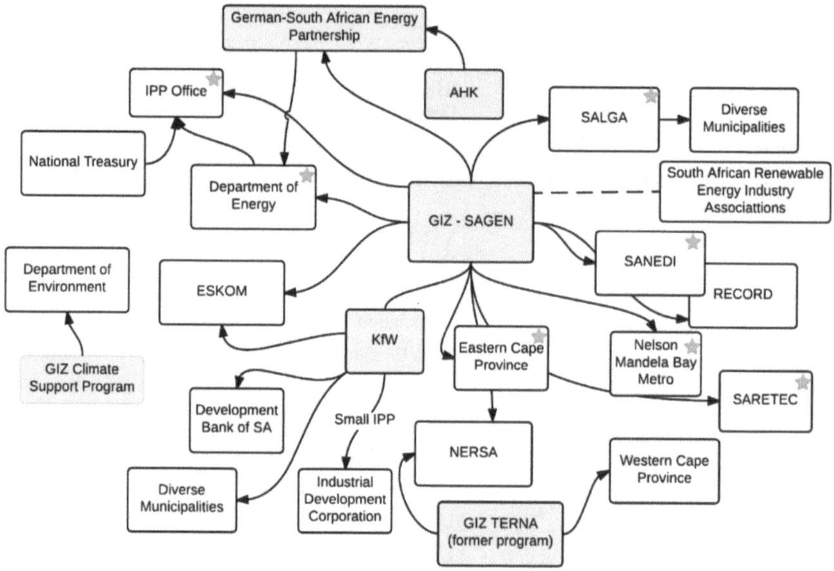

Figure 23: South African renewable energy policy network and main transfer channels. Colour: programs/agencies, star: German agent worked from within the organization. Graph by the author

Given its importance in volume and perceived impact, measures linked to bilateral cooperation through GIZ and KFW constitute the core of this chapter. The German embassy in Pretoria, political foundations, individual policy entrepreneurs, AHK (which operates a dedicated office for renewables and efficiency in Cape Town), researchers, and companies complete the variety of potential channels of communication between Germany and South Africa. The German Embassy in Pretoria has created a unit dedicated to environment, energy, and climate in the run-up to the 2011 COP conference. The embassy regularly publishes information about the Energiewende and Germany as a "green nation" on its website, and frequently hosts energy-related events (German Missions in South Africa, Lesotho, and Swaziland n.d.). For all the German political foundations present in South Africa and covered through interviews in this case study (Konrad Adenauer Foundation, Rosa Luxemburg Foundation, Heinrich Böll Foundation), energy and climate policy played an important, and growingly important, role. This was also linked to the availability of additional funds for work in these areas made available by BMZ for political foundations and other actors of civil society (S60). Beyond bilateral technical assistance, scientific cooperation plays a role between Germany and South Africa. The German-South African "Year of Science" 2012-2013 reinforced cooperation between scientific institutions, including in the field of climate change mitigation and adaptation (BMBF 2015).

7.1.7 Overview of field research in South Africa

Field research in South Africa was carried out from October to December 2014. Fifty-nine respondents were interviewed, most of them in person in Pretoria, Johannesburg, and Cape Town, and three background conversations were conducted. Interviews were in English and German, with German quotes in this chapter translated by the author. Interview partners were selected for their role in the development of South Africa's renewable energy policy, in particular REFIT and the REIPPPP. Surrounding policy developments, in particular renewable energy policy frameworks at the sub-national level and energy efficiency were taken into account as well.

As in the Moroccan case study, a focus of the interviews was on stakeholders involved both on the German and on the South African sides of bilateral cooperation projects in the field of renewable energy. Table 16 gives an overview of the affiliation of interview partners to particular sectors.

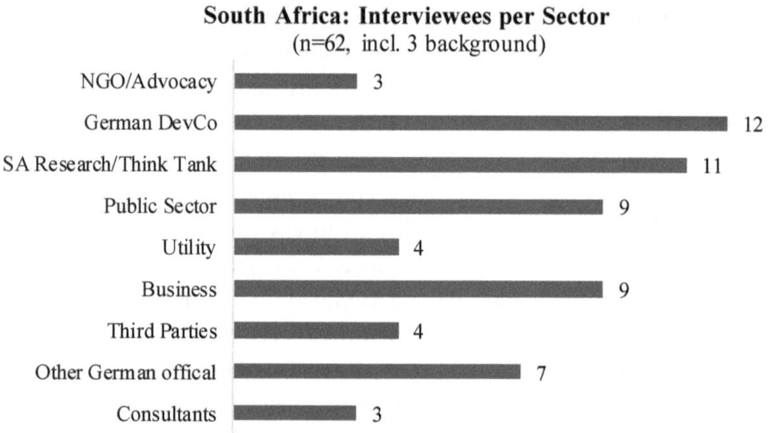

Figure 24: Number of interviewees per sector (South Africa)

Table 16: Interviewee affiliation by sector[73] (South Africa). DevCo: Development Cooperation

Identifier	Sector	Id.	Sector	Id.	Sector
S1	Public Sector	S22	Public Sector	S43	Public Sector
S2	German DevCo	S23	German others	S44	Consultants
S3	Consultants	S24	German others	S45	German DevCo
S4	Utility	S25	Consultants	S46	German DevCo
S5	German DevCo	S26	NGO/Advocacy	S47	Utility
S6	Public Sector	S27	Business	S48	German DevCo
S7	Research	S28	Business	S49	Utility
S9	German DevCo	S29	Business	S50	Public Sector
S10	Public Sector	S30	Business	S51	Business
S11	German DevCo	S32	Business	S52	Public Sector
S12	Research	S33	Research	S53	Business
S13	Research	S34	Research	S55	Public Sector
S14	Business	S35	Research	S56	German DevCo
S15	German DevCo	S36	German DevCo	S60	German others
S16	German DevCo	S37	Third Parties	S61	German others
S17	German DevCo	S38	Third Parties	S62	German others
S18	Third Parties	S39	Research	S63	German others
S19	Research	S40	Research	S65	Business
S20	Utility	S41	Third Parties	S66	Research
S21	NGO/Advocacy	S42	German others		

73 Missing numbers in interviewee identifiers were allocated to background conversations, second interviews with the same interviewee, and contacts that were not eventually followed up with an interview.

7.2 Laying the Ground for Renewable Energy Policy

7.2.1 ProBEC and support to pro-renewables pioneers

Cooperation between Germany and South Africa in renewable energy started with the ProBEC program on biomass conservation, which was implemented between 1996 and 2008 (Wentzel & Pouris 2007). The project was carried out in partnership with the Southern African Development Community (SADC), a regional organization of Southern African states, rather than on a bilateral level. It aimed at the conservation of energy from biomass, in particular through the deployment of more efficient cook stoves. Starting with field tests from 1996, different types of solar cook stoves were tested in South Africa through a cooperation agreement between GTZ and the South African Department of Minerals and Energy. The pilot program's results were positive with regard to energy, money, and time savings, but neither led to large-scale commercialization nor to the local production of solar cooking stoves (Wentzel & Pouris 2007), as initially expected. Results were however judged conclusive enough for BMZ to include solar cook stoves into its "standard" set of tools regularly proposed in policy advice activities to partner countries (GTZ & DME 2004: IV). The pilot field tests were run through a bilateral agreement between BMZ and the Department of Minerals and Energy, expressing the "will and commitment of both Governments to significantly contribute to solving the shortage of household energy, and more specifically the fuel wood problem, by coming up with a market oriented solution in South Africa" and to "once and for all [...] [show] that solar stoves are not only a niche solution" (GTZ & DME 2004: III). The project was described as being in line with the 2003 White Paper on Renewable Energy's aim of bringing "renewable energy into the mainstream energy economy of South Africa" (GTZ & DME 2004: III). Despite the link to broader efforts of promoting renewable energy, only one interview partner – who had been directly involved in the ProBEC project – mentioned it as an early example of German-South African cooperation in the energy field. Subsequent cooperation in the wind sector, through the TERNA project, was far more frequently cited at the start of German efforts in this domain in South Africa.

In parallel to the first bilateral cooperation between governments for renewable energy, German non-governmental organizations provided targeted support and capacity building to renewable energy advocates in South Africa. Heinrich

Böll Foundation, the political foundation close to the German Green party, played a particularly important role in this endeavor. South African researchers received support from the mid 1990s to conduct studies on the South African energy system and the potential of renewables in South Africa. Although this support "might [...] not have been incredible amounts of money", "if it wouldn't have been for them, we would not be standing here today" (S39), a researcher and consultant in the renewable energy field stated. Since 2011, Heinrich Böll Foundation has also been supporting annual training sessions on energy matters offered to members of the Portfolio Committee on Energy, a committee of the South African parliament. These one-day training sessions are carried out exclusively by South African experts because of the Foundation's concern of "work[ing] for local interests and local problems" (S63). Hans-Josef Fell, the Green Party parliamentarian and author of the 2000 German Renewable Energy Act EEG, visited South Africa on several occasions to advocate for renewable energy deployment and the introduction of feed-in tariffs in the country (S63, D13). Understanding the challenges linked to an energy transition was a particularly important takeaway from these meetings, according to an early South African renewables advocate: "I asked [Hans Josef Fell]: what is the magic trick to get renewables going? And he said 'there is no magic trick, it took us 30 years'. This was greatly encouraging for me." [S39].

7.2.2 *Supporting renewables at the provincial level: TERNA and the Western Cape*

TERNA was a global GTZ program promoting the uptake of wind energy in developing and emerging countries, including South Africa and Morocco. The project ran from 1997 to 2009 and was sponsored by BMZ and carried out by GTZ (GTZ 2009). The project's initial focus was on the assessment of wind energy potential through wind measurements, the evaluation of market potential, and the gathering of information about market conditions for wind energy. Cooperation with the German wind industry was a core aspect of the project, with an annual "Wind Energy and Development Dialogue" event (GTZ 2010). Two partner countries of this project – South Africa and China – received policy advice through GTZ in addition to the identification of wind energy potential and market studies. In this framework, activities in South Africa pursued "the broadest advisory approach of all the TERNA projects" (GTZ 2009: 5). In South Africa, this advice was decisive in bringing about a regulatory framework for wind energy in the Western Cape Province and, as a further consequence, "a feed-in tariff for renewable forms of energy according to the German model", resulting in "a rapid increase in planned wind energy projects, especially in the Western Cape Province" (GTZ 2009: 8).

Although the feed-in tariff was never effectively implemented, cooperation under the umbrella of TERNA was key in leveraging the pioneering efforts of pro-renewables agents in the Western Cape, and for translating the momentum to the national level. The Western Cape Provincial Department for Environmental Affairs approached GTZ to conduct "an extensive grid study for the Western Cape Province, a study for developing a business case for renewable energy (wind, solar and wave), and for support in generating a policy framework and support systems" (Essop 2008: 10). Provincial Minister for Environmental Affairs Tasneem Essop expressed her satisfaction with receiving support from the German government on this matter since "Germany has long-term experience in renewable energies (especially wind energy) and its exemplary renewable energy support system and policy framework can benefit us greatly" (Essop 2008: 10). The resulting renewables integration study was presented in 2009 and showed that 2,800 MW of wind energy capacity could be added without requiring major adjustments to the grid (DIgSILENT 2009). A first of its kind, the study was mentioned both by South African and by German officials as calming the heated debate on how much renewables Eskom's grid could absorb, thereby putting discussions on a more factual ground (S43, S45, S56, S17). Training sessions were also organized to accompany the study and to enable Eskom staff to perform renewable energy grid integration studies themselves (GTZ 2009: 7). Another study sponsored by GTZ and published in 2009 helped project developers navigate the regulatory steps toward project approval and highlighted barriers and opportunities for market entry in the Western Cape (Curren et al. 2009).

Support from GTZ, through studies and a study tour described in more detail below was seen as crucial in the development of a regulatory framework for renewables in the province and nationally, as well as in increasing acceptance for renewable energy among Eskom officials in particular (S43, S56). The main channels of how cooperation played out in favor of renewables were the empowerment of pro-renewables voices in the policy network through German support and the provision of evidence as a basis for arguments in favor of renewables. As one official put it, “all that would not have been possible, all of those things, if it was not for the German government. The capacity, the support, the technical knowledge. And I think it created what I would call a groundswell. Because it was really a fight. [...] I was told openly in meetings by top energy officials that solar would never work, that wind would never work in this country. And I needed a confident. I was so much alone; I was sending off stuff [...] without even knowing where the stuff is going. [...] And what really gave me a lot of confidence was learning lessons from GIZ and BMU and BMZ and a lot of interactions with German officials, technical consultants [...] we would get a factual, scientific basis for why this could happen” (S43). A third main channel of how German activities facilitated progress on renewables was by improving the connections between stakeholders from the energy space through meetings and study tours, one of which was to unfold particularly lasting effects.

7.3 Towards a National Renewable Energy Policy: the REFIT

7.3.1 An EEG study tour with wide-ranging impacts

In 2008, to further speed up the development of wind energy in the Western Cape province, officials requested GTZ’s advisor responsible for South Africa in the TERNA project to organize a renewables-centered study tour to Germany in 2008 (S56, S43). Participants in the tour included provincial and national government officials, staff from NERSA, Eskom, and private sector representatives. The one-week study tour included meetings with ministerial officials in Germany and site visits, in particular to wind farms. It equipped participants with hands-on experience and knowledge about renewables and examples to use in discussions: “I was able to show the pictures, to say we went there, we saw, we walked into a turbine. We saw the solar plants, touched them, we felt them, and it is happening” (S43).

The motivation for South African stakeholders to visit Germany specifically was directly linked to Germany’s unilateral leadership in renewable energy: “the country that was doing most of the strides and significant strides in the world [...] was Germany. So we thought, ‘Let’s go and see what Germany’s doing.’ Spain was also doing something but we said that we would rather have a study tour in this one that seems to be very organized” (S55). According to a German advisor involved, the motivation for the study tour was not to “export” a policy model, but to showcase how Germany as a forerunner dealt with renewables and, in particular, to "create a team" among South African energy stakeholders, "develop ideas" and "adapt" experience from Germany to the South African context (S46). The tour indeed provided a unique opportunity to energy stakeholders – both skeptics and early proponents of renewables – to meet and exchange. Opportunities for personal contact between different organizations on this subject in South Africa were judged insufficient also in the following years. A participant in a 2012 study tour recalled participants "even had a joke during that trip that it was easier to meet government in Germany than in South Africa" (S29). Creating a network for discussion and strengthening ties within the energy community, including the private sector, government, and Eskom, were seen as decisive achievements of the 2008 study tour as well as of tours in 2012, 2013 and 2014 (S29, S43, S28, S6, see Table 17).

Most importantly, the trip was judged fundamental in bringing about the REFIT as the first policy for renewables deployment at the national level in South Africa: "I was really fascinated by the learning and by Germany being a forerunner, in the context of renewable energy […] we basically, that group of people [participating in the study tour], changed the face of Southern Africa in terms of renewable energy. When we came back we wrote the first feed-in tariff for South Africa" (S43). As another interview partner put it "when we came back we said, 'Right. This is potentially the structure that we should follow'" (S28). Meetings with ministry officials in Germany and site visits in convinced the regulator that feed-in tariffs were a feasible and appropriate option for South Africa, leading to the unilateral announcement of such tariffs by NERSA only months later in 2009 (S55). NERSA leadership was already determined to act in favor of renewables prior to the study tour, with a view to finally implement the 2003 White Paper on Renewable Energy. The choice of a policy instrument and confidence in policy design were however decisively impacted through the trip to Germany (S55). Germany's policy experience was seen as particularly attractive for its ability to bring about renewable energy capacity fast and for its perceived strength in achieving "economic development […] distributed generation, and […] some development of manufacturing capabilities and facilities" (S55) through renewables. South African stakeholders who had not participated in the tour nevertheless spontaneously mentioned it as an example of the direct impact of German cooperation on the development of renewable energy policy in South

Africa (S3, S29, S39). The tour's impact on further policy developments in South Africa was not only confirmed by interview partners but was also mentioned in an impact report on TERNA published by GIZ: "A study visit at the start of the project focused on the German Renewable Energy Sources Act (EEG) and motivated participants from the national regulatory authority to introduce a feed-in tariff for renewable forms of energy according to the German model" (GTZ 2009: 8). Despite the clear intention of showcasing the German energy transition and the EEG in particular, participants in the tour did not feel pressured to adopt this particular model (S28, S55) – an impression in line with the general perception of Germany acting as an "honest broker" with a vague agenda (see case discussion).

The detailed design of the national REFIT policy in the months following the study tour was not accompanied by GTZ, although advice was offered, but was taken care of by a working group within NERSA. In the group's meetings, Germany was used directly as an example, according to a working group member (S26). A group leader described the REFIT development as a selective transfer process, where the group took "what's good from this and what's good from that and develop what's suitable for our country [...] we had considered what was being done in Germany, but we also always had the view and the feeling that we're not talking to the same kind of investors so we can't use all of the things that Germany has used. [...] But to a large extent, we looked at the German model, we customized it to fit our situation" (S55). Design features that differed from the German feed-in tariffs included planned program caps, which would have limited the amount of installations that could receive payments from a REFIT in any given year (S3) and a minimum project size of 1 MW, excluding smaller-scale installations (NERSA 2011, see also section 7.5.3). This choice of capping the program was explained by the provisions in the IRP on how much renewables capacity should be added and which the regulator could not exceed (S55). The automatic digression of tariffs in the German model was another design feature that did not appeal to the South African working group, and a more discretionary yearly review of prices was therefore introduced in the REFIT (NERSA 2011: 15, S55).

Germany was not the only country organizing energy-related study tours to Europe. Denmark organized tours on wind energy and local renewables similar to the German tours before German activities in the energy field took off (S37).

France's energy utility EDF invited South African stakeholders, including selected union representatives, to France in 2011 to showcase nuclear power – leading to strong criticism from the anti-nuclear Congress of South African Trade Unions (COSATU) (Faull 2011).

7.3.2 From REFIT to REIPPPP: Turf wars and preferences

Although it was never effectively implemented and tariffs were not paid out to producers, the adoption and publication of (German-inspired) feed-in tariffs by the regulator changed the dynamic of renewable energy policy in South Africa and facilitated the adoption and implementation of the REIPPPP. Interview partners, in particular from the private sector, agreed that because of NERSA's feed-in tariffs, developers were able to submit bids to the REIPPPP at short notice in 2011 (S29, S53, S28, S32). They had already identified promising sites for renewables, had developed financing concepts and the foreign firms among them had familiarized themselves with the South African context. Although project developers expressed their discontent regarding the policy uncertainty surrounding two years of postponed and eventually unsuccessful implementation of the REFIT, these early project developers were highly successful in round 1 of the REIPPPP (Eberhard et al. 2014: 8). The formerly proposed feed-in tariffs were used as ceiling prices in the first round of auctions and led to generous rates of return for the first successful REIPPPP projects.

Interview partners mentioned a number of different reasons for the sudden switch from the REFIT to the REIPPPP, including a turf war between NERSA and DoE as well as National Treasury (see also Pegels 2011). The struggle between NERSA on the one hand and the DoE and the National Treasury on the other orbited around NERSA's competencies in setting a tariff and the constitutionality of a public procurement program that was not based on competition over prices (Pegels 2011: 107). Eventually, a legal assessment commissioned by DoE and National Treasury concluded feed-in tariffs were unconstitutional in the South African context since they did not constitute a "competitive" procurement process (S25). Whether the constitutionality of feed-in tariffs in general was really an issue – and whether it was this concern that motivated DoE and National Treasury to take back ownership of renewables policy in South Africa – was strongly questioned by a range of interview partners from all sectors (S25, S26, S3, S55). For an interview partner from the NGO community "the Department of

Energy suddenly realized that its policy space was gonna be usurped by the regulator. So, [they] got into it" (S21).

As far as the use of other foreign examples in policy design is concerned, the experience of auction schemes elsewhere around the world was not considered a particularly supportive argument, but once the decision was taken to introduce the REIPPPP, other country examples such as Brazil were looked at by the "transaction advisors" (consultants) hired to design the program (S10, S50, S52). The German example did not get much attention in this process and no open debate on the choice of REFIT vs. auctions took place. Only one interview partner suggested the example might have been considered a threat by anti-renewables advocates: "there was a strong move against renewables and people felt, based on the German experience, that renewables could really take off if you put in a feed-in tariff" (S26). Interview partners were split over the question of whether the preference given to auctions was due to a conviction within DoE and NT that auctions were a more appropriate policy instrument for the South African context or whether the move was motivated by an aim to regain ownership of renewable energy policy-making. Overall, the move to the REIPPPP was in any case not based on negative lesson drawing from the German and other examples, but was motivated by domestic considerations. Auctions as a policy instrument did in any case allow for greater control on the amount of renewable energy capacity to be added to the grid within a given time frame (Baker et al. 2014) – including to ensure announcements of preferred bidders for the first REIPPPP window could be made in time for the COP17 in December 2011 (Eberhard et al. 2014: 15).

How much capacity is tendered in each renewable energy technology in the REIPPPP is defined by "ministerial determinations" on the basis of the IRP. The first phase of the REIPPPP, with five rounds of tenders between 2011 and 2016, will lead to the procurement of 6,725 MW of renewable electricity. An additional 6,300 MW are to be tendered until 2020 in a second program phase. Given the REIPPPP's success in bringing down PPA prices due to competition, South African Minister for Energy Tina Joemat-Pettersson announced an extra, expedited bidding round of 1,800 MW in 2015 (Joemat-Pattersson 2015). While the REIPPPP focuses on utility-scale installations, often at considerable distance from consumption centers, a program for "Small IPPs" of 1 to 5 MW capacity has been created (Fourie et al. 2015: 12), with strong support from Germany (South Africa-Germany Binational Commission 2014: 8, S16).

7.4 Technical Assistance and Incremental Transfer to Implement the REIPPPP

7.4.1 Germany as a learning partner

The switch from a feed-in tariff system, which German advisors were highly familiar with and had promoted, to an auction-based system required some changes to bilateral cooperation within the energy program SAGEN, which was launched in the same year as the REIPPPP. Several interview partners described the switch in policies as surprising and as requiring fundamental adjustments to be made to the initial work program of SAGEN in 2011 (S45, S46), while others said the program could be adjusted before its launch (S56, S17). In any case, one of the tasks foreseen in SAGEN's program proposal was the support of NERSA in implementing a feed-in tariff, drawing from experience in Germany (S17). With this task no longer relevant after the launch of the REIPPPP, GIZ found itself in a somewhat more marginalized position. Offers to provide advice on the design of the REIPPPP instead were declined by DoE, a fact interview partners from the German side said they somewhat regretted (S45, S36). DoE and National Treasury brought external "transaction advisors", from international law firms and private consultancies into the process. The Spanish government, through its business internationalization fund, sponsored a private Spanish consultancy firm, Novadays, with €436,000 to support the IPP unit as transaction advisors upon request from the South African government (S43). GIZ advisors only came to the process after the first bidding round and have since been advising the IPP office regarding monitoring and socio-economic development, as outlined below.

The South African switch to an auction system occurred at a time when GIZ's overall energy advice portfolio was seen as being in transition itself. Interview partners readily acknowledged feed-in tariffs were the dominant element in the agency's toolbox for energy policy advice around 2008-2010, based on the experience in Germany (S45, S56, M1, M13). This situation has since changed, with several important partner countries having introduced auction-based renewable energy systems, including with support from GIZ (Tietjen et al. 2015). A

2013 report commissioned by GIZ, for example, drew lessons from the experience with tender schemes in Brazil, Morocco, and Peru for the implementation of the South African program (Lovinfosse et al. 2013). Experience from South Africa also allowed GIZ to advise the Nigerian renewable energy program and a study tour to South Africa was organized in 2014.

7.4.2 Facilitating the REIPPPP: Monitoring and socio-economic development

Although German counterparts were neither involved in the design of the REIPPPP nor in the evaluation of bids, the implementation of the program was facilitated by GIZ and KfW. German policy promotion efforts in South Africa at the time of field research were globally targeted at perpetuating the change toward renewables brought about by the REIPPPP and at putting the nascent energy transition on firmer ground. Three German advisors worked from within the IPP office at the time of field research to provide legal and regulatory advice, assist with the development and implementation of program monitoring, and support the socio-economic development team. Advisors interviewed drew a direct connection between their German background in the Energiewende and advisory work in South Africa, in particular in the field of quality assurance, general project management, and the monitoring of IPP projects in terms of interconnection and safety (S9, S11). The presence of advisors within the IPP office clearly sets Germany apart as a partner, since no other foreign partners have placed advisors in this core unit of the REIPPPP (S50).

While the monitoring of socio-economic development objectives in the REIPPPP might appear as a marginal issue, the program's success in delivering job creation and local development are in fact decisive for its continuation. The REIPPPP is a particular renewable energy policy program, in which 30% of the decision to award a PPA to a specific project depend on efforts in areas such as job creation, local content, and support to the development of communities in the immediate vicinity of projects (Eberhard et al. 2014, WWF South Africa 2015). At the time of field research, anecdotes about project developers flying in foreign truck drivers to project sites and overall skepticism regarding the development impact of REIPPPP projects casted shadows over the program. Industry representatives highlighted the strong focus on job creation in South Africa: "government will support anything that creates jobs; they'll support anything. And if renewables don't do it, they don't care about the stories from Germany" (S28).

The importance of this objective needs to be kept in mind when assessing the impact of German-sponsored studies on expectations for job creation through renewables and on efforts to enhance the economic benefits of the REIPPPP. The South African development strategy "New Growth Path" sets a target of "300,000 additional direct jobs by 2020 to green the economy" (Department of Economic Development 2010: 12) – a number that coincides with the often-cited figure of 300,000 jobs created by the Energiewende in Germany. Although the origin of the identical number could not be established with certainty (interview partners could neither confirm nor refute the hypothesis of direct transfer of the number from Germany) Germanys' experience was used to justify the goal (DoE 2015: 23). Interview partners involved in its development also mentioned general trend to take into account the German job creation experience prior to the signature of the Green Economy Accord: "many of the studies from those guys [German research institutes] were used to back up many people's studies, and […] they said to government, 'If you do this, change these policies, you will create all of these jobs.' […] "Here's all the jobs you have to create – by 2020 – 300,000 jobs" (S28).

In the renewable energy sector more specifically, studies commissioned by GIZ evaluated the jobs potential for wind energy (Wiesegart et al. 2011: 64) and CSP (GIZ et al. 2013) in South Africa. But transferring the German example was considered dangerous in raising job creation expectations that could not be fulfilled in the South African context and could even threaten future extensions of the REIPPPP: "part of the problem was: there were also lots of studies done drawing on German examples of, like, how many jobs would be created. But the preparation for that means you need to have a structure as government to have renewable energy colleges or training centers" (S26).

Germany has taken various measures to facilitate the achievement of ambitious job creation objectives through the South African renewables program. The creation of the South African Renewable Energy Technology Centre (SARETEC) in the Western Cape province was cited as an example of a project that was initially pushed by Germany to this end (S6, S29). Germany had drafted a concept for the training of wind energy technicians based on the German and Chinese experience (Wiesegart et al. 2011) and argues it convinced the South

African government to invest ZAR105 million in the project (GIZ 2015b: 9). The center provides training to renewable energy technicians, including at a wind turbine sponsored by German manufacturer Nordex (SARETEC 2013). Through the intermediary of SANEDI, Germany partially funds the salary of SARETEC staff including a German senior advisor for renewables training, and repeatedly organized study trips both for technicians and trainers to Germany (S6, S17). A South African partner involved in the program, however, saw GIZ exaggerate their actual share in implementing the project: "okay, they organized [...] a study tour for German and Chinese experts to South Africa and this really was quite helpful to put the topic on to the agenda. But the idea certainly did not come from them. After the study tour they left us and came back with an expert when things were settled" (S19). Another interview partner, on the other hand, spontaneously mentioned SARETEC as a major contribution GIZ had made (S29), showing the diversity of perceptions on the issue. While the SARETEC project is supported by SAGEN, Germany has also established a separate "Skills for Green Jobs" program in 2011. The program is sponsored by BMZ and aims at developing curricula, training trainers and ensuring knowledge transfer to small and medium enterprises, including in renewable energies (GIZ 2015b: 13), to ensure the REIPPPP reaches the objectives of greatest relevance to South African policy-makers.

German political foundations working in South Africa are also active at the link between renewables, climate protection, and jobs. Rosa Luxemburg Foundation (RLS, connected to German left-wing party *Die Linke*), is a founding member of the "One Million Climate Jobs" campaign that calls for a just transition and a focus on socio-economic development in climate and renewable energy policies in South Africa (Bahadur et al. 2011). Heinrich Böll Foundation is a longtime sponsor of Earthlife, another NGO that is part of the campaign (S21, S63).

7.4.3 Appeasing concerns: renewable energy integration and Eskom capacity building

Facilitating the achievement of one of the top objectives of South African renewable energy policy, job creation, is one way in which Germany attempts to ensure the continuity of the REIPPPP and strengthen the program against skepticism. Another focus area is the integration of renewables to the grid and the management of the intermittency inherent to solar and wind power. To this end, GIZ, working in partnership with DoE and Eskom, conceptualized a range of studies on the integration of renewables into the South African grid (Table 17). Starting with the wind energy integration study for the Western Cape province (DIgSILENT 2009), these studies were seen as particularly useful in strengthening the arguments of pro-renewables advocates (S43, S56). They also contributed to appeasing concerns, in particular within Eskom, regarding the impact of renewables on grid stability (S20, S47). A study coordinated and sponsored by GIZ presented at the time of field research for example looked at the impact of different geographic locations of PV on grid management. The quality of the studies and their usefulness to South African partners were widely acclaimed, as well as GIZ's role as an "honest technical knowledge broker" in the commissioning and oversight of the studies. As an Eskom manager underlined, "GIZ […] helps a lot of people, also outside Eskom; government takes a lot of direction from them. Our network people got a lot of good studies. […] we are going to use them more for additional work that we don't have the capacity to do due to a lack of resources. […] [we] want, through these studies, to indicate to the government the importance of flexibility and to change the specifications of future generation [...] I haven't seen any kind of hidden agenda from them. They are honest and upfront and you can tell them what you think […] I think there is an underlying current of they like renewables, not so much nuclear (laughing). But we are asking them to look into renewables so..." (S20). This quote illustrates the role of lacking capacities and the utilization of transferred knowledge in renewable energy policy design and planning in South Africa. The strong self-reported ties that link GIZ advisors to the Energiewende in Germany and the use of German consultants for many of the studies presented in Table 17 justify looking at the knowledge they contain as a particular form of transfer.

In addition to these and other studies, tours and training sessions were organized for Eskom staff to visit power grid control centers in Germany and enhance their knowledge of dealing with renewables in the grid to appease concerns and mitigate opposition to renewable energy policy. These tours were aimed at countering a general feeling "from Eskom's side […] that, 'these renewables are unstable. They are going to make our grids to be uncontrollable.' So we said, 'Okay. Let's […] get the GIZ to assist us to get some of our controllers to go and sit with the controllers in Germany and see, yes the challenges are there, but see what is also being done" (S55). A manager from Eskom's renewables branch highlighted the immediate usefulness of German support for Eskom's wind farm project in Sere: "the guys that went, the technicians, for the wind farm […] the beauty of it is, they were exposed to training before we started construction, and now that we have started construction of the wind farm, and in fact we are now closer to commissioning, it is almost a revision of what they've learned and now, again, they are putting that learning into practice because they are actually doing it. Yeah, so it was very good" (S47). Eskom had started the construction of a wind park and a CSP plant as a consequence of conditionality from the World Bank and other donors, including KfW (S3, S16, S20, S47). Donors had agreed to contribute to the financing of Eskom's Medupi and Kusile coal-fired power plants of 4,800 MW *each,* but requested Eskom to push forward with the development of a 100 MW wind farm in Sere and a 100 MW CSP project in Upington, the latter co-financed by KfW with a €100 million loan (KfW 2015).

7.4.4 Strengthening institutions for sustainable energy in South Africa

Germany's support to sustainable energy in South Africa is strongly related to building capacity and reinforcing institutions in the energy field. These efforts change the structure and balance of power within the renewable energy policy network and are therefore likely to affect policy transfer. In this context, SANEDI holds a particular place. SANEDI is a state-owned entity in charge of promoting research and development in energy efficiency and energy and to promote the uptake of green technologies in South Africa. It has a central role in the dissemination of knowledge, including the studies produced or sponsored by GIZ. Interview partners estimated the share of GIZ financing in total SANEDI activities to be about one-fifth since the inception of SAGEN (S5), with a much larger share of over 50% for SANEDI's Renewable Energy Center of Research and Development (RECORD) (S5, S6). Ideas about how to support SANEDI as a strong voice in a transition to sustainable energy in South Africa were strongly influenced by the German model of *Länder* energy agencies, with an aim to translate the "pioneering role we saw in Germany to the South African market" (S5, also S6). A South African expert saw support from GIZ as "absolutely essential" and as "absolutely supporting South African strategic objectives […] they've really been driving the agenda of the South African government, if I can

Table 17: Examples of GIZ activities in the framework of SAGEN and TERNA.

Study Title	Year
Western Cape Regional Regulatory Action Plan	2009
Grid Integration of Wind Energy in the Western Cape	2009
Capacity Credit of Wind Generation in South Africa	2011
Costs and Benefits of Implementing Renewable Energy Policy in South Africa	2011
Green Skills Baseline Overview for renewable energy and EE	2011
Assessment of Training and Skills Needs for the Wind Industry in South Africa	2012
Support to the Design, Implementation And Evaluation Of Municipal Energy Efficiency Programs	2012
Overview and Assessment of the Energy Efficiency and Energy Conservation Policies and Initiatives of the Republic of South Africa	2012
Lessons for the Tendering System for Renewable Electricity in South Africa from International Experience in Brazil, Morocco and Peru	2013
White Paper: Net-metering concept for Small Scale Embedded Generation in South Africa	2013
Mapping of Provincial and Municipal Permitting and Authorization Processes for IPP Projects in the Eastern Cape	2013
REI4P Value Chain Analysis: Creating local opportunities in the renewable energy value chain: Value chain promotion and a guideline for IPPs in ZF Mgcawu District, Northern Cape Province	2013
RE Policy Mapping Study	2013
Enhancement of positive effects of REI4P Projects on Socio-Economic Development	2013
Analysis of options for the future allocation of PV farms in South Africa	2015
State of Renewable Energy in South Africa	2015
Solar Energy Technology Roadmap	2015

Study Tours	Year
EEG Study Tour	2008
Grid integration study tour	2012
Solar, wind & biomass energy training	2013
Grid integration study tour	2013
Energy Performance Certificates Study Tour	2014
Wind energy technician training	2013
PV, wind and bioenergy training sessions in Kassel	2014

put it that way. There hasn't been any German agenda that's been driving, from my perspective, at least" (S6). Study tours were again deemed particularly useful to build capacity since they allowed participants to see "things in action", an important experience given that "until you actually see it, and you see the scale, and you see how it works, you can't really comprehend it [...] You sort of read about all of these things and you think, 'Man, I don't know how we're gonna do this here'. And then when you see that it can actually be done, you think, 'Right. What do we need to do to get there?'" (S6). GIZ's support to the IPP office described above and the steady interaction with DoE – where a GIZ advisor also had an office and worked part-time – also fall into the category of supporting institutions and changing the policy network from the inside. In addition to the national level, GIZ is also supporting the South African Local Government Association (SALGA), with a GIZ advisor on energy placed within the association. Support mainly focuses on embedded generation (see section7.5.3). A German-sponsored advisor is also placed with the Eastern Cape Department of Economic Development, Environmental Affairs and Tourism, where she supports the implementation of the Eastern Cape's Sustainable Energy Strategy (Eastern Cape Sustainable Energy n.d.)[74]. Another German advisor works for the Nelson Mandela Bay Metropolitan Municipality (S2, S45), one of the pioneers in the field of small-scale generation from renewable electricity sources in South Africa. Support to institutions, especially by advisors placed with these institutions, was deemed necessary to ensure the first advancements in renewable energy in South Africa are institutionalized and translate into a longer-term transition. This type of support also addresses one of the most frequently cited challenges for renewable energy policy in South Africa, a lack of capacities: "South African institutions are under-resourced in terms of human capacity and financially. The lack of money and capacity is the major obstacle to our work here. Sometimes we didn't even have counterparts in the organization we were dealing with" (S2).

74 Cooperation with the Eastern Cape and Mpumalanga provinces is part of the South African-German partnership agreement (S45). The focus is motivated by the province's specific challenges, Eastern Cape being South Africa's poorest province and Mpumalanga being home to most of South Africa's coal mines.

7.5 Renewable Energy Policy: Design Choices and Agenda-Setting

7.5.1 Technology choice and the Integrated Resource Plan

Taking a closer look at the role of German agents in the choice of renewable energy technologies in South Africa is particularly interesting for matters of comparison with the Moroccan case. In Morocco, Germany's role in selecting CSP as a political choice to implement the Moroccan Solar Plan was clearly visible. Interestingly, both cases are very directly linked regarding this technology. A South African CSP industry stakeholder recalled having discovered the influential 2007 energy scenarios study GTZ had commissioned in Morocco (Roller et al. 2007, S53). The stakeholder then approached GIZ in South Africa asking specifically for a similar study to show the potential of CSP in terms of job creation and localization, following the example of the Moroccan study, with a view to strengthen the case for CSP in the IRP (S53, S2). Arguments from the study regarding the job and industrial potential of the technology for South Africa were then readily used in advocacy efforts to increase the share of CSP and broaden the knowledge base on this new technology: "So the point is, has the study been helpful? Yes. We wanted the study to help policymakers actually understand what is the value proposition of CSP. [...] the Department has been learning. That's why when you saw in the revised IRP that the CSP quantity had changed from the original scenario of 1,200 megawatts to where we're talking about the potential 3,300 megawatts for CSP. [...] so the bottom line is, has it been useful? Yes. Why? Because we asked for GIZ's support in order to have a document that we can use to engage policymakers and say it's been done by Ernst & Young, and Enolcon, and everyone involved is reputable" (S53). Although the study might have contributed to an increased allocation for CSP in the 2013 update of the IRP[75] (DoE 2013), several observers of the process did not see GIZ or Germany as pushing CSP in particular or German companies being particularly present in that technology at the time (S19, S41, S53): "you would never have found CSP in the IRP if it wasn't for the lobby. [...] it is ten times

75 The preparation of the 2013 IRP update followed the same process as the 2010 IRP, but while it is publicly available, the 2013 version has not been officially adopted by the ministry; this was because of political reasons, especially because it would have delayed the addition of nuclear capacity, as interview partners noted.

more than the cost of wind. It was more Spain and some American lobbyists" (S20).

A more direct impact on the IRP is linked to the work of a German professional working for Eskom. The expert had been working for an international consultancy firm on a project with Eskom and was subsequently hired by them to directly support its renewables unit as well as the unit in charge of modeling the IRP. The decision to hire was directly linked to his experience in evaluating the potential for solar PV and wind in Germany and his knowledge of the Energiewende (S4, S20, S47). The expert's contribution of insights into how the intermittency of solar and wind could be managed and of updated – more favorable – cost and learning data for PV, also based on the German experience, changed Eskom managers' view on the role renewables could play in the South African energy mix: "he's a PV guy [...] he kept on saying, 'PV will run, whether Eskom supports it or not, based on the learnings.' And I think, what was more exciting, he even installed PV on his own house in Germany. So I mean, showing us his annual revenues at this time, it was enlightening. It was really enlightening" (S47, also S49). The expert's influence on data input into the IRP was confirmed by another Eskom manager: "[he] was the single most influential person who got me so far as to even consider these things. [...] We had already done a public consultation for IRP, and [got] a lot of very negative comments on our cost assumptions, specifically the learning rates for PV. We knew we had to find better information. [...] Any future IRP will have far more PV. That definitely was totally due to them" (S20). While the expert personally drew his motivation and knowledge about the potential of renewables from the German Energiewende experience, he in practice often also referred to data from Spain to counter arguments according to which the German example was not applicable in South Africa due its lack of interconnectors with neighbors (S4, S49).

7.5.2 *Setting an agenda because "it makes sense": Cogeneration and biogas*

Most interview partners did not see any particular agenda for GIZ, except for the basic focus on renewables. In two areas, biogas and cogeneration[76], GIZ advisors

76 Cogeneration is the simultaneous generation of electricity and heat ("combined heat and power") to increase efficiency; in tri-generation, cooling is also provided in the same installation, in addition to heat and electricity.

did however feel a need to shape the agenda and push discussions further. GIZ staff saw an almost complete lack of knowledge about and familiarity with co-generation and tri-generation technologies in South Africa at the start of SAGEN (S46), a claim backed up by experts outside the agency (S44). The situation changed through GIZ's efforts of putting these items on the agenda, including two study tours to Germany and site visits (S46). A link was established between Germany's strength and policies in the sector and activities in South Africa, since given "the fact that cogeneration is big in Germany [it] is certainly not a coincidence that [a GIZ advisor] pushes for it" and "when they organize study tours it's to Germany, they show German technology examples" (S44). However, the interviewee also underlined GIZ's "primary mandate and interest is broad objectives, it's idealistic, helping to reduce emissions. And even if they are secondary interests in the background, it is not, and has never been in my perception, at the expense of the quality and sincerity of what they do" (S44). Another advisor cited the creation of a cogeneration facilitator unit within SANEDI as an example of direct transfer of experience in Germany to South Africa (S5). The unit's mandate is to facilitate the creation of a South African market for cogeneration and energy service companies (ESCOs). During its first year of operation from 2013-2014, it was managed by a German consultancy and the South African subsidiary of a major Germany electricity producer, STEAG (SANEDI 2014: 2).

Although the perceived priority objective of GIZ's work clearly was climate protection and development cooperation and GIZ advisors firmly rejected any aim of specifically facilitating market entry for German companies, the example of cogeneration certainly shows the inevitably blurred line between development cooperation, policy transfer, and industry in a technology-heavy field like energy. It also illustrates the challenges of balancing the need to bring in technical and regulatory know-how in a virtually non-existent sector with large GHG reduction potential and still conform to the "honest broker" reputation GIZ advisors in South Africa were particularly attached to. GIZ advisors described the aim as a "type of agenda-setting [that] has a positive connotation, if you say 'we have looked at the country and we understand the sector very well and we now bring in an idea that might not have been considered before'. Because we think it makes sense. Then we might have brought it onto the agenda and maybe had an agenda, but not for our own interests that have nothing to do with the country, but out of a conviction that this is the right thing for the country" (S56). Agenda-setting for cogeneration through frequent interaction with industry and government officials has resulted not only in the setup of the facilitator unit with SANEDI, in training sessions and publications, but also in policy results. A national 800 MW IPP program for cogeneration was launched in 2015, illustrating the effectiveness of GIZ's aim of "enhancing consciousness on the South African side that co-gen is an important topic" (S17).

A second example of a niche topic that became more prominent through GIZ is biogas. GIZ plays an instrumental role in this topic area as organizers of the National Biogas Platform, where needs for policy framework adjustments and other issues are discussed with a diverse group of stakeholders and regulators (Giljova 2015). Rather than being pushed to the agenda by GIZ, biogas emerged as an issue through direct contact with South African municipal officials participating in PV and wind training sessions in Germany and then asking whether "something could be done on biogas" in South Africa as well (S36). The logic here was described as looking at "What is our most feasible technology for municipalities? Which would have the most core benefits as well in terms of climate impact, in terms of job creation, localization, skill development, and everything? […] So that's where it came up […] I can really see a lot of interest for it" (S48). A South African consultant was surprised the focus had been on PV and wind for so long, since "Germany has so much biogas experience and now only is work being done on biogas. […] it does make sense to focus on those areas where there is expertise and where there are technologies" (S3). GIZ advisors confirmed that the question of whether their activities in the field of biogas served German industry interests was raised and openly discussed by South African counterparts (S48). Although the German biogas industry association regularly looked to enter in contact with the program (S36), GIZ advisors interviewed firmly refused the idea of taking steps in the direction of facilitating market entry for German companies as this was clearly seen as being outside GIZ's mission.

7.5.3 The development of a regulatory framework for embedded generation

South Africa's approach to renewable energy promotion is focused on large-scale plants. Against the background of persistently poor quality of electricity supply and rising electricity prices, it has become more attractive for households and businesses to consider producing electricity from renewables sources themselves: "You cannot stop people from doing that, and I think a tipping point has been reached" (S44).

A regulatory framework for this type of auto-production or "embedded" (i.e., decentral, small-scale) generation was still missing at the time of research, with the exception of pioneering schemes in metropolitan areas such as Nelson Mandela Bay. South African municipalities are confronted with a growing number of (illegally) connected solar rooftop panels or legal steps toward embedded generation like BMW's move to source electricity for its Rosslyn production plant from South Africa's first commercially viable biogas project (Odendaal 2015, also S16, S47, S49). This bottom-up development is worrying not only for Eskom, since it means a reduced demand for electricity generated by the utility, but in particular for municipalities who act as intermediaries between Eskom and about half of all final electricity consumers in South Africa (Bischof-Niemz 2015). Municipalities generate a considerable share of their income, sometimes more than 50%, from buying bulk electricity from Eskom and reselling it to consumers within their territory (Sustainable Energy Africa 2015).

The issue of rooftop solar could be a major area of knowledge transfer and policy transfer in the future, while the topic had not played a role early in cooperation activities: "Germany was moving toward rooftops for solar. At the time [2008], it was not something that we were seeing as being suitable for the country, mainly because of the costs. […] as we speak, we are looking at the German model now to expand our embedded generation […] We want to understand how you are dealing with the revenue shift from your traditional utilities to households and commercial entities, as well as your balancing […]. And we will take the good lessons and apply it at the end. And we're not going to go and look at how the Americans are doing it. We are fine with the German model." (S55). Contrary to the Moroccan case, the issue of small-scale generation was not heavily pushed by GIZ as an area of policy transfer at the national level, but organically emerged as an area for cooperation through existing working relationships and demand from municipalities who confronted a rising number of small-scale installations, often connected illegally (S17). GIZ, through the intermediary of SALGA and the Association of Municipal Electricity Utilities (AMEU), plays an important role as a facilitator in the development of a regulatory framework for embedded generation at the municipal level. Although rooftop solar is an area where "one can look at Germany, because it is very impressive" (S56), GIZ's advice and cooperation with municipalities goes in the direction of developing net-metering schemes rather than German-style feed-in tariffs. In net-metering schemes, consumers are allowed to feed electricity to the grid when they do not consume it themselves and receive a credit on their monthly bills in exchange. GIZ animated a working group on embedded generation and sponsored studies on international lessons for micro-generation as well as on net energy metering options for South Africa. Work on net energy metering resulted in a White Paper adopted by a working group consisting of SALGA, GIZ, Eskom, AMEU and other stakeholders (Pöller 2013). The German experience with small-scale renewables is fed into the process by the same German consultant who is also advising the Moroccan energy ministry, on behalf of GIZ, and who is "coming and telling [working group members] about the issues which were faced in Germany and how they solved it; [it] makes it a bit more approachable and reassuring. They think that, 'Okay, they also had issues and they actually managed it very well and didn't have huge problems'. I think that is something which was very well received by our colleagues here" (S48). In addition to regular working group meetings, a bi-annual "embedded generation forum" is hosted by GIZ and a tool has been developed to allow municipalities to calculate tariffs and prepare measures to accommodate the inevitable development of embedded generation (S36, S48). Certain design features of the preferred net-energy metering scheme for municipalities were influenced by input provided by the German consultant. For example, time-of-use[77] tariffs, which would put owners of embedded generation under a specific regime, were first preferred by municipal utilities. Evidence brought in by the German consultant then convinced municipal representatives that exchanging one KWh produced for less than one KWh consumed was a better choice (S17). Systems installed are thereby likely to be dimensioned as close as possible to actual consumption, a concern for municipalities and grid managers within Eskom.

Outside official bilateral cooperation and starting in 2013, a German head of unit at CSIR brought the German feed-in tariff experience to the debate in South

77 The term "time-of-use" means that producers of renewable electricity receive a compensation that depends on the time of the day, since demand for electricity differs greatly over the course of a day. Time-of-use tariffs can also be offered to consumers, e.g., when electricity is cheaper at night-time or at times of high solar generation.

Africa (Bischof-Niemz 2015). The proposed "NetFiT" is "at the end of the day, the German feed-in tariffs system with the sole difference that you have to compensate municipalities here" (S4). Both the regulator, GIZ, and industry stakeholders were aware of the proposal, but at the time of field research, skepticism regarding the likelihood of its implementation in South Africa prevailed, as municipalities preferred a net-metering system (S36, S55). To address the development of a framework for small-scale generation, NERSA launched a public consultation in February 2015. Interestingly, the final consultation document for this policy process does not refer to any foreign country experiences, while an earlier draft of the document from December 2014 describes the German program "which is the most advanced in the world" (NERSA 2014: 43) in detail as well as referring to other country experiences from Spain, the UK and the US. The topic of small-scale generation, in particular rooftop PV, has also been identified as a new focus area for the work of the renewable energy competence center of AHK. While German companies were seen as uncompetitive in utility-scale PV projects in the REIPPPP, small-scale projects were seen as a particular strength of German companies and therefore as an interesting market segment (S23, S24)

7.5.4 The South African nuclear debate: a role for the Energiewende model?

One of the most controversial, yet least openly discussed issues in South African energy policy is the role of nuclear in the future electricity mix. South Africa currently operates the only nuclear plant on the African continent, near Koeberg in the Western Cape province. The 2010 IRP foresees the addition of 9,600 MW of nuclear capacity, more than five times the current nuclear capacity in the country and more than the entire renewable energy capacity to be created through the REIPPPP until 2020. Eskom's financial situation, the lack of sufficient maintenance and operating capacities, and above all the technology's price have seriously put into question the appropriateness of new nuclear capacity in the country, but the utility nevertheless applied for licenses for two new nuclear sites in March 2016 (Roelf 2016). Numerous interview partners suggested political reasons behind the proposed increase in nuclear capacity, citing South African president Jacob Zuma's close ties with Russia, evidenced by the signature of a nuclear energy cooperation agreement between Rosatom, Russia's nuclear company, and South Africa (Rosatom 2014). Nuclear energy is a controversial topic, even within the leading party, the African National Congress (ANC), and

DoE has underlined the necessity to perform cost-benefit analyses before any final decision on nuclear could be taken (Paton 2015). The Economic Freedom Fighters, a revolutionary far-left party in South Africa, even urged the South African government to "strongly consider Germany's advice on nuclear energy" and to refrain from building new reactors (Economic Freedom Fighters 2015). But as in the case of Morocco, replacing or avoiding additional nuclear was nevertheless seen as the least important objective the government was seen as pursuing through renewable energy policy (see section 7.1.4).

German officials observed an increase in the negative use of the German Energiewende example by pro-nuclear advocates (S62). Among the most vocal of these nuclear proponents is Andrew Kenny, an engineer and advocate, for whom "the disaster of Germany's green energy should be carefully noted by South Africa", stating that thousands of "horribly unreliable", "gigantic wind machines now loom over [Germany's] beautiful countryside" (Kenny 2014b). Kenny, quite unsurprisingly, calls for negative lesson drawing from Germany's Energiewende experience: "Green electricity in Germany is bad for the environment, bad for the economy and a calamity for the poor. South Africa, please learn from Germany!" (Kenny 2014b). His perception of the German nuclear phase-out as a "fit of political madness" should lead "South Africa [to] be rational, and choose nuclear power as the safest and cleanest source of energy we know" (Kenny 2014a). South African pro-nuclear voices were not the only ones invoking the German example as a negative showcase for an energy transition. According to a report on an energy policy meeting of South Africa's trade union association COSATU, when "discussions at the meeting turned to the fact that Germany had scaled back its nuclear power generation plans since the Fukushima disaster, French nuclear lobbyists present at the meeting objected, arguing that Germany still imported nuclear power from France. It was at this point that [the meeting coordinator] discovered that the French lobbyists were not union members and asked them to leave" (Faull 2011).

German entities in South Africa see an increasing need to respond to negative reporting on the Energiewende from pro-nuclear (and anti-renewable) advocates. The German embassy in Pretoria is looking into creating more opportunities for journalists to come to Germany and gain first-hand insights into the Energiewende and held a training session for journalists on climate issues in Johannesburg prior to the COP21 (German Missions in South Africa, Lesotho, and

Swaziland 2015). An Energiewende-themed study tour for members of the Portfolio Committee on Energy in the South African parliament was organized in November 2014 (German Missions in South Africa, Lesotho, and Swaziland 2014, S62). Whereas GIZ focuses its work on cooperation with the renewables energy branch of DoE and is not involved in the nuclear energy debate directly, Heinrich Böll Foundation and Rosa Luxemburg Foundation strongly support anti-nuclear campaigns in South Africa (S36, S21, S60, S63). The risk of a nuclear renaissance in Southern Africa and its strong anti-nuclear stance motivated Heinrich Böll Foundation to publish a study on the issue in 2015 (Martin & Fig 2015), outlining costs and motivations behind nuclear on the African continent. The Foundation has been supporting South African NGO Earthlife in its fight against nuclear since 2008, notably by financing its "Nuclear Energy Costs the Earth" campaign (S63).

7.6 Case Discussion: South Africa

7.6.1 Honest broker or export agent? Germany's leadership agenda in South Africa

In addition to the success of its model, a sender's (perceived) motivation is likely to influence the legitimacy of its activities in the recipient country and hence transfer. As already discussed for the case of Morocco, understanding the agenda behind leadership is key in order to evaluate its effectiveness. German, South African, and third party interview partners were therefore asked to assess Germany's agenda for its involvement in the sustainable energy field in South Africa. As in the case of Morocco, interviewees drew a distinction between the general agenda pushing Germany to promote sustainable energy and the activities of GIZ agents on the ground in South Africa.

One interview partner expressed the difference between agents on the ground, within the policy network, and his ideas about Germany's general strategic interests in the following terms: "The people I am dealing with – I would say it's purely to provide support with no agenda. But: There is always an agenda. Trade. It has to be. […] I'm not saying they wouldn't do it anyway, and there is a development objective" (S3). The perception of Germany's not entirely uninterested agenda was rather linked to a general view on bilateral aid than concrete examples: "it's more from, yeah, from what people are saying. So it's hard to say exactly […] But in all these bilateral arrangements, the politicians will tell you, this bilateral thing is not about friendship, it's about trade" (S28). A South African official directly involved in German-South African bilateral governmental negotiations however said "the impression that I get from those discussions as well, every time I come out of one of them, is really that Germany is pushing South African government objectives. You know, they want to aid the South African government in achieving strategic objectives towards economic growth through the energy sector" (S6). In light of South Africa's high emissions, global climate protection was seen as an important motivation for general German leadership efforts, in addition to South Africa's central role as a major power on the African continent (S9, S11, S45, S46).

GIZ, the most important interface of German Energiewende leadership in terms of policy transfer, was generally perceived in South Africa as an "honest broker" of technical knowledge. This was linked to three main aspects. First, the perceived lack of GIZ's interest in promoting German technology; their demand-driven approach; and the legitimacy and technical knowledge stemming from Germany's own Energiewende experience, creating "natural synergies" and a perceived strong rationale for Germany to be present in South Africa. GIZ advisors were in fact much more worried about being perceived as opening doors for German companies than their South African counterparts. While they recognized Germany's economic interest in a promising market as one motivation for the country's involvement, they strongly viewed *GIZ*'s mission as being driven by demand from South African counterparts and the general goal of creating a sustainable energy sector. Striking a balance between the use of German industrial know-how for the benefit of GIZ's mission in South Africa and avoiding to be perceived as export promoters was a major concern for GIZ advisors: "We brought several hundred participants to Germany in different formats, in that regard we were an intermediary to German technology. But we try to be neutral and not be perceived as a door opener for German industry because being perceived as a neutral intermediary and broker is a very essential element of our credibility. That influences our credibility in general, but also technical standards, where it is not about pushing German standards or formulating them in a way so that German manufacturers have the best chances, but achieving the best solution. And that wouldn't be possible if one was perceived as a representative of German industry" (S17). As part of their mission, GIZ advisors also saw a goal of reducing concerns regarding renewables, in particular within DoE and Eskom, through capacity building, training and information (S45).

Areas where GIZ was perceived as having pushed the political agenda were cogeneration in particular and, to a lesser extent, biogas. These pushes were only partly associated with an interest from German industry and rather linked to advisors' personal preferences and their Energiewende-related background in Germany. The broader view among South African partners was that "they never pushed for a certain topic or technology" (S55) and that while "they of course expressed a preference since they only support renewables [...] they have been very careful to avoid any technological preference" (S29). GIZ advisors were seen by a utility executive as "very professional people [...] I haven't seen any kind of hidden agenda from them. They are honest and upfront and you can tell them what you think, but they will do what professionally they think is the right thing to do. I expect nothing else from people" (S20).

Germany's active leadership in South Africa was seen as "linked to experience and sharing that experience" related to the Energiewende since "Germany is known for world leadership in renewable energy. It is a shining example that it can be done. It is really a core area of German expertise" (S2). Focusing on renewables and specific issues within the field was seen as justified by Germany's experience. An external consultant therefore found "It's okay to say 'we will draw upon our expertise', that makes sense" (S3). These synergies with Germany's Energiewende experience were often – but not always – linked to global climate protection efforts. German embassy staff saw support to South African climate commitments and the country's role as a potential model and frontrunner on the African continent as decisive for German engagement (S61, S62). The COP17, which Germany had helped prepare, created opportunities for strengthening cooperation, with the establishment of full-time positions on energy and climate in the embassy and additional funds for political foundations from BMZ to do climate work (S62, S60). Germany's €2.2 million support to the South African International Renewable Energy conference SAIREC in 2015, as part of the series of international renewable energy conferences that started with *renewables 2004* in Bonn, is the latest example of cooperation between Germany and South Africa in the international renewable energy governance arena (South Africa-Germany Binational Commission 2014: 8).

Among the few critical voices, a civil society representative saw Germany's involvement and sharing its example in South Africa as being too much: "there's been so many Germans that have come down to South Africa to talk about it, that I've noticed now at policy meetings, people just roll their eyes to hear about the German transition again. They don't want to hear about it again. You know, it's just a, 'Yes, yes, fine. You're doing whatever in Germany. We're tired of hearing about this.' It's possibly been a bit too much, you know? And sometimes it can come across as a bit of a lecture, and you know, 'We've done it so well in Germany. You should do what we do.' Which by and large is true, we should be doing that kind of big transition Germany is. But that still doesn't mean that it doesn't come across badly at times" (S21).

In summary, Germany was seen as a leader whose intentions are "of a better kind" (S19), who is "serious about renewables and climate protection" and acts "very diplomatically" (S29). While being perceived as legitimate (a facilitating factor for transfer, see section 2.4.4), its agenda remained unconcrete for a number of interview partners. Even for a South African working frequently with GIZ, "there hasn't been any German agenda that's been driving" (S6). The lack of clarity is likely linked to GIZ's low profile – as an interview partner had also suggested in Morocco (M7) – and its focus on intense contact with South African counterparts in the style of consultancies. Data on ongoing work and outputs is not always easy to identify and access and several interview partners, including from other German organizations in South Africa, deplored a lack of transparency regarding SAGEN: "one of the biggest critiques that would be there, from my perspective, is that it [SAGEN] is really a closed project. It's between the two governments, so I don't think the civil society has had an impact. And I remember I talked to someone about this, you know, 'How is this project? Do you know anything about this project?' And not many civil society people who are active in energy know much about the content of that program. […] So in a way, it's government to government. We don't know whose interests are met. Is it Germans'? Is it the South Africans'?" (S63). As discussed in Chapter 9 as a policy lesson, clarity about the objectives leaders pursue is likely to be an essential component of credibility and legitimacy in the long run. While Germany's "good intentions" in South Africa were widely accepted, the perceived lack of transparency among civil society actors indicates an area for concern. As in Morocco, some stakeholders were so puzzled by Germany's apparently altruistic approach that they expressed a preference for a clear-cut strategy of promoting German industry. A South African industry representative suggested the numerous studies that were carried out should be replaced with implementation efforts by simply bringing German companies to South Africa instead: “There is nothing like a free lunch in this world. You don’t do aid and development just to say, ‘Okay. South Africans, we like you, here’s 500 million in Euros paid.’ No. […] Don’t write me another study or report. Say, ‘No, okay. This is how we’re gonna do it. […] you lead the way and you get your industries to be there. And that’s straight forward. So if you can get me to the top of GIZ, I can give them that strategy for free” (S53).

7.6.2 Hypothesis I: Unilateral leadership and mediated consideration

Theory shows that transfer can result from unilateral leadership, that is to say, the mere existence of an example stakeholders and decision-makers in the follower country may draw from (Bennett 1991: 35). As stated in Hypothesis I, Germany’s unilateral Energiewende leadership is likely to be considered by stakeholders in South Africa if the example is known, available/observable and considered successful in the German context. The prominence of Germany as a sender, another likely driver for consideration, is clearly established for the group of stakeholders interviewed in South Africa: “Germany made a very big decision, and I believe lesson-drawing is possible from that decision” (S40). The Energiewende as a direct source of lessons was considered by South African stakeholders in the case of the REFIT, and is making a comeback in recent discussions on embedded small-scale renewable electricity generation, but the effect of active leadership or “mediated consideration” of the Energiewende experience through the presence of and interaction with German transfer agents clearly prevails in importance. Few South African interview partners reported they regularly informed themselves about the Energiewende beyond a general interest in the issue and the context of “the whole world looking at it” (S19). Views on the Energiewende *in* Germany were generally positive, especially among stakeholders who had participated in one of the numerous study tours to Germany the country had sponsored. It is important to note that a critical view on the Energiewende, mainly related to cost and sometimes to the challenges of managing the grid, did not preclude recognition of leadership. In comparison to the California case study in particular, few interview partners (S28, S44, S62, S63) reported a negative use of the German example by others in policy debates. This is also due to an overall less important role for direct lesson drawing from the Energiewende example, except for the REFIT and embedded generation. Negative views of the German case focused on the consequences of nuclear phase out (among a very small group of vocal pro-nuclear advocates) and the fate of utilities in Germany. An interviewee reported the different views stakeholders had of the same policy model: “One is: ‘Germany has done so well, we can also do it since we even have sunshine’. The other one is: ‘Look, RWE and others now have to ask for capacity payments otherwise they need to switch off their plants and security of supply will suffer.’ And people in South Africa jump on that” (S28). Overall, interview partners described a general reluctance of South African decision-makers to consider any foreign policy models and experience due to the country’s colonial past, its apartheid-era history of isolation, and its very specific geographical and social situation. An NGO representative insisted on his organization – despite receiving funding from a German source – not being “an extension of some external power. We like it that way and we want to keep it that way. [...] things like environment, climate, emissions, energy emissions, you know, this must be a debate internal to South Africa. South Africa then makes up its own mind about what it must do. If it’s seen as being something that is coming from outside the country, and in particular from Europe, that will have extreme negative consequence” (S21).

Whereas the general view on the Energiewende was positive and unilateral leadership was recognized, important perceived differences between Germany and South Africa limited the likelihood of direct consideration of the model. The lack of interconnections with neighbors (as opposed to Germany’s integration in the European electricity market), the lower level of economic development in South Africa, and Germany’s higher ability to pay were mentioned as limiting the consideration of the German example (S14). An interview partner whose opinion about the manageability of renewables changed through contacts with a German expert stressed that initial opposition to renewables was in part due to South Africa’s isolated location: “The argument at the time and still is that we are not interconnected to France – you have to bail other people out in fact here. And we didn't think PV and wind could do that reliably” (S20). Another interviewee confirmed this view, adding that the sheer degree of ambition (80% renewables in Germany) was “the part that’s crazy” about the Energiewende, but Germany could “afford to be on the extreme side because they’ve got very good interconnectors" and "electrons from France" (S47). The absolute need for South Africa to create jobs through the renewables program was mentioned as a particularity: "That's very unique to South Africa and probably from a German perspective... you might think you need to get the best price, always be economically efficient. But no! It has to be something uniquely South African. The guys who work for GIZ are really aware of that and the local conditions" (S3). South Africa's need for additional capacity, as opposed to Germany's move of replacing existing capacity, was seen as a difference as well, albeit one that could also facilitate renewables uptake in South Africa.

The empowering potential of the Energiewende for pro-renewables advocates was recognized despite these differences (S6, S29, S39). An interview partner from academia saw "awareness that it is not perfect, the Energiewende", but felt that being able to "[show] that a very high share of renewables is possible in one of the most significant economies is very important" (S40). The comparative lack of solar resources in Germany reinforced the appeal of the example, as a former participant in study tours to Germany noted: "I looked at Germany and I thought, 'Wow. How did they do this and in such a short […] time. […] It's crazy and it all works. […] that sort of converted me, also the study tours, you know, the fact that it can work and in a place like Germany […]. The solar resource that Germany has compared to the solar resource that Africa has […]. So I mean, if Germany's making it work, imagine what we can do here" (S6). Despite a globally positive view on the German example, consideration without active leadership from Germany was not enough to trigger the use of knowledge and its reflection in policy decisions. Even for the example of the REFIT, which comes closest to a case where consideration of a policy alternative led to policy change, participation in a study tour and interaction with advisors played an important role. The importance of "mediated consideration" as a mechanism can therefore be seen for the South African context as well as for Morocco. A lack of capacities to perform searches for information was however less often cited as a reason for this, and differences in systems, objectives, and a general perceived lack of similarities can rather be seen as obstacles to deeper consideration, beyond awareness of the mere existence of the Energiewende model. Interestingly, the argument of a "cultural reference group" of leadership that played a role for transfer in Morocco and California could not be observed in South Africa, which is in line with the low rank leadership as an objective received in the ranking exercises; only decentralization and avoiding nuclear power ranked lower.

7.6.3 Hypothesis II: Utilization of transferred knowledge and active leadership

As in the case of Morocco, the question of utilization of Energiewende-related transferred knowledge in South Africa first requires addressing Germany's active leadership and its link to the Energiewende. This is important since transfer agents are members of the renewable energy policy network in South Africa (as well as in Morocco) and are hence also to be considered as stakeholders who utilize transferred knowledge in the recipient context.

The volume and approach of German activities in the renewable energy field in South Africa were widely seen as outstanding, because "German experience and expertise is highly valued in South Africa. You need to see that GIZ has people placed *in* the departments. Other donors don't have these structures" (S2). Representatives of other donor-related agencies confirmed the particular status Germany had in the South African energy community, especially since the launch of SAGEN in 2011 (S18, S37, S38, S41): "GIZ is really *huge* in South Africa, it's very impressive. Actually we are jealous about the number of people and the resources they have in South Africa. And they seem to have good access to the policy making level" (S38). Germany is indeed the only donor with advisors working from within the IPP office, the nodal point of implementation of the REIPPPP. One advisor also worked part-time from within the DoE, where only Denmark was also sponsoring a (Ghanaian) expert as an in-house advisor on renewable energy at the time of field research (S37). As shown in Figure 23, advisors also work from within other organizations at different levels of government. South African stakeholders underlined that the German approach stood out, not only with regard to the size of the programs and financial means invested, but also regarding Germany's approach of preferring to place on-the-ground, long-term consultants in the partner country over short term missions: "I think GIZ, over time, has really managed to build up very close links, also with government. They have people in the IPP office […] they do have a strong sphere of influence, it's a bit as if a big guy was standing next to you to look at what you do, help with monitoring" (S29). The longer-term approach (SAGEN has a planned duration of 3x3 years) and frequent cooperation with South African consultants was contrasted for example with the Swiss and Japanese approaches (S44). GIZ advisors also saw the long-term perspective of their approach and the ability to draw from "very specific technical know-how", based on the Energiewende in particular, as main points distinguishing their approach (S17, S36, S45). As far as the use of transferred knowledge is concerned, three variants can be identified in South Africa.

The first type of utilization was the use of the German example by pro-renewables advocates, especially business representatives to underpin their position. South African pro-renewables advocates said they used the German Energiewende as an example of how the share of renewables could be increased substantially (S28, S29), but nevertheless rather carefully since "first thing, you must never say 'feed-in tariff'. All right? 'Feed-in tariff', this is a German thing. This is a Swedish thing. This is a British thing. This is a whatever, a Danish thing. This is for the colonials again" (S28). Others said the German example was widely used by pro-renewables players, but statements regarding that use remained mostly general, such as "everyone would refer to Germany. Germany did this, Germany did that, Germany did whatever. Germany has so much renewables" (S26). The above-mentioned CSP study sponsored by GIZ upon the request of a South African industry stakeholder is an example of the more concrete and strategic use of information by a pro-transfer stakeholder (S53, section 7.5.1). Another veteran of the South African renewable energy industry also described the direct use of the German example: "I do use Germany as an example when I do presentations in South Africa. [...] It shows that variability is not really a problem. But unfortunately, people are predisposed; they do not change their mind with new evidence. But it nevertheless helps to build an argument" (S29). Publicly available presentation material confirms the use of the German example by industry representatives, like the CEO of the South African Wind Energy Association (SAWEA), who referred to "Germany and what we can aspire to" in a presentation to the Fossil Fuel Foundation (Van den Berg 2013).

The second type is an instrumental use of transferred knowledge for the design of policies, especially the REFIT. Knowledge about the Energiewende and information gathered through a study tour and interaction with German advisors was used by the regulator in the feed-in tariff design process. Although the use of transferred knowledge was not reported for the REIPPPP, it was important for modeling in the IRP and within Eskom's renewables department in particular.

Knowledge about quality control and project management based on experience in the German energy sector was fed into the work of the IPP office by German transfer agents and used within the office (S9, S11, S52).

The third type of use of transferred knowledge was to increase the pool of available information and thereby mitigate concerns over renewables. Table 17 shows a selection of studies produced to bring new knowledge to the South African renewable energy policy network. Based on their perceived high level of legitimacy, linked to the Energiewende, German transfer agents extensively used these studies, as well as study tours, workshops, training sessions, and bilateral working meetings to increase knowledge and mitigate concerns regarding the feasibility of renewables in South Africa, especially among ministerial and utility representatives. For example, a "flexibility study was done based on discussions with a high-level Eskom manager, who is in charge of capacity planning, and who said [...] it was not made sufficiently clear that different plants were needed in the future, not only base load like nuclear or inflexible coal plants. He had made this argument for a long time, but was not listened to and therefore just wanted some international input in order to bring a different type of credibility to the discussion" (S17). South African partners underlined GIZ's ability to create information regarding renewables and efficiency that could then be used to facilitate programs: "One of the main contributions I guess is they created more information, about the potential and for the sector. They were definitely more influential than the Danes. It's true that the Danes gave a lot of money into climate change issues, but since 6 or 7 years it has really been the Germans that were more influential, certainly" (S29). This was also seen as applicable for the case of municipalities and embedded generation, where a German consultant's sharing of experience made issues related to embedded generation "more approachable and reassuring" (S48). The work of a German expert within Eskom and his bringing in knowledge about the Energiewende and the management of renewables also had a lasting impact on the perception of renewable technologies for managers interviewed. The expert's role within the unit was seen as "an impetus for change, to challenge the thinking and rock the boat, really rock the boat, which he did. [...] his influence there was enormous and what he achieved, given the organization that we are, he achieved really a lot" (S20). Another Eskom manager independently confirmed this, underlining that the expert brought in decisive knowledge about German utilities and negative lessons that could be drawn from their resistance to embracing renewables: "based on the learnings of what happened out of the borders, [he was] helping us to shift the thinking [away from a] 'coal mentality'" (S47).

7.6.4 Hypothesis III: Selective transfer and implementation

Much of Germany's leadership efforts in South Africa take place in the framework of continuous, long-term working relationships with the DoE, Eskom and municipalities, generally outside the public eye. Foreign influence is a particularly delicate issue in South Africa and GIZ's discrete approach might thus be a strategic advantage. From a methodological point of view, however, the South African case strongly shows the need of conducting interviews with stakeholders involved in cooperation and potential transfer agents to gain an understanding of transfer processes. In this case study, Germany's aim for leadership can be summarized as facilitating the uptake of renewable energy in South Africa (be it for climate or other reasons). Several observations indicate substantial impact of German leadership efforts in this direction. The support to the Western Cape province as an early pioneer in renewable energy and a resulting study tour to Germany were essential in the adoption of a REFIT in South Africa, which reflects positive lessons from the Energiewende and adaptation with regard to specific design features. The program was decisive in prompting the development of the REIPPPP and in preparing the ground for project developers and the now successful (although overall small) uptake of renewables in South Africa (see section 7.3.2). The abandoning of the REFIT was not due to negative lesson drawing, but internal consideration of constitutionality and/or competition between state-level institutions. Much more important than direct transfers of the German Energiewende model, including instruments, were measures to facilitate the transition toward renewables while taking into account specific South African constraints. These activities still drew from the Energiewende as a pool of knowledge and legitimacy for GIZ advisors. German transfer agents within and outside GIZ were seen as having decisively contributed to a high share of PV and CSP in the IRP (S20, S47, S49) and were also seen as leading efforts in pushing for the establishment of a cogeneration IPP program (S44). For biogas, efforts and demand for support and know-how have led to the creation of a national biogas platform, but concrete regulatory developments were still missing (Giljova 2015). The biogas platform is an example of the structural changes to the renewable energy policy network supported by German transfer agents. In addition to the reinforcement of capacities through training sessions and the placement of in-house advisors, the creation of a cogeneration facilitator unit within SANEDI (SANEDI 2014), and support to the industry association SASTELA are other examples for this. Study tours and meetings organized by GIZ were appreciated as rare occasions for stakeholders from different backgrounds to come together: "They really did develop the core capacity here, a network of people. Otherwise it could not have happened. We could not have done any work without their support, wouldn't have gotten this thing going" (S39).

In summary, cognitive leadership and capacity building clearly outweigh the effect of structural incentives or conditionality in the South African context. An exception is the request from the World Bank and other donors, in the framework of loans for new coal-fired power plants in Medupi and Kusile, for Eskom to also build a wind and a CSP plant (S16, S47). In addition to GIZ's support to actors "within" the system, German political foundations have supported challengers to South Africa's current energy policy, especially with regard to nuclear. Cooperation with South Africa required Germany to be a learning leader. The switch to an auction-based renewable energy promotion system required adjustments to the advice offered and these are linked to a global rethinking of GIZ's energy policy toolbox. The invitation of a South African energy policy expert to share his experience on auction-design with BMZ in 2014 (Eberhard 2014) is a sign of reverse transfer from South Africa back to Germany. Even more remarkably, the South African experience with auctioning renewables figures as one of two foreign examples (the other one being France) in the German government's official report on its solar pilot auctions (Bundesregierung 2016b).

8 Case Study: California

8.1 Introduction to the Case Study

Germany and California, the world's fourth and eight largest economies (Sisney & Garosi 2015), stand out globally as leaders in the deployment of renewable energy technologies for electricity generation. With more than one quarter of their electricity now being supplied by wind, solar, biomass, geothermal, or hydro-energy, the question of whether and how policy transfer took place in the field of sustainable energy policy is a relevant one (Galiteva & Moss 2014). A comparison of California's and Germany's energy and climate policy goals and indicators reveals a high degree of similarity in terms of ambitions, as shown in Table 18.

Table 18: Key energy and climate goals in California and Germany. Adapted from Galiteva & Moss (2014: 34–35)

	California	Germany
Renewable electricity share	25% (2014, excl. hydro)	27.4% (2014, incl. hydro)
CO_2 Reduction targets	80% by 2050	80-95% by 2050
Renewable electricity targets	30% by 2020 50% by 2030	35% by 2020 50% by 2030 65% by 2040 min. 80% by 2050
Energy Storage	1,325 MW of storage to be procured by utilities by 2020	No quantified target. Incentives program
Zero-Emissions Vehicles Target	1.5 million vehicles by 2025	1 million vehicles by 2020

Despite these similarities in ambition and general direction, Germany's and California's energy challenges nevertheless differ. While close to 40% of California's GHG emissions are from the transport sector, this sector is responsible for only 16% in Germany, where, in turn, industrial emissions are much higher (Galiteva & Moss 2014: 37). Interview partners underlined that about one-fifth of California's electricity demand was caused by irrigation and water transport needs, a share that rose sharply as a result of the draught crisis (C43). While the

K. Steinbacher, *Exporting the Energiewende*, Energiepolitik und Klimaschutz.
Energy Policy and Climate Protection, https://doi.org/10.1007/978-3-658-22496-7_8

share of solar and wind capacity in Germany is roughly equal, California's share of solar capacity is about one-third larger than wind installations in the state and is set to increase further (CEC 2015: 3).

The focus in this chapter is on the mechanisms through which policies, ideas, and positive and negative lessons in the field of renewable energy promotion travelled from Germany to California, even though it used to be the other way around. In the late 1970s, German policy-makers closely observed the policy framework for electricity production from renewable energy that was emerging in California. Debates surrounding Germany's first electricity feed-in law adopted in 1990 made reference to California's experience with "qualifying facilities" (e.g., Bundesregierung 1988; see also online material[78] for relevant passages from parliamentary proceedings). While California's experience was a source of lessons in the past, attention to Germany's approach only started to grow among a broader group of stakeholders in California in the mid 2000s. Germany's expressed interest in gaining followers to its energy transition justifies the focus on transfer in the direction from Germany to California in this chapter and was also confirmed by interview results, but further research into bidirectional transfer appears promising. With California being a "most likely case" for transfer from Germany due to similar economic conditions, high environmental capacities and shared overarching (climate) goals, California is a test case for the effectiveness of Germany's aim to exercise leadership through transfer.

8.1.1 *Overview of field research in California*

Findings presented in this chapter are based on interviews with 59 interview partners and one background interview. The interviews were carried out in California during two months of field research in March and April 2015. Interview partners were selected for their direct involvement in key renewable energy policy debates and decision-making in California, from the Renewable Portfolio Standard (RPS[79]) to the Million Solar Roofs Program/California Solar Initiative in the mid-2000s, to the 2007-2009 debates on the introduction of a feed-in tariff

78 To access the book's appendix, please follow the URL given in the imprint or visit www.springer.com and search for the author's name.

79 A renewable portfolio standard (RPS) requires utilities to source a certain percentage of the electricity they sell from renewables. Producers of renewable electricity compete for contracts with utilities.

to the ongoing debate on future adjustments to net energy metering[80]. Interviews were carried out in Sacramento, Los Angeles, Palo Alto, San Francisco, and the Bay Area (Oakland, Berkeley). Figure 25 provides an overview of interviewees per sector, while Table 19 shows organizations covered through interviews and Table 20 provides information on the sectorial affiliation of each interviewee.

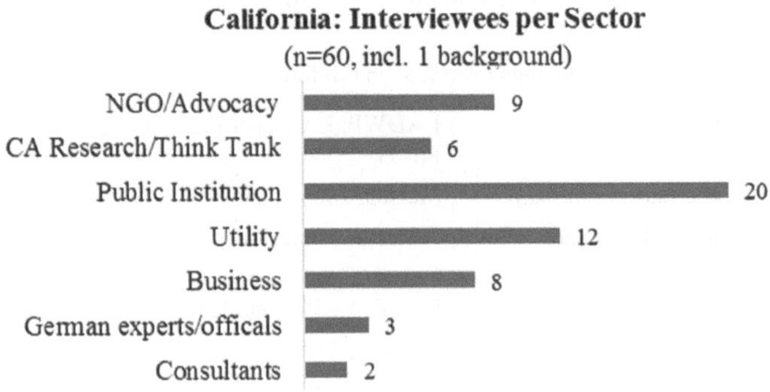

Figure 25: Number of interviewees per sector (California)

A special focus was put on four categories of organizations of particular importance to renewable energy policy-making. While energy policy initiatives emanate from the Governor and the bicameral state legislature, actual policy-design takes place to a large extent within the California Public Utilities Commission (CPUC) and its "proceedings" (negotiations/consultations on a given piece of regulation). Important energy policy impulses also come from the California Energy Commission (CEC), which is involved in the implementation of programs such as the California Solar Initiative, the RPS, and the permitting of larger plants.

The CPUC is in charge of regulating investor-owned utilities operating on California territory, including the three most vocal ones in the debates studied,

80 Net energy metering is a framework that allows individuals or companies to cover part of their electricity demand by on-site renewable energy generation. Electricity produced is netted against electricity taken from the grid, for example at nighttime, allowing for a significant or even total reduction of the electricity bill. Different from feed-in tariffs, net-metering schemes do not generally incentivize electricity production beyond their own consumption.

Pacific Gas and Electric (PG&E), Southern California Edison (SCE), and San Diego Gas and Electric (SDGE). Given the strong voice of these companies in renewable energy policy discussions, stakeholders from these utilities, as well as from municipal utilities (Sacramento, Los Angeles and Palo Alto), constitute the second important group covered. The third group is solar companies of different sizes and renewable energy industry associations. In California's open political culture, the role of environmental, consumer groups, and renewables advocacy

Table 19: Organizations covered through interviews in California.

Renewables 100 Institute	LADWP: LA Department of Water and Power
AMS: Advanced Microgrid Solutions	Meister Consultants
BECI: Berkeley Energy & Climate Institute	Office of Public Accountability, Los Angeles
Borderstep Institute	PG&E: Pacific Gas & Electric
CEC: California Energy Commission	Palo Alto Utilities
CAISO: California Independent System Operator	Run On Sun
California Institute For Energy and Environment	SMUD: Sacramento Municipal Utility District
CPUC: California Public Utilities Commission	SDGE: San Diego Gas & Electric
California State Assembly Committee on Utilities and Commerce	San Francisco U.S. Export Assistance Center
CALSEIA: California Solar Energy Industry Association	Schwarzenegger Institute, Univ. of Southern Calif.
CALWEA: California Wind Energy Association	Sierra Club
CinnamonSolar	SEIA: Solar Energy Industry Association
Clean Coalition	SCE: Southern California Edison
ClimateResolve	SunlightandPower
German Consulate General San Francisco	TURN: The Utility Reform Network
AHK, San Francisco office	University of California Berkeley
GTAI: Germany Trade and Invest	Vote Solar
Governor's Office	Windworks
LABC: Los Angeles Business Council	

groups merits particular attention. Representatives of these groups constitute the fourth focus of the interviews. Interviews with researchers, some of them directly involved in policy advice, were highly valuable, also in putting the California discussion in a global picture. Compared to the two other cases in this chapter, the number of German "transfer agents" interviewed is substantially lower due to far less institutionalized transfer channels. Interviews in Germany that also touched upon the California case more specifically (D5, D6, D9, D16) complete the picture presented in this chapter. Approximately half of the interviewees were identified prior to field research through lists of attendees at relevant hearing and meetings, with the rest of interviewees identified through snowball-method recommendations.

Table 20: Interviewee affiliation by sector[81] (California), Inst: Institution

Identifier	Sector	Id.	Sector	Id.	Sector
C1	Utilities	C23	NGO/Advocacy	C44	Utilities
C2	Public Inst.	C24	Public Inst.	C45	NGO/Advocacy
C3	German officials	C25	Research	C46	Utilities
C4	Utilities	C26	German officials	C47	Utilities
C5	Utilities	C27	Business	C48	Public Inst.
C6	Business	C28	NGO/Advocacy	C49	NGO/Advocacy
C7	German officials	C29	Public Inst.	C51	Business
C8	Research	C30	NGO/Advocacy	C52	Public Inst.
C10	NGO/Advocacy	C31	Public Inst.	C53	Research
C12	NGO/Advocacy	C32	Business	C54	Research
C13	Public Inst.	C33	Public Inst.	C55	Utilities
C14	Business	C34	Public Inst.	C56	Public Inst.
C15	Public Inst.	C36	Public Inst.	C57	Utilities
C16	Public Inst.	C37	Public Inst.	C58	Public Inst.
C17	Public Inst.	C38	Utilities	C59	German officials
C18	Business	C39	Utilities	C60	Business
C19	Utilities	C40	Business	C61	Consultants
C20	Public Inst.	C41	NGO/Advocacy	C62	Public Inst.
C21	Consultants	C42	Public Inst.	C63	Public Inst.
C22	NGO/Advocacy	C43	Research	C64	Utilities

8.1.2 Germany-California transfer channels

The focus of Germany's active policy promotion efforts is on building capacities in developing countries, and targeted at avoiding lock-ins into fossil-based elec-

81 Missing numbers in interviewee identifiers were allocated to background conversations, second interviews with the same interviewee, and contacts that were not eventually followed up with an interview.

tricity systems. Compared to permanent structures like bilateral energy partnerships or longstanding development cooperation programs in place in Morocco and South Africa, the main channels for cooperation and exchange in renewable energy between Germany and California appear less institutionalized and rely more strongly on individual initiative.

The Transatlantic Climate Bridge, which was established by Germany's BMUB, the German Federal Foreign Office and the US, offers a framework and limited funding for study visits (German Missions in the United States 2015). For example, short-term visits to California from a German researcher and an official from KfW on the Energiewende in 2012 and workshops in California, e.g., on demand-response mechanisms (Bayer et al. 2014), were supported (C7). In 2011, the German Consulate in San Francisco created the position of a "science liaison officer", who – because of the topic's relevance and her personal interest – has put much emphasis on clean energy in her work connecting German and California experts and facilitating events sponsored by the Transatlantic Climate bridge. For the German-American Chamber of Commerce (AHK) and Germany Trade and Invest (GTAI), Germany's investment agency, who share offices in San Francisco, energy is a priority area and renewable energy market studies as well as business delegation visits have offered opportunities for learning and exchange in the past (C26, C58).

As evidenced by the interviews conducted in California, individuals with a special interest in Germany, with personal connections to German stakeholders or Germans living in California have been crucial to enabling information flows and transfer. For example, regular visits from Hermann Scheer, spurred the creation of the Renewables 100 Institute. Renewables 100 – whose Californian co-founders are in regular contact with German stakeholders – has repeatedly created and used opportunities such as conferences, study visits, and press releases to make information about Germany's approach to renewable energy deployment available in California. This has been done with an aim to push higher California's level of ambitions. In 2014, the institute organized a "learning and collaboration tour" for high-level officials from California to Germany to enable exchange between the two frontrunners (Galiteva & Moss 2014).

Other renewable energy policy advocates from Germany, such as Hans-Josef Fell, representatives of Heinrich Böll Foundation, Eicke Weber (a former UC Berkeley professor and director of the Fraunhofer ISE institute), officials from BMU(B) and BMWi and Harry Lehmann, (a Head of Division with the German Federal Environment Agency *Umweltbundesamt*, UBA) were quoted as "ambassadors" for German renewable energy policy, by those interviewed. Delegation visits by BMU/BMWi[82] officials to CPUC and CEC have taken place approximately once a year according to interview partners. Staff from the CPUC and consumer advocacy group The Utility Reform Network (TURN) has visited Germany as fellows through scholarships for one year or more to work on specific energy-related issues and have brought knowledge back to their organizations (C10, C15, C29, C56).

With regard to German-style feed-in tariffs in particular, California advocacy groups and individuals were instrumental in bringing the idea to California. Paul Gipe, a wind energy specialist and contributor to the Ontario feed-in tariff program who had become familiar with European renewable energy policies in the 1980s, frequently communicated on the German experience in newsletters, speeches and articles (Gipe 2015). The "CLEAN coalition" (formerly the "FiT Coalition"), is another example of an advocacy group bringing knowledge about the German example to debates in California (Wang 2013). The importance of direct communication and bilateral visits between companies and grid operators created information flows on the more technical aspects of enabling the integration of large shares of renewable energy in the system. Although at first sight, these information flows might rather resemble technology transfer than policy transfer, knowledge about "ways of doing things" in Germany created a demand for similar policies in California. German utilities entertained contacts with California counterparts, and also appeared publicly, such as in a 2014 CPUC "Thought Leader Series" event (Weale 2014).

Although California is only a state, national-level German actors have carried out the majority of cooperation and exchange activities regarding the Energiewende. Interview partners on both sides clearly saw California at eye-level with Germany, a perception that is justified by the fact that the state's competences in setting energy policy goals and instruments almost equals those of a nation state (Elliott 2013, Litz 2008:14), especially compared to a member stated

82 The unit in charge of exchange with California shifted from BMU to BMWi in 2013, when all renewables-related topics were brought to BMWi. Staff and leadership remained the same despite the change in departmental affiliation.

in the European Union. Nevertheless, California is also highly involved in organizing climate action at the subnational level, e.g., by linking its cap and trade system to the one of Quebec. Cooperation between California and German *Länder* in the fields of climate, environment, and sustainable development, has a long history, including through formalized partnerships, as analyzed in great detail by Ralston (2013). In 2015, a milestone of German-California cooperation at the subnational level was reached when the *Land* of Baden-Württemberg and California agreed to reinforce cooperation in climate protection and launched the "Under 2 MOU" movement in the run-up to the Paris COP21 climate conference (Under 2 MOU 2015, Governor's Press Office 2015).

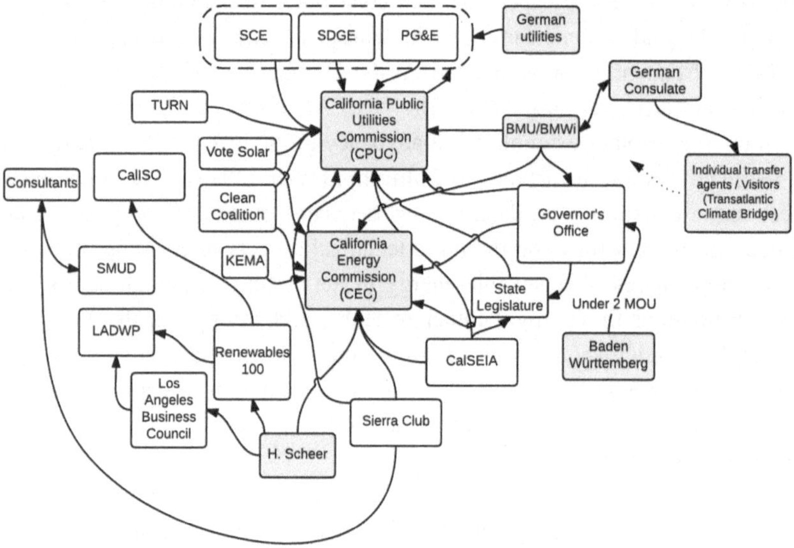

Figure 26: California renewable energy policy network and main transfer channels. Graph by the author

More important for the transmission of information than institutionalized transfer channels were direct contact with German stakeholders, individual visits to Germany, encounters at conferences, active research on the German example, and studies that brought information about the Energiewende to the Californian debate, especially studies commissioned by the CEC. Another example is the United States Solar Energy Industry Association (SEIA), which, in 2014, also published a study providing "a closer look" at German support for solar energy (Weiss 2014). The aim was to counter claims from utilities that the model "was a disaster" and California "had to make sure we don't do it that way" (C6), in particular regarding the high share of small-scale renewables installations in the country.

8.2 Tracing Renewable Energy Policy Transfer from Germany to California

8.2.1 Tracing Energiewende transfer to California

No strict selection of specific policy decisions and debates to be traced was made before conducting the interviews. Interviewees were instead asked to point out at what point since the introduction of the RPS (which marks the beginning of the "modern era" of renewable energy development in California), the German example was present in debates, how it was used, and whether it was reflected in policy outcomes. Figure 27 shows the main processes and events considered in this chapter.

The period of 2006-2010 emerged in the interviews as the time where the German example figured most prominently in the debate in California. The introduction of a feed-in tariff was discussed at the state level since the RPS had not yet delivered expected results. Concurrently, renewable energy deployment in Germany was accelerating year after year. Both proponents and opponents of feed-in tariffs as a policy instrument for California heavily relied on the example of Germany to underpin their respective arguments. This phase therefore constitutes one main observation point.

In addition to the period of discussions on a state-level feed-in tariff, several other policy discussions emerged as observations within the broader case study through the interviews: the Renewables Portfolio Standard, the Million Solar Roofs Program, feed-in tariffs at the local level (Los Angeles, Sacramento and Palo Alto), permitting procedures for renewable energy installations, smart inverters, and debates on the reform of California's net energy metering system. The German example was considered to very different degrees in each of these policy debates, from virtually no consideration in the RPS – also given the lack

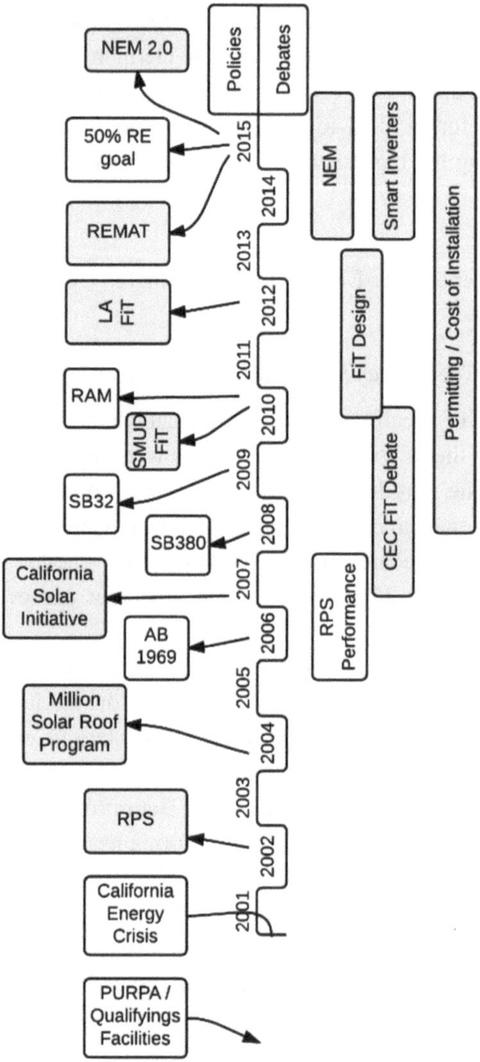

Figure 27: Timeline of California renewable energy policy initiatives and debates. Graph by the author

of conclusive evidence from Germany at the time – to being a dominant theme in the debate on state-level feed-in tariffs.

8.2.2 *Absence of the German example in the development of the RPS*

The history of renewable energy policy in California dates back to well before the introduction of Germany's first electricity feed-in law. California's interpretation of the 1978 federal Public Utility Regulatory Policies Act (PURPA) led to a "gold rush" for renewable energy installations that were considered "qualifying facilities" (QF, see also section 5.1.4). The experience with QFs in the 1980s, in particular the high cost of the program that resembled a feed-in tariff, was considered negative by interviewees (C2, C15, C20). The negative experience with QFs motivated the choice of a fundamentally different program in the aftermath of the 2000-2001 California energy crisis, when the decision was taken to promote renewable electricity as an alternative to volatile fossil fuels (Kristov & Keehn 2013). As confirmed by an attorney directly involved in the negotiations on the RPS, no alternatives to a portfolio standard were seriously considered (C22): "As [the RPS] moved forward, there were no major competing alternatives. No one was talking about a feed-in tariff as an alternative. It was really all about competitive solicitations, fixed price resources, under long-term contracts." The RPS entered into force at the state level in 2002 and obliges investor-owned utilities to procure a certain percentage of their electricity sales from renewable sources, without fixing a price or quotas for specific technologies (CEC 2016).

Germany's approach to promoting renewable energy through feed-in tariffs only entered the debate in California around the middle of the 2000s. However, interviewees close to California Governors Arnold Schwarzenegger (Republican, 2003-2011) and Edmund G. (Jerry) Brown Jr. (Democrat, 1975-1983, 2011-present), saw the German example as relevant as the level of ambition in their state's RPS increased, from 20% to 33% and 50%[83] (C13, C31, C45). Germany's experience with managing higher shares of intermittent renewables was seen as salient in this specific context (C14). In addition, the presence of Germany as another global leader was seen as reassuring, despite differences in policy choic-

83 The share of renewables to be bought by California investor-owned utilities has been ratcheted up over time: Senate Bill 1078 (SB 1078, 2002) set a goal of 20% by 2017, SB107 (2006) accelerated the goal to 2010; Governor Schwarzenegger's Executive Order S-14-08 (2008), codified by Senate Bill X1-2 (2011), pushed the goal to 33% by 2020 for all utilities in the state (CEC 2016). In 2015, the California legislature, following a proposal by Governor Brown adopted SB350 to scale up the RPS goal to 50% renewables by 2030.

es between Germany and California: “He [Gov. Brown] sees a country [Germany] that has been very determined, very persistent, overcame obstacles, moved the market, really changed things, produced a dramatic amount of [renewable] electricity. […] that’s very, very serious about clean energy, about climate change, about renewables and has done a great deal. So it has been a positive example” (C45).

8.2.3 *Analogy rather than transfer: The Million Solar Roofs Program*

Entering office in 2003, Governor Arnold Schwarzenegger’s unconventional background shaped his perception of California as a “global leader that happens to be located in the United States” (C59), which also meant comparing the success and drawbacks of California’s energy policies to other global leaders: “While California is leading in the U.S., we are building on the work of the European countries who have led the way up until now and have done extraordinary work. [...] Germany has pioneered solar” (Schwarzenegger 2007). In 2003, Governor Schwarzenegger had announced his vision of having 1 million “solar rooftops”[84] in California as part of his campaign for Governor of California. One of his closest advisors on environmental policy, Tarry Tamminen (who was to be appointed by Schwarzenegger as the head of California’s Environmental Protection Agency in 2003), is being quoted in Johnstone (2011: 2609–2610) as saying “Arnold is a competitive bodybuilder […] If you can do ten pounds, he can do a hundred. It was an embarrassment to him that a state with so much sun was lagging behind [Germany]”. Based on interviews with Schwarzenegger’s advisors, Johnstone (2011: 2612) goes on to explain that Germany’s solar policies acted as a motivation because of competition rather than as a concrete inspiration for policy design since Germany’s feed-in tariffs at the time had not shown their full potential in accelerating the growth of small solar installations. This assessment is in line with interviews carried out (C2, C31, C63): “From Arnold's standpoint, he'd hear people in the utility sector, the business community say ‘We can't do

84 The term “solar rooftop” signifies the installation of PV solar panels on the roofs of – generally residential – buildings. These installations can be owned by home owners, or as is often the case in California, by solar leasing companies (SunRun, SolarCity etc.) which own the panels and lease them to home-owners.

that.' and he would point to Germany. And say 'Come on, why could they do that and we can't - what makes it easier for them than for us?'" (C31).

The resemblance in the programs' logic between Governor Schwarzenegger's Million Solar Roofs vision and Germany's "100,000 solar rooftops" program is hard to ignore (BMU 2003, Eddy 2015). But even though the rhetoric hints at competition being a potential driver for California's program, two former direct advisors to Governor Schwarzenegger underlined the disconnection between rhetoric and "real" drivers for the program (C9, C59), namely climate protection and implementing Schwarzenegger's electoral campaign program.

Several interview partners emphasized that the Japanese example of promoting solar rooftops through rebates and direct financial incentives was much better known at the time in California (C12, C31, C51), and that Germany became relevant later on: "originally with the California solar program, the original model was based on Japan, because in the early 2000s, Japan really got going before Germany. But then, once Germany got going as big as it did, it was helpful in all the policy conversations" (C31). Inter-temporal transfer was mentioned as another source of lessons for the California Solar Initiative, which was an incentive program offering investment subsidies for solar rooftops to home owners with a view to reach the one million solar rooftop goal (CPUC 2016a). An official involved in the process mentioned a solar-water heater program established in the early 1980s as a source of lessons: "When we were trying to scale up the CSI [California Solar Initiative] and the talk of solar roofs started, we tried to get people to go back and look at that earlier experience to see what had gone wrong [...] and what planning considerations were taken into account" (C24).

If transfer from Germany took place with regard to the Million Solar Roofs Program and incentives proposed to homeowners through the California Solar Initiative, it did so on the level of general inspiration and a sense of belonging to a group of global leaders rather than at the level of policy design. One interview partner went as far as to affirm that the California Solar Initiative program used to implement the Million Solar Roofs vision "was built around the idea of avoiding high cost by leading the decline [in technology prices] rather than tracking it, contrary to the German feed-in tariff" (C48). Most interview partners also confirmed they had become aware of Germany's approach to promoting renewable energy only around the mid-2000s, when "Germany really took the torch from

Japan […], really stepped up. It was interesting to look at why it was that Germany was so bold and visionary on renewable energy" (C31).

8.2.4 *The German example enters the debate: Feed-in tariffs at the CEC*

The period of 2006-2010 clearly emerged in the interviews as the time when the German example was utilized most intensively in broad energy policy debates in California although interviewees felt the utilization of the German experience has been again intensifying since around 2013. From the mid 2000s, the idea of introducing a comprehensive German-style feed-in tariff program at the state level in California emerged, was championed by the CEC; triggered heated discussions; encountered resistance from utilities, ratepayer advocates, and the CPUC; and finally resulted in the adoption of the Renewable Energy Market Adjusting Tariff (ReMAT), a type of tariff with a "totally different philosophy" (C27). Germany was the primary example used by proponents and opponents of feed-in tariffs in California to underpin their arguments. This phase therefore is of central importance to assess how different stakeholders in California used evidence from the German case.

Channels through which interviewees first came in contact with the German model were as diverse as their backgrounds. An interviewee from the renewable energy industry recalled having "read about it in the media […] when it started back in 2003, 2004" and said he "got some information from my solar module suppliers who were starting to say that sales in Germany were taking up very quickly as a result of a feed-in tariff, and that's when I did more investigation" (C14). Another industry stakeholder even "went to Aachen [in 2005, 2006] to meet with the people who started the feed-in tariff program there to learn about how it was working. The industry at that time in California was really puzzled and fascinated by the crazy low prices for solar installations compared to here. So I went there to do some research and crunched some figures and did an internal report on the differences, on the various cost items" (C27). A consultant and environmental advocate learned about the German model from the book "'Who owns the sun?' from the 1990s, where the German example is explained. I found the financial model really convincing, I come from a family farm and this model appealed to me" (C21, see Berman 1996). The most important entry point for the German experience to the California debate, however, was through the initiative of the CEC.

A window of opportunity for discussions on a complement to or a reform of the existing RPS system opened when, by the middle of the decade, the RPS had not delivered expected results in terms of renewable capacity added to the grid: "every year until 2007 our renewable percentage declined […] we keep talking RPS and we don't see much come online" (C2). This lack of visible success prompted CEC officials to consider alternatives and reforms, whereas utilities and the CPUC did not see a need to act at this point in time. Personal encounters played an important role in shifting the CEC's focus to the German experience. A high-level decision-maker within the institution had been invited to speak at the Bonn *renewables 2004* conference. Then, through his involvement in different renewable energy fora, especially the American Council on Renewable Energy (ACORE), the interviewee met Hermann Scheer: "I can't remember exactly when, it may have been 2005, that I was speaking at an ACORE event in New York […] I was describing the California RPS. And Hermann, who I didn't know at the time, stood up in the audience and said 'Great, but why hasn't California had the nerve to establish a feed-in tariff?' And I said 'because our utilities wouldn't allow us to'. And I went on at some length to describe how once before we had tried something like that[85] […] he and I spent a great deal of time at the conference and later talking about our experiences" (C24).

Based on these encounters and strong personal contacts ("we both shared an anti-nuclear background. Actually we became quite good friends" C2) the interviewee immediately saw the relevance of Germany's experience for California (C2, C24). What rendered the German model particularly attractive, against the background of "frustration" with the RPS at this point, was the speed of renewable energy deployment; factors such as the share of decentralized generation were of secondary importance from a CEC perspective (C2, C20, C24).

On an institutional level, as underlined by one interview partner involved in California energy discussions for close to two decades, bringing thought-provoking and controversial topics like German-style feed-in tariffs into the debate also allowed the CEC to return to the center stage of policy discussions, leading the interviewee to conclude that "one of the reasons the Energy Commis-

---

85 The "something like that" experience the interviewee referred to being California's experience with Qualifying Facilities, from which negative lessons were drawn for future policy developments.

sion pushed it, is because they felt left out of the conversation" (C22). Once the interest of CEC leadership in the German example was established, its staff "started raising in the different workshops that we held: 'wouldn't it be a lot better if we simply embrace the approach of the Europeans?' And we said Europeans, but we know Germany was the instigator, but it broadened our political base by saying 'the Europeans'. Arnold Schwarzenegger was the Governor here at the time and most people confuse Austria and Germany" (C24). Despite the rhetorical trick of referring sometimes to the German and sometimes to the "European experience" (both were extensively used), interviewees were very clear about the discussion being about the German experience specifically and "Germany was presented as the model of how they were expecting this would work" (C30).

The CEC's 2007 Integrated Energy Policy Report – the institution's main, biannual publication – strongly recommended the CPUC work with the CEC to establish feed-in tariffs that "incorporate features of the most successful European feed-in tariffs" (CEC 2007: 6) and described the German feed-in tariff system – "one of the most successful"– in considerably greater detail than any other example (CEC 2007: 113). At that time, a very limited feed-in tariff program already existed in California. In 2006, Assembly Bill 1969 had introduced time-of-use feed-in tariffs for small renewable energy installations linked to wastewater treatment facilities with an overall program cap of 250MW. The cap was expanded to 478 MW and all types of installations by the CPUC in 2007 (Rickerson, Bennhold et al. 2008: 4). As described in more detail below, this decision cannot be seen as a sign of the CPUC embracing the idea of feed-in tariffs beyond a small niche.

The recommendations formulated in the 2007 IEPR of introducing a broader feed-in tariff program constituted the beginning of a phase of intense discussions within the renewable energy community in California, where both proponents and opponents of the instrument used the German example. The prominence of this example in workshops and debates following the publication of the 2007 IEPR not only made observers ask "Is California the next Germany?" (Rickerson, Baker et al. 2008), but also led an exasperated high-level energy policy official at the State Capitol to put a note on her door informing stakeholders that they were "not allowed to discuss Germany in this room" (C21), according to one interviewee.

Subsequent steps taken at the CEC to discuss the idea of feed-in tariff further were a series of workshops, and most importantly, the commissioning of studies (Grace et al. 2008, KEMA 2008) explicitly reviewing the German – and Spanish – feed-in tariff experience and lessons for California. Even for interviewees deeply involved in renewable energy policy-making, the "KEMA studies" (KEMA 2008, KEMA et al. 2010, KEMA 2011) were the first opportunity to consider the German experience more in-depth: "obviously [we] were aware of what was going on in Germany, but really for me at least, that KEMA report was really the first in-depth introduction to what was going on in Germany, and how that may or may not translate into California. So I think for a lot of us, at least, that KEMA report was kind of the starting point for that, that drawing the connections between Germany and California" (C16).

The studies led to the presentation of six different "policy paths", three of which "resemble the German approach" (Grace et al. 2008: 51). These policy paths were introduced at workshops (Grace 2008) as "Full German-Style Tariffs" (Policy Path1) and "German-Style for [installations] under 20 MW" (Policy Path 6). The main consultants responsible for the reports had previously worked on projects involving German stakeholders and had regular direct communication with German officials during the drafting process (KEMA 2008: 59; C2, C20, C61).

According to a member of the project team overseeing the studies within CEC, the institution's aim, through the studies, was to "point to Germany to show they were doing well and that it was all a question of design […] that was the main thing we wanted to say, we could design it [a feed-in tariff] so it would be successful" (C2). Spain's experience with feed-in tariffs soon became much less attractive to California stakeholders in favor of the instrument, due to design flaws and financing problems in the Spanish policy. The importance of Germany's example therefore grew even further over the years: "The feed-in tariff has been mismanaged pretty much everywhere around the world, although Germany had figured it out. How can you get it that wrong? Portugal, Spain, Czechoslovakia [sic], Italy… it's absolutely crazy how they managed to get the feed-in tariff so wrong" (C27).

8.2.5 Differentiated use of the German example: Politics of transfer

Throughout the process of discussing an alternative or improvement to the RPS and also net-energy metering, the German feed-in tariff example was used both positively and negatively by members of the policy network, pointing to the differentiated use of transferred knowledge assumed in Hypothesis II.

Environmental NGOs, smaller solar companies, CEC staff, and pro-renewables advocates were among the proponents of German-style feed-in tariffs. Utilities and ratepayer advocates from TURN, in an unlikely coalition, expressed a clearly negative opinion on the appropriateness of the German model for California, deeming it "right for Germany, wrong for California" (Hawiger 2008). A consumer advocate stressed in particular the resemblance between German-style feed-in tariffs and the unsuccessful experience with QFs of the 1980s as a reason for resistance to transfer, in addition to the model's complexity: "if someone tells you to try something like what you experienced in the 1980s again, we were very resistant. And it seemed completely unnecessary. It was clear to us if you set prices, you had to set many prices as in the German model. There were dozens and dozens of different prices in Germany, it seemed like an incredibly complicated exercise" (C22).

Opponents, both from the pro-renewables (but anti-FiT) community and from utilities, also pointed to the cost of the program; for a longtime advocate of solar in California, the level of the German feed-in tariff at the time "was so beyond the pale unthinkable in the United States that [...] it made solar look really bad, frankly. We needed to make a case for [...] positive returns on investment [...] it's not that I have anything against the feed-in tariff system, it's really a question of pragmatism" (C12). In a striking example of the "politics of transfer", some interview partners' opposition to a transfer of Germany's feed-in tariff experience was not due to any negative lessons drawn from the German example, but rather to a fear of seeing "Pandora's box" of renewable energy regulation opened in California. These interviewees worried about losing ground in overall renewable energy development by replacing the RPS and/or net energy metering by a feed-in tariff with potentially weaker goals or any other policy instrument that would mean leaving the nascent path dependency for renewable energy created by the RPS. Some doubted feed-in tariffs or any other form of administrative price setting would ever be politically feasible in California, also due to the perception that "Germany has a different political culture around that, people can absorb that" (C31).

Another feature of the German feed-in tariff that was deemed ill-fitted for the California context was the lack of a program cap, meaning that, in principle, an unlimited amount of renewables could be added every year: "you know, one of the most beautiful aspects of the German feed-in tariff was that it was not capped. Anybody could do it. In the U.S. context, that is absolutely impossible. There is no way to have a program with a completely unlimited upside. So then, once you start to have program caps, everything changes. You know, the pure ease of participation under the feed-in tariff, under the German feed-in tariff is lost" (C12). Hence, while being "very supportive of the German feed-in tariff in Germany", this group of interviewees expected "a lot of pushback against feed-in tariffs because of rate impacts" in California and therefore "wanted to support whatever policy is the most politically feasible, and net energy metering turned out to be it" (C31, also C12, C14).

Around 2007, members of the Sierra Club, one of California's main environmental NGOs, also became familiar with Germany's more and more visible success in bringing high volumes of renewable energy capacity online in a decentralized manner. A member of the Sierra Club's energy committee recalled he "heard about the feed- in tariffs in Germany and […] thought 'wow, this is great'" and stated that "probably my most important advocacy work was for feed-in tariffs based on the German model" (C49). Several members of the Sierra Club's energy and climate committee became advocates of the idea of German-style feed-in tariffs within their organization and the Club's official comments to policy discussions and proceedings soon reflected both a strong preference for German-style feed-in tariffs and a strongly positive view on Germany's experience overall (Sierra Club California & Metropulos 2009: 3). To the contrary of CEC and industry representatives, the "democratic" aspect of a policy instrument that allowed anyone to become an electricity producer appealed to members of the Sierra Club as much as its ability to speed up renewable energy deployment (C21). Given the positive view within the Sierra Club of the German model, the organization advocated for "Policy Path 6", the item from the six-proposal menu developed by the CEC's consultants that most resembled the German model (Grace et al. 2008: 51, Sierra Club 2008). The main policy design features supported were the inclusion of all types of renewable energy sources; the calculation of rates based on costs plus reasonable profits; long-term contracts (Sierra Club recommending 20 years like in the German model); technology-differentiated tariffs; and, above all the absence of a market cap, which would have yielded a program of unlimited size as in Germany (Sierra Club 2008). Among the reasons the Sierra Club cited for its support to an unlimited program was the "competitive battle with other states and countries for technological leadership in renewables [...] such as Germany, Spain and Denmark that are already far ahead of the US and California in these technologies and manufacturing capacity" (Sierra Club 2008: 2), an argument the organization hoped would appeal to a broad range of actors (C49).

8.2.6 Opposition to German-style feed-in tariffs at the CPUC

The CPUC was the key institution in terms of formulating the details of policy design in the California renewable energy policy network. As explained by stakeholders, including CPUC officials interviewed, the institution's dominant view on feed-in tariffs was – and is – that of a policy instrument appropriate for a certain niche, and which should be designed very differently from the German example due to "more differences than similarities between the two systems" (C48). It therefore comes as no surprise that German-style feed-in tariffs as an instrument were never fully embraced by the CPUC, which "steered away from a feed-in tariff in the light of the example of Spain in particular" (C12), but also "because of the lessons people have taken from Germany, correct or not correct" (C51). Another reason for the CPUC's critical view, mentioned by an observer was linked to capacities, namely an "inherent lack of confidence on the part of the public utilities commission staff that they could set the price right. [...] And that their process is so cumbersome that it would take forever to fix it" (C24). As mentioned above, a limited feed-in tariff for small installations of less than 1.5 MW had been in place since 2006.[86] The tariffs set by the CPUC in the program were based on utilities' market price referents and time-of-day adjustments and hence fundamentally different from the German approach. Its overall program cap of close to 500 MW needs to be put in comparison to overall system capacity of about 65,000 MW in California in 2008 (Rickerson, Bennhold et al. 2008: 4).

86 AB1969, extended in 2007 to installations outside wastewater treatment facilities through SB380.

Pressure coming out of the feed-in tariff discussion at the CEC and from a coalition of actors led by the California Solar Energy Industries Association (CalSEIA) led to the adoption of Senate Bill 32 (SB32) by the California legislature in 2009. The bill required the CPUC to design a feed-in tariff program with tariffs taking into account "current and anticipated environmental compliance cost" (California State Legislature 2009) in addition to a market price referent. The provisions in SB32 were seen as a slight improvement compared to the former feed-in tariff program by proponents of the German model, but the actual program designed by the CPUC to implement SB 32 reflects the institution's opposition to administratively set prices as in the German example. A CEC official confirmed his institution, in alliance with other groups, had "pushed, and pushed, and pushed, and got a lot of push back from the [C]PUC" (C24). As a longtime advocate of feed-in tariffs in California put it, "to do anything in politics in places like California you need institutions like the California Energy Commission saying, 'Oh. This is a very good idea. We like this.' That's extremely important. [But] that's not enough. […] They have some political power but they don't have the power of law. And so that's the responsibility of the California Public Utility Commission which [said] 'Nah. We don't care. You guys talk all you want over there in Sacramento. We do what we want over here'" (C28).

Negative lesson drawing from the German experience was one reason for the CPUC's opposition to the idea of feed-in tariffs. As a high-level official from the CPUC put it, "the fixed-price approach in Germany was a problem, costs went up. This was the single takeaway from my visits there. So right from the beginning I knew this was not for us, looking at the prices there" (C48). More fundamentally, the German approach was perceived as a poor fit in terms of regulatory tradition, and even culture. In particular, the element of setting prices administratively was used by different groups to emphasize the conflict between feed-in tariffs and the American or Californian "way of doing things". As an interview partner reported, "those people who are opposed to feed-in tariffs, whether in the coal industry, whether in the natural gas industry, whether it's in the electric utility industry, whether they're in the solar industry, or the wind industry, will use this ideological debate to say, 'We can't do it here because we're Americans.' And you know, 'The socialists over there in Europe can do that kind of stuff'. […] Sometimes you can almost hear the shock in the audience. 'What? The government is setting the price?'" (C28).

In the ReMAT program resulting from SB32 and introduced in 2013, tariffs are set through auctions, a fundamental difference compared to the German approach (DSIRE 2015). Also, the CPUC's interpretation of SB32's program cap of 750 MW was to include the capacity awarded or offered in the earlier feed-in tariffs program under AB1969. This leads to a small incremental addition of capacity under the ReMAT program. These policy design decisions proved to an industry stakeholder interviewed "that the feed-in tariff concept doesn't work with the CPUC, that they are not capable of making it work [and] not willing to make it work" (C27).

During the lengthy development phase of the ReMAT program, between 2009 and 2012, some stakeholders still pointed to the German example in order to achieve adjustments such as a higher project size cap of 20MW instead of 3 MW. The Sierra Club stressed in its comments to the CPUC's feed-in tariff proposal that "[it] is well known that the German feed-in tariffs have been very successful in achieving this goal under the Renewable Energy Law (EEG) established in 2000. In just 6 years renewable energy doubled from approximately 35 billion kWh to over 70 billion kWh" (Sierra Club California & Metropulos 2009: 3). A session of the CPUC's "21st Century Thought Leaders in Essential Industries Speaker Series" on 27 August 2009 also featured an "international panel of experts" to "discuss feed-in tariffs", with two out of four participants being German, one Spanish and one from the US (CPUC 2009).

Governor Jerry Brown's push for *decentralized* renewable energy generation as part of his "Clean Energy Jobs Plan" gave a new impetus to proponents of feed-in tariffs and/or decentralized renewable electricity generation in California. The plan, announced during his electoral campaign in 2010 called for the development of 12,000MW of decentralized renewable electricity by 2020 and for the CPUC to "implement a system of carefully calibrated renewable power payments (commonly called feed in tariffs)" (Brown 2010: 3). Shortly after the announcement of the plan, a one-day conference on "Feed-in Tariffs: A Time for Real Action on Renewable Energy" in July 2010 was co-sponsored by Deutsche Bank Climate Research and Heinrich Böll Foundation. They invited a number of German experts to California to speak again about feed-in tariffs, including Hans-Josef Fell, Eicke Weber, and Georg Stryi-Hipp from Fraunhofer Institute ISE (Pacific Environment 2010). Several other speakers also had close links to Germany, including Paul Gipe and Angelina Galiteva from Renewables 100. An "educational briefing" on feed-in tariffs organized by the FIT Coalition (now the Clean Coalition) at the State Capitol in Sacramento in the same month also featured Hans-Josef Fell, who spoke about the German experience with the policy instrument in front of a "standing room only crowd" (Bernhardt 2010). A 2011 "Governor's conference on local renewable energy", organized to promote the Governor's decentralized generation plan, featured Harry Lehmann from the German Environment Agency as the only speaker from outside the US on a "Best Practices Panel" on the question of "How is the deployment of localized renewables working elsewhere?" (Office of Governor Edmund G. Brown 2011).

Despite an ongoing push for German-style feed-in tariffs by a part of the environmental and renewable energy community (except for solar leasing companies), the adoption of the ReMAT as California's state-level feed-in program is a sign of highly selective transfer and major adaptation. Nevertheless, a strong proponent of German-style feed-in tariffs in California stressed that the exact contribution of transfer in policy decisions was hard to evaluate: "The attorneys that I worked with said 'you know there is a dial needle and the forces of evil pull the needle in one direction and we pull it back further in the other direction'. Sometimes you can't say because we did this they made that change, that's often not the case. Often our influence is just change it to a better outcome overall than it would have been because we made some strong arguments" (C49).

8.2.7 The German example and feed-in tariffs at the municipal level

Municipalities and municipal utilities are of importance to renewable energy developments in California. Three municipal utilities, the Los Angeles Department of Water and Power (LADWP), the Sacramento Municipal Utility District (SMUD), and Palo Alto utilities – which introduced feed-in tariff programs that go beyond legal requirements – were considered and representatives from the three utilities were interviewed.

In the case of Los Angeles, interview partners mentioned that a trip to Germany strengthened Los Angeles Mayor Antonio Villaraigosa's interest in going beyond the 75 MW feed-in tariff program the utility was required to implement by SB32 (C39, C47). The city council approved a 150 MW feed-in tariff program in 2012. This amount of renewable energy capacity might appear modest compared to the German case, but was still referred to as "largest-in-the-nation" (Los Angeles Business Council 2013). During Mayor Villaraigosa's 2009 trip to

Berlin, he stated "Germany has been a leader in renewable energy. The example set by Germany is a model for the rest of the world" and that he hoped "to implement a similar system in Los Angeles in the coming years to stimulate the local economy and create jobs" (KPCC Wire Services 2009). Along the same lines, an interview partner close to the former mayor also recalled him saying that "nobody understands why I am doing this [LA feed-in tariff] program, but I went out to Germany, [...] I saw how many jobs it created. I understand this is a jobs program, I am not confused" (C39).

The Los Angeles Business Council (LABC) very soon recognized the instrument's potential of bringing renewable energy projects – and therefore economic opportunities – to the Los Angeles Basin and became the most fervent proponent of feed-in tariffs in Los Angeles. Both Mayor Villaraigosa's and LABC's enthusiasm for the German experience is hence linked to congruence between objectives deemed important (job creation, local business opportunities) and the perceived track record of the German model, in line with expectations formulated in hypothesis II. The Los Angeles Business Council commissioned several studies on the benefits of feed-in tariffs (DeShazo & Matulka 2009, DeShazo & Matulka 2010a, DeShazo & Matulka 2010b), only one of which discussed the German feed-in tariff example explicitly, noting that "Germany has led the world in annual solar installations" and that this "steady growth was due to a national feed-in tariff" (DeShazo & Matulka 2010b: 15). Asked whether not referring to the German example extensively was a deliberate decision, an interviewee involved in the process acknowledged that although they spent "a lot of time looking at the German model", the idea needed to be "sold to the LA city council", "with something they could compare to, we tried to do as much 'apples to apples' as we could" (C39). US-based examples, above all the city of Gainesville, Florida, were therefore preferred.

Just like at the state level, German stakeholders supported the idea of feed-in tariffs in Los Angeles. Hermann Scheer's visit to a reception in Malibu in 2007 not only led to the creation of Renewables 100 Institute, but also familiarized a range of stakeholders, including LABC's president, with the feed-in tariff idea (Renewables 100 Institute 2007, C23, C40, C45). In workshops surrounding the introduction of the program in LA, the German example was discussed vividly, according to two of the program's designers within LADWP (C60, C46), along with other local feed-in tariff examples in the US: "a lot of the research that was done definitely cites Germany as one of the largest and most successful feed-in tariff programs internationally. […] There were a couple of other utilities that launched, one being SMUD, Gainesville[87] in Florida. […] We kind of took some lessons learned from them. Definitely, a lot of the research that was done and the advocates group cited the German example as a model" (C46).

LADWP had also invited German utility representatives and experts on a study visit in 2010 to "learn from their experience – what worked and what didn't, potential problems to avoid and effective policies that perhaps we can apply here in Los Angeles" (LADWP 2010). This openness to the German example is in some contrast to LADWP's earlier position in CEC discussions on the introduction of a state-wide feed-in tariff: "While the information presented by CEC staff on the German and Spanish programs was insightful […] LADWP believes that significant […] initiatives already approved […] provide a robust course of action. [...] There are significant differences between the State of California, the Federal Republic of Germany, and the Kingdom of Spain in terms of infrastructure, regulation, environmental controls, rate structures, climate, and topography that must be considered" (LADWP 2008: 2). While the general idea of implementing a larger feed-in tariff program than required by state law – although much smaller than the 600MW initially advocated for by LABC – has certainly been encouraged by Germany's positive example and learning, transfer has been partial when it comes to design features. Importantly, the LA program has a low overall program cap and excludes very small projects. The contract duration of up to 20 years, guaranteed grid access and pre-set prices do however constitute design features both programs have in common. As two of the program designers underlined in interviews, although they closely looked at the German example, a tailor-made approach for Los Angeles was necessary (C46, C64).

87 As an interesting side-note, the Gainesville, FL, feed-in tariff – frequently cited as a prominent example of feed-in tariffs California stakeholders looked at, was directly transferred from Germany. A German teacher who had migrated to Florida brought up the idea in a city council meeting and in 2008, a representative of the Gainesville utility participated in a renewables study-tour to Germany. Following this "mind-boggling" experience, the city decided to introduce one of the first municipal-level feed-in tariff on the American continent (Johnstone 2011: 2892).

In California's state capital, SMUD had been pioneering renewable electricity programs for years when the decision was taken to introduce a feed-in tariff program of 100 MW. The feed-in tariff was introduced prior to the entry into force of SB32, and going beyond the quantity required by the Bill (SMUD was allocated 33.5 MW). A team of SMUD's feed-in tariff program designers interviewed stated that the idea of a feed-in tariff had been brought to the utility by the company's representative within the Solar Electric Power Association (SEPA), and that SMUD was motivated from the start to be proactive on the issue, given the company's innovative culture as a citizen-owned utility (C4). The company official involved in SEPA "kind of read the tea leaves that this was both an opportunity for solar leadership, and an opportunity from success in Germany and Spain in getting lots of solar on the grid quickly. And [...] there were draft bills in the legislature [...] floating down [...] That's a pretty standard thing for us, that if we hear enough about it that we think that may be becoming the trend, [we] come out with a design for something that is how we would like to do it and try and lead the policy in that way" (C4, C39, C57). A consultant advising SMUD on the establishment of a feed-in tariff recommended the team look at the experience of German cities with feed-in tariffs, in particular Aachen, as "examples of showing them how in Germany local government really took on the leadership in sustainable energy" (C21).

While the team in Sacramento considered the German example, it is not reflected in detail in the design of SMUD's program. Although the success of the model in bringing large amounts of renewables online in Germany was recognized by the program designers, SMUD's main motivation was to test the ability of different policy instruments, including feed-in tariffs, to deliver renewables cost effectively (C4, C39, C57). Specific design features of the program, such as the small overall program cap (100 MW) and its limitation to large projects were motivated by a desire to contain costs, contrary to what program designers perceived to be the case in Germany: "we were aware of that [the German example], and aware of how expensive that was to the customers in Germany, it seemed. And we didn't want to follow that model, I don't think. We wanted a feed-in tariff that provides a good deal of certainty to the developers as to, 'This is what I'm going to get.' But wanted it based more on the value of the power to us. And we actually won an award from an organization called the Clean Energy States Alliance for the feed-in tariff design that we did" (C57).

Interviewees involved in the design of the Palo Alto feed-in tariff were also aware of the German example and considered it a positive encouragement, but it was not considered in a detailed manner in the process either (C1, C41). A renewables advocate from the CLEAN coalition stated he kept himself informed on developments in Germany and then translated the evidence in a more generic way to the discussion in Palo Alto (C41). News reports about the tariff also drew a connection to Germany, referring to the instrument as what "Germany, Italy, Gainesville, Florida, Sacramento, California and Palo Alto, California have in common" (Wesoff 2012).

8.2.8 Unexpected support for German-style tariffs in the net energy metering debate

While the CPUC's opposition to introducing German-style feed-in tariffs on a large scale in California persisted at the time of field research, some stakeholders seemed to have fundamentally changed their view on the instrument as compared to the peak of discussions in 2008/09. The growing share of decentralized generation in small-scale renewable energy installations had created a new landscape and challenges. Utilities in particular considered the "invisible[88]"electricity generated and consumed on-site in net metering arrangements a problem. By 2014, a debate on a reform of California's net-energy metering scheme had started, since the existing system was scheduled to be reviewed by the end of 2015 and a "successor tariff" be developed. In this debate, utilities and consumer advocates appeared to see a feed-in tariff-like policy as a more attractive alternative, an important evolution compared to their earlier stance on the issue. In comments to the CPUC, Southern California Edison for example suggested a model in which "any power exported to the grid would be compensated at a feed-in tariff rate. In SCE's opinion, this structure is principled and fair to all customers, whether they are participating customer-generators or not" (Walsh et al. 2014: 19–20). It should be noted that by the time of the discussion on a successor net energy metering tariff, the average retail electricity rate in California (US Energy Information Administration 2015), i.e., what a net metering consumer saves by producing 1 kWh of electricity, was about one third higher than the feed-in tariff for

88 This electricity is "invisible" insofar as it is not fed into the grid, making it more complex for utilities to know the actual demand and production at a given point in time.

small residential installations in Germany. This is a different situation that at the time of the CEC feed-in tariff debate, when German feed-in tariffs where still higher than average residential electricity rates in California and net metering therefore comparatively cheaper.

Even ratepayer advocates, who were extremely critical of feed-in tariffs in the past, were acknowledging that a system, in which electricity production and consumption are taken into account separately and paid at different prices, would now be a good option, instead of allowing solar rooftop owners to discount one kWh from their bills for each kWh produced at any time of the day. Asked what his organization's preference for the new scheme was, a ratepayer advocate and former opponent to the German model stated, laughing, it was to "adopt the German feed-in tariff" and "more seriously" to pay "everyone who produces solar on their rooftop a fair price for their solar output", which would be "very similar to a feed-in tariff. Here in the States they've called it a value of solar tariff" (C30). An environmental advocate's hope that "by the end of the year [2015] in California [we] will have some kind of a structure to incentivize people to do exactly what is happening in Germany" (C40) therefore appeared more realistic than two or three years earlier.

Still opposing the introduction of a feed-in tariff were parts of the solar industry, in particular big solar leasing companies like SunRun or Solar City, the world's biggest solar installer. These companies were considered "new utilities" (C43) by interviewees, who would see their business model fundamentally put into question in the absence of full-retail rate net energy metering (C23, C24, C27, C28, C36). The opposition by solar firms to feed-in tariffs, which had led to a tremendous growth of the market in Germany, might seem counterintuitive. But, as an interview partner explained, "you and I would say, 'But, but, but, but, but…the market would be so much bigger, you would make more money.' No, no, no. Because the market would be much bigger, but it would be a different market, and because it's a different market, there may be new competition. And no business, German or American, wants competition" (C28). Solar leasing companies had already made their voices heard against feed-in tariffs earlier on, including within industry associations. A former CalSEIA official remembered the case of a bill that would have introduced feed-in tariffs for projects up to 20 MW in 2008, that "at that time, the large financial [i.e., solar leasing] companies started to work against CalSEIA. [...] So, as soon as we introduced the bill, SB

1714, they began telling that feed-in tariffs are bad for growth, that they were costing jobs. And the utilities also tried to keep the feed-in tariffs down. So the bill died“ (C36). SunRun, Solar City, and others spoke out publicly again German-style feed-in tariffs already two years before the net energy metering successor tariff decision was to be taken (Shahan 2013). The US-wide solar industry association SEIA, where solar leasing companies have a strong voice (C27), also advocated for the “successor tariff” to remain close to the existing system and firmly opposed “other distributed generation tariffs, such as feed-in- tariffs or value of solar tariffs” (SEIA 2014: 2). The reasons provided by interviewees for this opposition to an instrument that was very much defended by the renewable energy industry in Germany was that solar leasing companies had already tailored their business model and split the market based on a net energy metering scheme. Also, the companies’ identification with the goals of solar energy, beyond business interests, was put into question: “They are not solar firms, but financial firms, Solar City and the like. They have nothing to do with solar. If you got incentives for putting your grandmother on your roof, they would develop a grandmother-leasing model” (C36). The lack of support for feed-in tariffs from this important part of the renewable energy industry was seen by a proponent of the German model as a major reason it was only very selectively transferred to California: “very few things get done in either California politics or US politics for conceptual reasons. It requires an engaged profit-driven business interest to actually succeed and [for] both the solar and the wind industries [there] wasn't as much an interest in scaling up the technology or even scaling up the market so much as scaling up my business” (C24, also C28).

The eventual CPUC decision regarding the net energy metering successor tariff was taken after the completion of field research, in January 2016, and maintains the principle of the current net-energy metering system (CPUC 2016b) The re-emergence of German-style feed-in tariffs as an idea in the debate on net energy metering reform is nevertheless interesting as Germany’s own model was already undergoing fundamental changes at that time (BMWi 2015c).

8.2.9 Distributed renewables: transfer of incremental policy adjustments

While many interviews focused on policy instruments to promote the uptake of renewable energy, the question of how renewables can best be integrated into the electricity system is an area where transfer has occurred and will likely occur in the future between Germany and California. A recent CPUC white paper on the integration of renewables for example extensively reviews the German experience and acknowledges, “Germany has not had to face serious difficulties with grid integration yet” (CPUC 2015a). Starting from the fact that “when we talk about distributed generation and high shares of it, we talk about the German example” (C13), learning took place among stakeholders in California, especially regarding so-called “smart inverters[89]” and permitting.

General interest in ways of integrating distributed generation had intensified with Governor Brown’s 12,000MW goal for small-scale installations by 2020 in the state (Brown 2010). The California Energy Commission – which had championed the idea of German-style feed-in tariffs earlier on – commissioned a report in 2011 on lessons learned from electricity markets in Germany and Spain with regard to distributed electricity generation (KEMA 2011). A participant at a presentation of the study said the document convincingly showed these countries could manage their grid without major problems and that the presence of a German consultant in the study team and at the workshop made it difficult for California utilities to argue renewables “destabilize the grid, cost a fortune and cause a lot of problems” (C49). Even though CEC staff from the distributed energy team interviewed had “obviously” been aware of developments in Germany, the KEMA reports were essential in deepening that knowledge (C16, C20).

Puzzled by the low prices for solar installations in Germany compared to the United States, two interview partners from the solar industry travelled to Germany around 2005/2006 to understand the difference in the cost of installation, especially for residential PV panels. Much simpler permitting procedures[90] in Germany were identified as a main reason for the difference in price. An in-

89 Simply put, smart inverters are technical components that are either integrated or retrofitted to renewable energy installations and allow for the installations to be controlled remotely or react automatically to extraordinary situations in the grid, in addition to inverters’ basic function of converting direct current (DC) to alternating current (AC). For example, unusual or unexpected spikes in solar output may create a need for grid operators to curtail production for a certain amount of time, to maintain grid stability. This can be achieved more easily through so-called smart inverters. See for example CPUC (2015a: 79).

90 These permitting procedures cover all steps building-owners or owners of free spaces need to go through in order to be able to construct and operate a renewable energy installation on their premises. These may include checks regarding the compliance with building or safety codes, requirements for the interconnection of systems to the grid, or compliance with technical standards.

terview partner involved in the subsequent discussions on Assembly Bill 2188 on permitting processes in California stressed that the German example was "at the background of discussions and everyone was aware of it", and that it "forced discussions into the right direction" (C27). The respondent did not personally use the German case in his communication outreach activities due to the fundamental adjustments a change toward the German permitting approach would have required. This is a case of transferred information hitting the obstacles of regulatory traditions: "It's just such a different system. Applying that example to the US would have meant getting rid of our National Electric Code, which is not imaginable" (C27). Advocates from the Sierra Club, however, used the example when sponsoring a bill to cap permitting fees for solar installations (SB1222) and in their support to AB2188: "One of the points [...] where we cite Germany even today and we did continuously is the installed cost of PV in Germany is half what it is in California, it's half! [...] we used that [...] when Sierra Club was sponsoring the bill to cap the permitting fees. [...] We were successful in getting this bill passed [...] We also used that argument where we supported a bill that was passed 2 years ago to streamline the permitting process. The cost of panels are pretty much the same in Germany as they are in California, it's all about the system cost. And in Germany, also because of the feed-in tariff, it is so much lower" (C49). A survey carried out in 2012 and 2013 by a team around a German researcher at the Lawrence Berkeley Lab on the price difference between solar installations in Germany and the United States delivered further arguments in favor of reducing permitting fees and streamlining procedures as evidenced in the case of Germany (Seel et al. 2014).

California utilities themselves became more proactive in the face of growing shares of distributed generation. A utility representative who took part in a 2012 study tour to Germany subsequently pushed for the introduction of "smart inverters" in California, as a consequence of Germany's negative experience with retrofitting these appliances to existing systems. As stressed by a participant of the study tour and member of Western Electric Industry Leaders (WEIL), this industry association used the German example as the main argument to push for the installation of smart inverters in California (C5). A working group has since then been installed at the level of the CPUC to implement rules for the introduction of smart inverters (C37, C42). Numerous German studies are being discussed in the group as technical reference materials (CPUC 2015b). Utility representatives interviewed said they were in contact with German utilities E.ON and RWE to continue dialogue on the issue and learn from Germany's experience (C5, C55). The German experience with retrofitting smart inverters made utility representatives push to "get those inverters to California and incorporate that sooner rather than later to be prepared. That's certainly one of the lessons we took away from RWE and E.ON, that you needed to be prepared and proactive" (C5). Respondents from other utilities made congruent statements, with regard to regulations for smart inverters, saying that "California is moving aggressively because of the lessons with Germany" (C37) and, more generally, that "Germany is a great example, they were able to bring such a large share of renewables online and it's great to be able to learn" (C44).

On a different issue, a renewables advocate pointed out that he had become aware of a German government program that subsidized the combined installation of storage and renewables in private homes. The interviewee said he brought the concept to utilities: "I have spoken with a senior person from SCE, the largest IOU [Investor-Owned Utility] in the state, here just recently. And I didn't mention Germany but there's where I got the idea from. 'You all have this rooftop solar, it's intermittent and it's variable [...] why don't you [...] provide subsidies to the home owner and let them get some storage for their system and it will create a much more stable grid'" (C49). This incident exemplifies the reality of "hidden transfer processes" that happen in informal conversations, and with the original source of policy ideas getting diluted in the process. The choice of methods attempts to capture these processes, but it can be assumed that many instances of transfer can still not be accounted for.

8.3 Case Discussion: California

8.3.1 Active policy promotion and the German agenda

As in the case of Morocco and South Africa, the extent and agenda of Germany's leadership efforts in the case of California is to be taken into account when discussing its effectiveness.

As the mission of initiatives such as the Transatlantic Climate Bridge illustrates (German Missions in the United States 2015), the overarching aim of German outreach to California appears to be strengthening common leadership for the benefit of global climate protection. But more specifically, numerous instances of German transfer agencies advertising Germany's particular approach of promoting renewables were reported. To support efforts in the direction of a feed-in tariff in California, German policy advocates regularly gave speeches at industry events in California shedding a positive light on the instrument. At Intersolar North America in 2008, the German initiator of the conference and former UC Berkeley Professor Eicke Weber reportedly asked his audience to let him "again make [his] pledge why the [German model] works" calling the alternative used in California, net energy metering, "the wrong way" (Gunther 2008). Events dedicated to feed-in tariffs also took place throughout 2010 and 2011 prior to the adoption of the ReMAT (Office of Governor Edmund G. Brown 2011, Pacific Environment 2010). Perceptions of how active German advocates actually were and to what extent their aim was to promote feed-in tariffs in California differed widely among interviewees. Whereas some saw "German delegations come over several times a year to explain to us the virtues of feed-in tariffs" (C31, also C29), others "haven't heard much from German advocates" (C27). Interview partners also deplored the fact that outreach from Germany wasn't more active, a finding of relevance for Germany's leadership efforts (e.g., C36). Interview partners from state-level institutions CEC, CPUC and the Governor's office all estimated the average number of visits from official German ministerial delegations to California (excluding visits from individual policy advocates) to about one per year. An interview partner involved in renewable energy for years, both in the public and private sector, stated "We had trade groups, companies, German officials, delegations organized by environmental groups. I don't think I ever had a year where I didn't meet with German delegations and every time they were always talking about a feed-in tariff. But you can't just transpose it here; the circumstances are different in every location" (C31). Another interview partner agreed: "the German American Chamber of Commerce would hold a lot of forums and discussions. I mean BSW [*Bundesverband Solarwirtschaft*, German Solar Industry Association] was here all the time, Gerhard Stryi-Hipp, I got to know quite well. You know, Eicke Weber at the Fraunhofer Institute [...] there were a lot of Germans who just couldn't quite understand why everybody didn't, you know, want to replicate exactly what Germany had done" (C12).

Nevertheless, Germany's information strategy regarding the Energiewende was seen as insufficient in the light of misinformation about blackouts and grid instability in Germany: “if there’s one thing Germany could do better it’s making more information available, making the information clear and providing it in English. They should clearly inform about the cost and reliability on a website that is easily accessible. […] The solar industry should really carry these messages from Germany to other countries, and the German government as well” (C35). With the focus of attention shifting to technical renewables integration issues, exchange between utilities was also seen as key for the future. A session on Europe and Germany at the annual conference of California’s transmission grid operator CalISO in October 2015 was mentioned by a number of interview partners as a sign of intensifying cooperation and a likely source of future, more concrete knowledge transfers between grid operators (C13, C23) – in both directions. At the event, which happened after the end of field research in California, a representative of a German transmission system operator gave the dinner keynote speech and two international panels with German participation were organized (California ISO 2015).

8.3.2 Hypothesis I: Strong consideration of the German example

Among the three cases studied in this book, California stands out as the case where the direct, demand-driven, consideration of the Energiewende as it is implemented in Germany played the strongest role. Most interview partners agreed they had become aware of Germany’s renewable energy policy around the middle of the 2000s, through diverse channels described earlier, from personal interaction and meetings at conferences, to self-organized study visits to Germany, business contacts, publications, and the media. The amplification of the feed-in tariff debate in California, initiated by the CEC, increased awareness for the German example.

Regarding the question of what Energiewende elements were considered by respondents, feed-in tariffs and their specific design features in Germany were obviously of relevance. Among these particular features, the absence of an overall program cap on how much capacity can be supported and of caps on the size of installations, the important role of small-scale, residential installations, simple interconnection procedures and guaranteed grid access were most frequently cited. General aspects of the Energiewende seen as successful were the speed of renewable energy deployment in Germany; its general level of ambition to achieve high renewable energy and climate targets; democratization of electricity production; low installation cost for residential solar; as well as regulations related to renewables integration. The aspect of "democratization" of electricity supply in the Energiewende through individual small-scale generation was, however, a minority concern: "there are advocates for democratization in California. They have not prevailed and the reasons they haven't prevailed are largely because there is a perception, whether it's real or not that centralization gives greater economies at scale and allows for more, more to be done quickly" (C56). This statement points to the importance of cost efficiency as an objective of (renewable energy) policies in California.

Nuclear phase-out did not play a role in transfer since the end of nuclear in California is de facto implemented through a moratorium on new installations and there is only one plant still in operation (CEC n.d.). From a technology point of view, solar energy played the most important role in discussions, as it is the fastest growing segment in California's electricity mix and California has excellent solar resources (CEC 2015). Steep increases in solar capacity in Germany in the mid and late 2000s therefore triggered particular attention in California. Other sustainable energy issues including efficiency, and demand-side management were occasionally mentioned as areas where cooperation could be reinforced for the benefits of both California and Germany, but Germany's experience in the renewable energy space was clearly considered more strongly (C56).

Overall, findings from California confirm the relevance of the prominence and perceived success of senders for their policy models to be considered. Additional factors that determine consideration of the Energiewende are California's and Germany's similarity in terms of economic development and leadership aim, as well as the availability of sufficient capacities to search for information. Germany was seen by far as the most relevant example, and a "key reference point" for officials and industry stakeholders in California (C58). A public sector official confirmed: "We definitely look at other states as well. Though the examples don't usually feel relevant, I do feel like we're sort of on the forward edge of most of the clean energy stuff in the states. [...] But Germany has always been like, on renewables – demand side is a different discussion – but on renewables, it's always been the test case. We're always watching to see what would happen" (C56). Germany's example was seen as effective in achieving the goals it set out to reach domestically, despite critical views regarding the cost of the program for German consumers, and, more recently, of grid management issues (C24). The country was referred to as a "trailblazer" (C20), as "leading the world in renewables and climate policy" (C13), and as "taking great leadership for the world and for the European Union in developing and implementing clean energy policy" (C13). The Energiewende's contribution to bringing down the cost of solar technologies through its policy-driven domestic market was recognized in particular: "the German feed-in tariff built the global market. It did the world a complete favor" (C12) and "Let's say Germany had done nothing, where would we be today? And I think the answer is nowhere near as far along" (C31). This effect of German policies was recognized as being intended by Germany, which was seen as willing to "[subsidize] solar power to the rest of the world" (C56). Germany's leadership was also recognized with regard to its role in the European context, both in terms of unilateral and active leadership: "Germany in my mind had and has its greatest influence as an example setter within Europe. It very early on, and I attribute much of this to Hermann [Scheer], captured the staff of the European Commission from a policy standpoint [...] I think that is purely attributable, primarily to the power of example but also to some very effective behind the scenes exercise of soft power, a lot of persuasion" (C24).

At the municipal level as well, the German case was considered for its perceived success. Again, the renewable energy program's sheer size was a main argument in its favor: "The other examples that were mentioned [...] were Gainesville, Florida, SMUD, and Vermont, also Indiana. But these were small, capped programs, some of whom under-delivered and then moved to a competitive system. Los Angeles wanted a bigger program to experiment" (C17).

Across the different policy debates, consideration can be seen as mainly driven by a motivation to improve policies that were considered sub-optimal (RPS/FiT), and to increase the knowledge base for policy action. Consideration, importantly, was not hampered by whether the example would likely affect stakeholders positively or negatively: knowing about the Energiewende was seen as relevant for stakeholders from any group interviewed to increase information. As for other drivers for consideration, it is challenging to clearly separate between the factors "competition" and "prominence/success", since the two are strongly linked and, according to interview partners, there was hardly a way to avoid considering the German experience (C13, C17, C20, C29, C42 etc.). Although economic competition regarding the market for renewable energy technologies (particularly solar) could appear as a plausible driver for consideration, this mechanism was of marginal importance. With much of global manufacturing of solar panels having shifted to China by the time of field research, several interview partners, including trade representatives, saw regulatory competition for economic reasons in the field of renewable energy as irrelevant (C18, C34). California's role as a pioneer, in environmental policy in particular, was seen as part of the state's very identity, resulting in a sense of fruitful competition over reputation and leadership with Germany: "the example of Germany doing more than California is important because California always wants to be number one. [...] So we have to at least keep up, right? Germany is a great example [...] leading the way and making us embarrassed so that we have to do more" (C20). Germany's progress despite its lack of sunshine was often referred to as an additional argument: "There is a wonderful dynamic with renewables where everybody draws inspiration from each other and starts to think bigger because you see it being done elsewhere. [...] the fact that Germany was ahead of us even though they don't have as much sun or wind, it really makes the case, [...] we have the better renewable resources so why don't we do it? And so I think it was very important" (C31). Others were more reluctant to speak about open competition for a reputation as a green leader: "[there is] maybe a little bit of a race to the top, but I wouldn't put it in a competitive way, it's more of a synergistic benefit that other countries are doing this. It makes it seem more possible, more feasible" (C51).

While opinions on the transferability of the German model to California diverged, a vast majority of interview partners recognized Germany's leadership in renewable energy and the value of dialogue among leaders, confirming "prominence" and the model's effectiveness in its original context as factors for consideration. Even among those interview partners who were skeptical about the transferability of Germany's model to California, the value of Germany as a "sparring partner" and a source of lessons was recognized. Knowing "there's someone out there you can talk with" (C13) was described as highly valuable by an interview partner with an otherwise critical view on transferability. Information about the German experience gained through study visits to Germany also appeased concerns, although different participants considered information in different ways: "they were there and learned a lot. Sometimes people learn different things when they see something. [...] And it also might have given them just more courage in saying 'that's what I want'. I noticed that the ISO [Independent System Operator] was much more comfortable afterwards in saying we could do 40% renewables if the Germans can do it. [...] It did, it did really encourage people and they would say 'They did it, why wouldn't we do it as well?' [...]" (C23).

Consideration of the example was somewhat promoted by active German outreach in favor of Germany's approach, as described above. But interview partners (C55, C56) also mentioned a 2014 presentation by a representative of German utility RWE before the CPUC (Weale 2014), for shedding a negative light on the German Energiewende's cost, burden on traditional utilities and grid management issues: "I attended a meeting with RWE this year and last year. This Mister Graham… and they explained how renewables are depressing wholesale prices and generators are in big trouble" (C55).

The generally close consideration of the German example and its particular status as the key reference also meant, according to a public sector official interviewed, that negative developments in Germany were overly amplified in the press: "What we see now in terms of the tension between existing generation in the market and the blip up in carbon […] these things get totally amplified in the US press […] and people really take a disproportionate amount of interest in these little data points about what's happening in Germany and they try to weave a story out, but that will temper the US's progress in this regard" (C56).

Interviewees' knowledge of the Energiewende was detailed and the example was seen as the most relevant source of lessons. Strong consideration in the case of California can certainly also be seen as linked to the capacities available for the search and processing of information, with access to information and media attention ("there have been multiple articles on the Energiewende, on Germany's leadership role in the world in terms of tackling climate change", C26), regular trips to Germany (e.g., Galiteva & Moss 2014), and participation in workshops and events. Also, studies were carried out specifically to understand the German example, including the "KEMA" studies mentioned above and a study initiated by SEIA to defy misconceptions about the German experience (Weiss 2014). Nevertheless, interviewees addressed the issue of gaps in Germany's Energiewende communication that could increase the knowledge base further.

In summary, while California would certainly have ambitious renewable energy and climate goals in the absence of the German energy transition, the existence of this prominent example was deemed very helpful for renewable energy advocates. No interview partner stated that the Energiewende example was irrelevant or that German active leadership was intrusive, despite differing views on how transferable the model was seen, confirming the relevance of the Energiewende as a reservoir of lessons, including negative ones, that were seen as facilitating the transition: "The first ones, leaders, don't always get it perfect but it's always valuable because then the people coming next can try to avoid the same problem" (C20).

8.3.3 Hypothesis II: Strategic utilization of transferred knowledge

Going well beyond mere awareness of the model, Germany's renewable energy policy example was extensively used in California policy debates. A few overarching conclusions on utilization are of particular relevance, also from a theoretical point of view.

A first striking aspect of renewable energy policy transfer between Germany and California is the instrumentalization of the German model by advocates of more rapid and ambitious renewable energy expansion, in particular based on decentralized generation, but also by skeptics of such an approach.

Members of environmental organizations in particular said "Germany was often the example that encouraged us to move forward" and that Germany's example had "done a lot for advocates in California" (C21). Interview partners found it particularly important to be able to "point to Germany and say [California] can do at least as much" (C20), as an answer to utilities who argued high shares of renewables and distributed generation in particular were not feasible: "generally I would say it was often somebody at a utility that said [...] these [German] policies are bringing in resources that are too costly" (C57, also C49). In addition to the cost of the Energiewende program, perceived grid instability in Germany (loop-flows to Poland, over-generation) was the most frequent negative argument used in debates: "this issue of grid instability. If I got a dollar every time I heard that I'd be a wealthy person!" (C23). An interview partner also criticized the fact that stakeholders who had been to Germany and "were told there are absolutely no problems with reliability", returned to California "tell[ing] everyone you should keep an eye on that because Germany has reliability issues. That's conscious misinformation" (C35). Despite this use of the Energiewende as a cautionary tale, the size and resilience of the German economy made the

Energiewende a “much more difficult example to dismiss” (C51), also compared to smaller European countries.

In terms of how the Energiewende example was used internally, staff from the CEC, the CPUC, and an energy policy advisor to the governor underlined the importance of knowing someone else was engaged in a comparable transformation of the electricity sector, in particular for ambitious policy proposals such as the increase of the RPS to 50%: “people on the one side use Germany as a punching bag to say there's problems. Others say it's an example of what can be done. I think for Governor Brown [...] he feels like it's a partner in leadership. [...] I think it has helped him be willing to be more proactive because there is another country with a very large and successful economy moving forward with this policy. [...]. It makes him feel like California has partners and it's not isolated” (C51). Officials reported they utilized evidence about Germany to counter utilities’ arguments against large shares of renewables: “You hear ‘it's gonna be a problem’ but the Germans are still able to manage the grid, with some circuits of 50 or 60 percent solar in Bavaria, and you don't hear about blackouts or transformers exploding. […] we tell them [utilities] to look at the German system: ‘look, they are managing their system with much higher penetrations than we have and they are able to do it and don't have any problems so you should be able to do that too’” (C15). But a utility representative as well recognized that it was natural for California to look to Germany: “California is a leader so it’s clear you look to other leaders to learn. Germany is a great example to learn what the experience is and also to avoid mistakes and learn what works well” (C55).

In terms of integrating renewables to the grid, knowledge about the regulatory framework in Germany was used to devise rules in California, in particular with regard to permitting and smart inverters: “Absolutely, the German example is what we pointed out originally. When we talked to the CEC and CPUC in 2011, we pointed to Germany who is retrofitting this inverters, why don’t we? And we could take advantage of their experience and specifications” (C5). But lesson drawing also concerned planning more generally: “We did a very early study with KEMA working on the German example concerning the distribution system, that provided some pretty good lessons as we are getting into the new proceeding at the PUC on distribution planning. Basically that's been healthy lessons from Germany which are reflected in the report” (C58).

As a longstanding advocate of renewable energy in California pointed out, the degree of detail to which the German example was utilized by stakeholders directly depended on their role within the policy network: "it depends on how sophisticated and how involved the person is in actual policymaking, you know? So the ones who are more sophisticated or who really live this as their work [...] they're going to be looking for more specific examples of, 'Okay, so, what can we actually apply here?' As opposed to, 'Well, that's great. Glad to know you can do it'" (C45). Another interesting aspect of utilization is the change of perception on the usefulness of the German example among utilities and consumer advocates in the net energy metering successor tariff debate. Features of the model that were seen as negative around 2008-2009 appeared as more advantageous in comparison to net energy metering by 2015 (C22, C30).

In summary, utilization of the German model in California is a particularly strong example of differentiated use according to stakeholders' position, objectives pursued and views on feasibility. These factors need not necessarily point in the same direction of negative or positive use, as the example of the group of interviewees who saw feed-in tariffs favorably, but politically unfeasible, shows (C12, C27, C31). The expectation that transferred knowledge is utilized when it can fill epistemic gaps or bring new insights was confirmed, in particular for instances related to permitting, smart inverters, and distribution grid planning. Compared to the other cases, the additional impact of active leadership on utilization, in particular *via* German transfer agents within the policy system, was limited in California.

8.3.4 Hypothesis III: Selective reflection of transferred knowledge in policy outcomes

Evaluating the outcome of renewable energy policy transfer from Germany to California requires taking into account several layers of policy, from general objectives to instruments to incremental and regulatory adjustments. At the intermediary level of policy instruments, German-style feed-in tariffs were discussed in detail, but reflection in policy outcomes is greatly limited. Several reasons for this can be put forward, in line with expectations expressed in Hypothesis III. Actors at critical positions within the policy network, in particular within the CPUC, and powerful groups (utilities, solar leasing companies, consumer advocates) were opposed to transferring feed-in tariffs to the Californian context. While staff and decision-makers interviewed at the CPUC shared arguments in line with concerns over "regulatory fit" and negative lessons from the German experience in terms of the cost of the program, utilities and solar leasing companies saw their business interests challenged by the German model of privately owned, small-scale electricity generation. A sub-group of respondents in favor of decentralized renewables had a positive view of the example in Germany, but saw a risk in bringing in such a disruptive model to the conversation, despite German efforts to promote it: "At the end of the day it's really about what is politically practical" (C31). For this group, putting into question the existing RPS program could have led to more fundamental questioning from other actors of renewable energy policies and possibly led to a less favorable outcome overall. Other arguments brought forward to explain the very selective reflection of the widely and intensively considered Energiewende model in policy outcomes in California have less to do with interests. The perceived complexity of the feed-in tariff system and a certain mistrust regarding the appropriateness of institutional structures and capacities in California to implement such a policy (lengthy decision-making process at the CPUC, perceived close ties between the CPUC and utilities) were mentioned as reasons. In summary, California's ReMAT as a policy outcome is the result of a contentious and lengthy battle over ideas, in which the German example served as a main reference point for both opponents and proponents of a larger feed-in tariff program. At the local level as well, lessons from the German program were reflected selectively, in particular in the choice to implement a program cap in the Sacramento, Palo Alto, and Los Angeles policies (C4, C46, C57).

The German experience is more strongly reflected in policy outcomes at the level of incremental, but important, adjustments, such as the introduction of smart inverters, with stronger consensus on the German experience across a broader range of stakeholders. This is in line with expectations from the transfer and diffusion literature regarding the more likely transfer of items that are "technical" rather than disruptive or redistributive.

On the level of overarching sustainable energy and climate goals, knowing about the German energy transition has comforted California's decisions to pursue an equally ambitious path of energy transition, but did not trigger them. Asked whether the German example played a role in the recent announcement of an increase of ambitions in the RPS to 50% renewables by 2030, an advisor to the governor confirmed that “yes, it was important, it was significant. Seeing that it can be done and that others were doing it too made the decision a lot easier” (C52).

Even though the Energiewende model is only selectively reflected in policy outcomes, the value of German leadership for the development of policies and general policy direction similar to Germany’s was recognized, pointing to “effective” leadership: “you know when you are starting something new, there will always be some things that are extremely successful and some things that are not as extremely successful. That's what progress is, making mistakes and being able to correct them. That is a whole lot better for everybody than being too afraid to do anything at all. And instead of focusing on what cannot be done, Germany's attitude, correctly was to focus on what can be done” (C23).

9 Cross-Case Discussion

9.1 Findings: Determinants for Effective Leadership through Transfer

9.1.1 Patterns of leadership through transfer

This chapter identifies common patterns and cross-cutting findings from the case studies presented in Chapters 6, 7, and 8, with regard to the book' main research question: What determines the effectiveness of German Energiewende leadership in promoting renewable energy policy transfer to Morocco, South Africa, and California?

The transfer processes studied in this book all need to be seen in the context of fundamental changes and challenges to energy systems. While California already had a renewable energy policy framework in place when the German example started to be considered, Morocco and South Africa developed their very first frameworks for renewable energy promotion while transfer was occurring.

In all cases, policy transfer from Germany fell into (and contributed to) periods of major policy change in a central sector of the economy, which is also closely tied to politics. Against this background, a first and essential common finding is that non-convergence in terms of policy instruments cannot be equated to "non-transfer" or the ineffectiveness of Germany's leadership efforts. Morocco, South Africa, and California were selected, among other criteria, as cases for their rejection of the most iconic instrument Germany used to build its leadership in renewable energy, the feed-in tariff. But the case studies clearly show that transfer can occur in the absence of convergence between leader and recipients in terms of policy instruments chosen. Assessing the effectiveness of leadership through transfer hence requires a differentiated look at transfer in its temporal context, at different levels of policy, and along the three categories for transfer – consideration, use, and reflection in policy outcomes – used to structure the case studies and cross-case comparison.

Direct consideration of the Energiewende was strongest in California, where the example was used extensively as a direct reference point and as a source of lessons in policy debates. Evidence from the Energiewende was politicized and

K. Steinbacher, *Exporting the Energiewende*, Energiepolitik und Klimaschutz.
Energy Policy and Climate Protection, https://doi.org/10.1007/978-3-658-22496-7_9

instrumentalized to steer discussions and introduce new knowledge to the policy-making process. Unilateral leadership also unfolded an empowering effect on pro-renewables advocates in the state, who were able to point to Germany's experience to underpin their arguments. The German experience was by far the most important example considered by stakeholders and triggered discussions, informed debates and spurred policy action, with concrete outcomes at the level of adjustments to regulations, as well as on a meta-level of validating California's move towards a renewables-based energy system. The importance of looking at transfer as a multi-layered process also became clear in Morocco and South Africa. Whereas stakeholders were aware of the Energiewende example and Germany's leadership was as widely recognized, independently from stakeholders' views on transferability, the German experience was more often translated into the recipient context by German transfer agents. Transferred knowledge was reflected in policy outcomes at the level of general policy orientations towards more renewables in Morocco and South Africa, and in adjustments to regulatory frameworks as well as with regard to program implementation. The interactions between transfer and the membership and structure of policy networks, which evolved with transfer, are an important aspect to consider here. Germany's Energiewende leadership goes well beyond the provision of a pioneering policy model, and contains an important part of preparing the ground for transfer in recipient states.

Results from the three cases clearly show that different levels of leadership and transfer are interlinked, as illustrated in Figure 28. Individual and structural filters affect how information about foreign policy models is received, used, and reflected in policy outcomes. But these were not static and transfer contributed to the evolution of individual and structural filters in the three cases studied. The building of necessary capacities to process and use transferred knowledge, targeted support to pro-transfer agents within the policy network and, very importantly in Morocco and South Africa, the deployment of transfer agents *into* policy networks, are factors that were conducive to transfer and are likely to shape future transfer outcomes. Active leadership is closely linked to the Energiewende as a domestic policy example, justifying the integrated study of these two aspects of leadership. The results open avenues for further research that grant an important place to the study of pioneering countries as active senders in transfer.

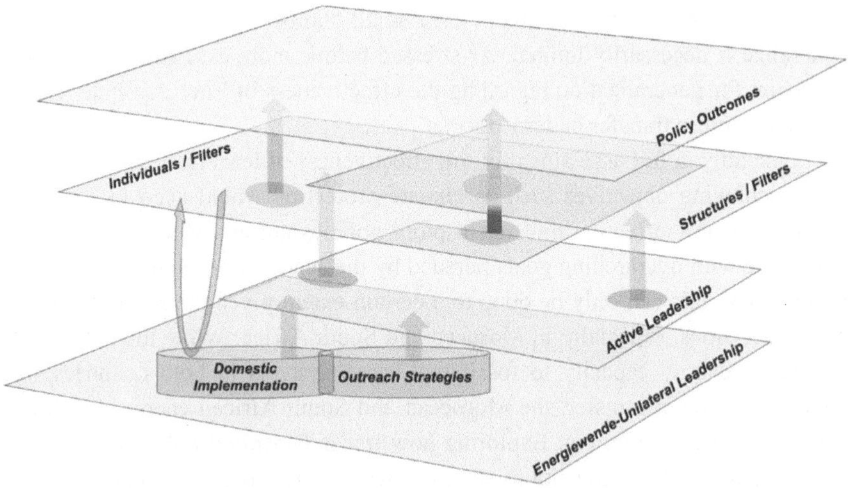

Figure 28: Visualization of Energiewende leadership and transfer

Across the cases, the perceived appropriateness of transferring the German energy transition approach, given its potentially system-changing effects and challenge to regulatory traditions and utility business models, varied greatly between different groups. While this is unsurprising from a political economy view on regulation and policy instrument choice, the politics of policy transfer are still a vastly under-researched aspect in this literature. The finding also points to a need to consider the specificities of the policy field in which transfer occurs. The Energiewende as a transfer item challenges the structure of a major sector of the economy and beyond, and transfer in this field is likely to be different from areas where less organized, less powerful incumbents are concerned. The relative power of stakeholders in favor or opposed to the transfer of German experience, as well as regulatory traditions and policy objectives eventually determined the limited reflection of transfer in policy outcomes at the level of instruments in all cases.

9.1.2 Research limitations

Several important limitations to the study need to be emphasized to put findings in context. First, the research aim was to explore a novel field by gaining in-depth insights into a small number of cases. Although cases were selected to cover a range of different channels, their small number means that the ability to generalize is necessarily limited. As stressed below, more case studies are needed to aim for generalization regarding the effectiveness of Energiewende leadership and explore transfer patterns further.

Secondly, a full assessment of the effectiveness of leadership through transfer in achieving objectives such as climate protection would need to cover the question of how effective transfer-inspired policies are and whether the results are in line with overarching goals pursued by the sender. For the policies studied in this book, this can only be done to a certain extent given the recent nature of policy decisions, especially in Morocco and South Africa. While the addition of renewable energy capacity to fossil fuel-based systems in both countries was facilitated through transfer, the Moroccan and South African energy transitions are only at their beginning. Exploring how transfer-inspired policies are implemented and how these transitions stabilize over a longer period of time is thus an area for future research.

Thirdly, while an effort was made to contrast Germany's role with other donors and international institutions, including through interviews with these agencies, the book has a focus on German efforts and the German Energiewende model. While the findings presented in the case studies very strongly suggest that Germany played an outstanding role in the three cases studied, single case studies covering all potential sources of influence in the same depth could provide additional insights.

Fourth, the transfer processes described in this book and Germany's very aim of inspiring policy action abroad – regardless of the legitimacy of its motives – can be seen in a critical light. The book does not include an explicit critical discussion of the very concept of policy transfer. This is due on the one hand to the focus on exploring a novel empirical field and generating knowledge about how transfer and leadership actually unfold, preparing the ground for future critiques. On the other hand, interview results and stakeholder reactions to Germany's outreach, as well as the strong adaptation of transferred knowledge to local objectives, did not indicate an immediate need for a fundamental critique of transfer along the lines of knowledge being simply "delivered" into a recipient context (Ellerman 2000) or even forced upon agents there. In the particular cases and for the particular policy field studied in this book, knowledge was often co-created and utilized with and by local stakeholders and transfer processes were largely driven by local considerations, but the question of the ethics of leadership through transfer is still a relevant one.

Finally, the choice to study the debates and decision-making processes presented in the case chapters through the lens of transfer inevitably means that other perspectives on the same processes were excluded. For example, a comparative policy approach, contrasting the different energy transitions, could have been adopted or the Morocco and South Africa cases could have been embedded in a more development-focused framework. Also, the focus on Germany's efforts was very much on sustainable energy, while broader considerations of Germany's foreign policy and role on the global stage were excluded.

With these limitations in mind, the following sections identify what determined the effectiveness of German Energiewende leadership through transfer in Morocco, South Africa, and California.

9.1.3 Unilateral leadership and consideration: Capacities and mediation

Literature on unilateral leadership is concerned with the question of whether first-movers send strong enough signals for followers to adopt similar policy choices. The question of whether Germany's Energiewende was able to trigger a conscious consideration of the model by foreign stakeholders is therefore a first fundamental question to be answered. Hypothesis I expressed the expectation that unilateral leadership, that is to say, the implementation of the Energiewende and an aim to find followers, is more likely to lead to consideration under the following circumstances:

Hypothesis I: Consideration of the Energiewende in recipient contexts
The Energiewende and related ideas are more likely to be considered by stakeholders in a recipient con-stituency, the more information about them is **available**, the more **capacities** there are in the recipient context to search for and process information, and the more the German example is perceived as promi-nent and **successful** with regard to its implementation in Germany. These factors can be impacted by active leadership through communication (esp. linking transferred information to local objectives / "issue-framing"), reinforcing capacities, and the creation of transfer channels by Germany.

In the three cases analyzed, the Energiewende appeared as a compelling enough example to trigger consideration. The prominence of Germany as a country and of the Energiewende as a policy model was a particularly important driver for consideration (Finnemore & Sikkink 1998: 906). Interviewees were almost unanimous to recognize Germany's leadership in the field of sustainable energy and were aware, at the very least in general terms, of the German approach to renewable energy promotion. Germany was perceived as an outstanding partner in all three cases, due to its domestic example, distinguishing it from other donors and other sources of transfer. More precisely, this prominence was linked to the degree of ambition of the Energiewende in transforming the German energy system, Germany's technological leadership, the dramatic increase in renewable energy capacity in Germany (despite a lack of sunshine, as many interview partners would specify), and sustained economic stability. Other arguments underlining Germany's prominence were related to the country's general industrial strength in areas such as the automotive industry. Negative elements, like the downturn in the German solar industry, high consumer prices, and, sometimes, perceived grid management issues, were mentioned, but did not overshadow prominence as a determinant for consideration. These elements rather played out when it comes to the utilization of the model and its impact on policy decisions. The prominence of the Energiewende as an example and of Germany as a country appeared closely intertwined across the three cases, underlining the need to take a model's source into account when analyzing transfer (Saint-Martin 2010).

The extent to which unilateral leadership alone led to a consideration of the Energiewende depended on local capacities. This finding confirms an expectation from the transfer literature (Jänicke 1997, Kern et al. 2001: 8, Tews & Jänicke 2005: 40, Weidner et al. 2002). Across the case studies, purely demand-driven transfer was strongest where capacities were strongest, namely in California. This is not surprising insofar as engaging in active research for information and screening the scattered sources of information on the Energiewende requires time, resources, and personnel or, in other words capacities. But even in California, stakeholders regretted that information on the Energiewende was not more easily accessible and asked where they could find reliable first-hand information on Germany's policy. The depth of knowledge about the Energiewende varied greatly between stakeholders who extensively used the example for their work and those who inevitably came in contact with it since it was a very present theme in policy debates. A differentiated degree of knowledge about a transferred model is in line with findings in other transfer studies (e.g., Biedenkopf

2011: 214). The determinant of availability of information or "visibility" (Heiden & Strebel 2012) for consideration thus appears relevant regardless of how extensive capacities are. Transfer studies might generally overestimate how much time and resources stakeholders in recipient contexts are ready and able to invest in order to engage in international knowledge transfer, in addition to their day-to-day activities.

In Morocco and South Africa, purely demand-driven consideration, was less likely to occur beyond a small group of well-informed specialists, some with personal ties to Germany. Resources and capacities in these countries were frequently reported as stretched to their limits. These capacities were, however, not static and Germany has decisively contributed to enhancing capacities in recipient networks, from strengthening institutions to providing training. German transfer agents also played a crucial role in translating Energiewende-related knowledge to the local context. The communicated information was often no longer labeled as "the Energiewende", but translated general principles such as "more renewables" or "decentralization" into the local context. For example, consultants in Morocco and South Africa were asked to sketch out high-renewables scenarios while taking into account assumptions and experience from Germany. Work by GIZ on cogeneration in South Africa and on decentralized electricity generation in South Africa and Morocco also drew directly from experience in Germany.

Although this was not publicly labeled as a transfer *of* the Energiewende per se, there was no doubt among recipient-side stakeholders and German transfer agents that the activities of the latter were immediately linked to either their own experience in Germany or Germany's track record in renewables. This process of sender-driven transfer of a policy model to the recipient context could be labeled "mediated transfer". In this, the distinction between unilateral leadership and active leadership fully disappears: while unilateral Energiewende leadership provides the basis and motivation for outreach to potential followers, active leadership can increase the receptiveness to the transferred model and spur consideration.

It is a truism that demand-driven transfer without mediation is more likely to take place where fewer mediators are active. In turn, where more transfer agents are present, the Energiewende model is more likely to be communicated through them. It therefore becomes impossible to assess how important recipient-driven demand for information about the transfer item would be in the absence of these transfer agents. What this also means is that the activities and agendas of individual transfer agents on the ground are absolutely decisive in shaping Energiewende leadership, especially in the absence of a consistent Energiewende outreach strategy. Transfer agents interviewed had a generally positive view on the Energiewende experience, and used it intensively as a source of evidence and lessons, including through study tours and hiring consultants from Germany. Many of these interview partners also had work experience in an Energiewende-related position in Germany prior to their jobs in partner countries. Mediated transfer remained close to the principles of the Energiewende, even though knowledge was adapted to the recipient context and framed to suit policy objectives of particular relevance to the recipient (rapid increase of capacity, job creation).

Although capacities are an important factor for consideration, in addition to the even more crucial elements of prominence and perceived success, demand-driven transfer is not automatically more likely at all times when capacities and prior knowledge are more abundant. Prior knowledge and experience can in fact act as barriers to transfer since they may constitute "mental path dependencies", i.e., strong beliefs about what constitutes appropriate and effective policies and reluctance to update those in the light of new information. This phenomenon is discussed in the theory as "cognitive shortcuts" and anchoring (Weyland 2004a), which leads to a prevalence of earlier experience over new information. A certain general reluctance to look across borders given the strong domestic experience was seen as a consequence of California's self-perception of leadership according to interviewees there. The prominence of Germany as a sender made it difficult to dismiss the Energiewende as a model, but as can be seen from the discussion of transfer outcomes, regulatory path dependencies played an important role. The tendency to rely on existing regulatory frameworks appears to be present in Germany as well, where the country's strong sense of leadership also resulted in a slow opening to input from abroad (Bundesregierung 2016b, Kopp et al. 2013).

In summary, Hypothesis I can be confirmed for the three cases, but needs to be extended with an essential additional element, mediation. Transfer agents who are related to the sender can translate a leading model into the local context, thereby making consideration far more likely. Across the cases, the availability of capacities and easy access to information about the Energiewende appeared as important drivers for consideration, but Germany's prominence and the particular nature of the Energiewende as an outstanding example were even more important. This is especially confirmed by the fact that leadership was recognized and the German experience was at least considered regardless of individuals' positions regarding transferability. Table 21 provides a summary of outcomes for Hypothesis I per case.

9.1.4 Utilization of transferred knowledge: Politics, strategy, and new agents

Beyond the mere consideration of the Energiewende and translated principles, Hypothesis II proposes conditions that render the utilization of transferred knowledge more likely:

Hypothesis II: Utilization of transferred knowledge
Transferred information is more likely to be used by members of the policy network when it is perceived as serving **objectives supported** and as **feasible** in the local context, when it can **underpin a personal or institutional position** and when it **closes epistemic gaps** that impede change. **Active leadership** can affect utilization through the entry of new agents to the network, incentives, the targeted provision of infor-mation to specific stakeholders ("empowerment"), and by increasing the spectrum of what policies appear feasible.

A first observation with regard to the intentional, and action-oriented utilization of transferred knowledge (Evans 2009b: 245, Bennett 1991) is the political nature of the process, in all three cases. Whether a piece of evidence linked to the Energiewende or reported by German transfer agents was perceived as positive and desirable in the recipient context essentially depended on how it would affect a stakeholder's position, in economic terms, but also in terms of status, opinion leadership, and beliefs about what constitutes "good" energy policy for the recipient state. This is an important finding given the overwhelmingly positive general view interviewees had of the Energiewende as it is implemented *in* Germany. A mechanical view on policy diffusion and transfer, based on fully rational learning or emulation, would suggest that a positive view of a model implemented elsewhere would almost automatically result in the transfer of such a model to the recipient context. But by distinguishing only between "normative

Table 21: Summary of Hypothesis I by case study

Case	**Evaluation of Hypothesis I**
Morocco	Energiewende considered, especially by stakeholders with personal ties to Germany (studies, prior visits); strong interest, knowledge about current developments; Mediation of unilateral leadership through transfer agents; emphasis on issues directly related to specificities of German model (decentralized generation, solar); Transferred and translated knowledge strongly considered in policy network, linked to prominence and close working relationships between transfer agents and Moroccan stakeholders; Strong presence of German transfer agents and frequent (daily) interaction reduces need for demand-driven knowledge searches; also constrained by lack of capacities.
South Africa	Energiewende more closely considered by specialists, especially industry representatives; example considered leading and prominent, but far away and not directly translatable into the South African context for most interview partners; Mediation of unilateral leadership, some of it more directly linked to the Energiewende (cogeneration, biogas; anti-nuclear by political foundations), some of it adapted (net-energy metering for decentralized generation). Early communication with South African renewable energy pioneers concerned Energiewende as implemented in Germany, in particular feed-in tariffs. Comparatively late creation of a dedicated South African – German renewables program (2011), which started operations after decision by South Africa to choose a different policy (auctions); this changes the focus of mediated transfer, which is less directly linked to Energiewende experience (Germany as a learning leader).
California	Very strong interest and direct consideration of the German example, which was the most important policy example considered in renewable energy debates; Clear recognition of leadership and success of the Energiewende in Germany, and German prominence, but limits to transferability for some actors (see hypothesis II); Even though capacities are plentiful, a lack of availability of information on the Energiewende was critiqued; Communication by German transfer agents was directly focused on the Energiewende strongest promotion of the German feed-in tariff model among all cases.

emulation [that] is driven by a desire to belong to a symbolic community" and learning or lesson drawing as a "systematic search for the most effective solution" (Torney 2015: 193), many transfer processes are likely to be excluded from transfer studies. A range of factors influenced how transferred knowledge was actually used in policy debates in the three cases, far from automatisms and clear-cut categories of either norm-driven or rational/functional responses. A second important overarching finding, which applies in particular to Morocco and South Africa, was the importance of the question of *who* uses transferred knowledge. Here, the presence of German transfer agents as full members of recipient transfer networks is a particularly relevant result to discuss.

Among determinants outlined in Hypothesis II, utilization first of all depended on the perceived congruence between features of the Energiewende and objectives deemed important by particular stakeholders or for the recipient context in general. These objectives were explored through ranking exercise and interviews. Policy drivers for renewable energy appeared most similar to Germany in California, where climate protection, but also technological and industrial leadership ranked high on the agenda. However, the particular importance of cost efficiency as an objective for California decision-makers meant that numerous interviewees judged the Energiewende inspirational and highly relevant in general, but saw a need to deviate from it in terms of policy instruments and therefore used it selectively. The German approach of feed-in tariffs and decentralized generation was seen as not cost efficient enough, and objectives that were only of secondary importance to many interviewees in California, in particular the democratization of electricity supply, did not outweigh this conflict between local priorities and the German experience. This resulted in a partially negative use of the German example, in particular by utilities and consumer advocates, but also within the CPUC. In Morocco and South Africa, energy security (and independence, in the case of Morocco) ranked highest on the agenda, and the ability to control the energy system in a top-down manner was a crucial implicit objective for state-owned utilities and decision-makers alike. This also led to an adaptation of German outreach to the local preference for policies that allow for a control of capacity additions, namely auctions. In South Africa, utilization of the German model was immediate as far as the introduction of feed-in tariffs in the REFIT program is concerned, but eventually a program more in line with the aim to control renewable energy additions was chosen.

In terms of objectives as determinants for utilization, the numerous facets of the German Energiewende model emerged as an advantage for transfer. Stakeholders were able to see job creation, industrial policy, or rapid capacity increases, as objectives that were also served by Energiewende policies in Germany. German active leadership was also adjusted to local objectives over time. Although policy promotion was first clearly targeted at feed-in tariffs, the support offered was adjusted to accommodate different policy choices, increasing the likelihood of utilization of German-sponsored knowledge in Morocco and South Africa. Given the importance of socio-economic objectives in this context, cooperation programs linking job creation, local economic development, and renewables were designed. As for the contribution to security of supply, German-sponsored studies showing how fast renewables could make a contribution to electricity capacity and that their intermittency could be managed were relevant. It is therefore important to stress, as discussed in more detail in Steinbacher (2015) that differing objectives are not an obstacle to transfer per se (Mossberger & Wolman 2003). They, however, shaped transfer outcomes, in that basic principles ("more renewables") and incremental adjustments to policies were more readily utilized than policy instruments of the Energiewende, which were seen as not well adapted to local needs by a considerable number of interviewees in all cases.

The question of feasibility – in the sense of whether transferred policies were seen as implementable in the local context – was answered differently for the three cases, and evolved in all of them. There are two main aspects regarding feasibility as a determinant for utilization. One is political feasibility, which overlaps to a certain extent with the question of policy objectives as determinants for transfer. The other one is technical feasibility, in the sense of seeing an approach similar to the German one as implementable at all in the recipient context. With regard to the first dimension, the perception of feed-in tariffs as a central element of the Energiewende that does not "fit" local needs was pronounced in Morocco, and especially in California, albeit with tremendous variance among respondents. A particular concern in California was the fit with regulatory traditions and even "culture". Some stakeholders therefore chose to "unlabel" transferred knowledge and present it in general terms or referred to other examples from Northern America – some of which also inspired by Germany – while still drawing heavily from the German experience. Also, some stakeholders were favorable in principle to a transfer of German policy instruments, but were worried that the fundamental policy change this would cause could lead to unwanted adjustments to other aspects of renewable energy policy as well.

With regard to the dimension of technical feasibility, active German leadership contributed to broadening the scope of what was considered implementable in Morocco and South Africa, but also in California. In the latter case, it was German leadership by example that pushed the boundaries of what was seen as feasible, e.g., in terms of high shares of decentralized solar electricity generation, in the minds of numerous interviewees. In Morocco and South Africa, studies developed in cooperation with local actors, the presence of advisors within institutions, training sessions for incumbents and challengers, and the aim to "objectify" the debate, expanded the realm of what share of renewable energy was considered technically feasible.

In addition to "mediated consideration", "mediated utilization" was also identified as a transfer mechanism. Evidence produced by German transfer agents or commissioned by German organizations, in particular GIZ, was utilized extensively in Morocco and South Africa. The provision of this evidence was tailored to the concerns of renewable energy opponents, and was able to fill decisive epistemic gaps regarding grid operations, the renewables potential, and socio-economic results to be expected. The 2007 energy scenarios study on how Morocco could shift toward an energy system with high shares of renewables (Roller et al. 2007) is a compelling example for this. In South Africa, a study showing how 2,800 MW of wind capacity could be interconnected to the grid in one region (DIgSILENT 2009) also reduced uncertainty about the technical feasibility of renewables, together with numerous other publications, training sessions, study tours to Germany, and site visits described in more detail in the case chapters.

Active leadership amplified utilization in South Africa and Morocco. It not only produced evidence that could be used by stakeholders from the recipient side, but also brought new agents into the local policy network, blurring the lines between utilization of transferred knowledge by local and by sender-related agents. A common observation across all cases is that Germany's active leadership efforts clearly stand out in comparison to other donors and partners. This is not only due to the legitimacy and pool of knowledge the Energiewende provides, but also to several aspects of Germany's approach that were perceived as outstanding by interviewees. In Morocco and South Africa, the fact that German advisors worked directly from within nodal points of energy policy-making and implementation, such as the Moroccan energy ministry and the South African IPP office, set its leadership efforts apart. A second particular feature, which was mentioned in the case of California as well, is the longstanding nature of exchange, dating back to the 1980s in the case of California and Morocco and the 1990s in South Africa. Cooperation on pilot projects through the PERG electrification program in Morocco and on biomass conservation and provincial wind policy in South Africa laid a fertile ground for future cooperation. Interviewees often referred to the special working relationships they had with German transfer agents from GIZ, who usually stay in the country for extended periods of time, contrary to other donors' approaches. By "blending" into the energy policy landscape, it was possible for GIZ advisors in Morocco and South Africa to feed in information and steer processes in cooperation with local pro-transfer/pro-renewables agents. In California, stakeholders transmitted information they learned through encounters with German renewable energy experts, personal research and study visits, to other members of the transfer network.

The findings clearly show that different actors, who hold different beliefs about how energy policy should look like in the future, strategically utilize transferred knowledge. The fact that different actors looked at piece of evidence related to the Energiewende in greatly different ways is unsurprising given the characteristics of the model. The Energiewende is a disruptive and redistributive policy model for the energy sector, and its full transfer would inevitably produce losers and winners. In South Africa and California, evidence from the Energiewende was also strategically used by institutions to devise proposals that resulted in temporary control over policy debates.

Although the strategic utilization of transferred information is in line with expectations on the role of knowledge in politics and "political learning" Radaelli (2009: 1157), the politics of transfer and the role of interests in transfer are rarely addressed in the literature so far (Garrett & Jansa 2015). The relevance of who benefits from transfer (Dolowitz & Marsh 1996: 355) and how transferred knowledge is used differently by different groups and individuals is therefore a promising area for further research and a main finding.

Table 22: Summary of Hypothesis II by case study

Case	Evaluation of Hypothesis II
Morocco	Knowledge produced/sponsored by German transfer agents is strategically utilized by Moroccan policy entrepreneurs to make the case for more renewables in the system; Active leadership first focused on the identification of renewable energy potential and then at the targeted mitigation of epistemic knowledge gaps to reduce uncertainty about how a high-renewables scenario could be implemented; Longstanding and close cooperation contributes to enlarging the scope of what is deemed feasible, with a particular impact on the incumbent energy utility; Objectives to be reached through renewables are perceived as different from German priorities (focus on independence, jobs and industrial policy rather than on climate and nuclear phase-out), but a shared sense of belonging to a group of leaders means these differences do not preclude transfer (but impact transfer outcomes).
South Africa	Utilization by the regulator of information about the Energiewende and feed-in tariffs leads to announcement of such tariffs in South Africa – example seen as in line with personal and institutional objective of rapidly promoting renewables after years of stalemate; Transferred knowledge used in implementation of subsequent REIPPPP program, esp. regarding socio-economic objectives and grid management/stability, which are central objectives. Leadership activities are targeted at the demonstration of technical feasibility, enlarging the scope of what is considered implementable; pro-renewables advocates use the evidence generated in their argumentation; Capacity building and training also increase the receptiveness of incumbents to transferred knowledge; facilitate the use of such knowledge in implementing the renewables program REIPPPP.
California	Strategic and purposeful use of transferred knowledge, both publicly and to devise positions without public reference to the German example; Strategic negative use of the German example by utilities and consumer representatives, along the lines of expected losses, if the German example was to be transferred; view on German example drastically changing among these groups; Perceived utilization of transferred knowledge to strengthen the position of an institution (CEC) in the policy network; Importance of active German leadership more limited, bilateral visits and study tours to Germany lead to knowledge about precise technical regulations, which was then used.

9.1.5 Policy outcomes: Path dependencies and windows of opportunity

Germany's approach to leadership by transfer ticks all theoretical boxes, from a domestic model that is perceived as legitimate and as a valuable source of positive and negative lessons, to policy promotion efforts. How these efforts led to transfer in the "classic" sense of impacting policy decisions was nevertheless a question of the decision-making structures and the positions held by powerful members of policy networks, and of when "windows of opportunity" (Kingdon 2011: 166) for policy change opened, as outlined in Hypothesis III. The Hypothesis also reflects the assumption that similar dynamics apply to policy-making with and without transfer, but that transfer can bring important ingredients to the process that might alter dynamics within the policy network.

Hypothesis III: Effects of Transfer on Policy Outcomes
The likelihood of Energiewende-related transfer impacting policy outcomes (in the sense of leading to a specific policy design or policy change) depends on the relative **power of stakeholders in the policy network** and their **respective position with regard to transferred policies**. Whether transferred knowledge affects policy outcomes also depends on the **presence of windows of opportunity** that ask for a politi-cal/regulatory action. Active leadership can impact these factors by attaching incentives to policy choices (mitigating losses, enhancing gains for particular groups or agents), and contributing to the recognition of windows of opportunity.

A first common finding across the three cases is that incumbent energy utilities were most opposed to the transfer of experience from the Energiewende, both in terms of concrete policy instruments and the general idea of a high-renewables energy system as embodied by the Energiewende. Their opposition to the transfer of knowledge was mitigated in Morocco through political support from King Mohamed VI for renewables. The opposition by incumbent ONEE and local electricity distributors played out more effectively regarding the opening of the low-voltage level to small-scale electricity generation facilities. ONEE in particular emphasized the importance of a controllable increase of renewables and of dispatchable capacities, hence also preferring CSP to PV initially. Increasing demand from Moroccan citizens and businesses willing to benefit from lower prices on solar modules has led to pressure for change that started to materialize around the time of field research and translated into a law for small-scale renewables generation at the end of 2015. Opposition to feed-in tariffs was widespread, out of concern for rising electricity prices, the perceived need to control capacity additions to the grid, and differences between Morocc and Germany. In particular, Germany's own solar industry and consumers' willingness and ability to pay top-ups on electricity prices were seen as reasons for the country to rely on feed-in tariffs, as opposed to what was seen as sensible in the Moroccan context. German transfer agents adjusted to this preference, supporting the Moroccan renewable energy strategy and the implementation of its renewable energy tender programs.

Pressure for change is also a main reason South African policy-makers are currently looking into ways to open the electricity generation sector to households and businesses in a way that does not threaten the income of municipalities, but opposition from utilities remains high. At the national political level, the turf war between the regulator and ministries led to the non-transfer of a feed-in tariff policy inspired by the German model, but, eventually, prompted renewable energy policy action nevertheless. The success of the current renewables program, the REIPPPP is eroding opposition to renewables further: prices for renewable energy capacity have fallen bidding round after bidding round and much-needed capacity has come online on schedule – a stark contrast to the cost overruns and delays in the construction of new coal-fired power plants. Germany supports the implementation of the REIPPPP, especially with regard to the integration of renewables to the grid and concerning the very important socio-economic dimension of the program.

In California, resistance to the German feed-in tariff model from utilities, solar leasing companies, consumer groups, and most importantly, the institution that is in charge of designing such regulations (CPUC) led to the non-adoption of this policy instrument in a large-scale program at the state level at the end of the 2000s. A particularly interesting development in California is how traditional utilities, consumers, and "new utilities", that is to say, companies who own, install and lease solar rooftops, joined forces against the German model of feed-in tariffs. Challengers to the status quo heavily relied on the Energiewende example, but could not overcome path dependencies, including a perception that feed-in tariffs were incompatible with the political and general culture in California. The pre-existence of a regulatory framework for renewables – and hence of regulatory path dependencies – at the time when the German example was considered is a point that differentiates California from the other cases studied. The picture is again more nuanced when different policy layers are considered, as proposed by Knill et al. (2010: 417) for the evaluation of transfer outcomes. The German model produced an impact on the level of technical regulations as well as in terms of validating the move towards an energy system with high shares of renewables. The finding that transfer outcomes depend on the policy layer concerned is consistent across the three cases and an important insight for transfer studies.

In all three cases, the full impact of the presence of transferred knowledge unfolded when windows of opportunity opened. Knowledge that often lay dormant was activated and utilized when external drivers prompted action, suggesting that a "transfer stream" in a "multiple streams model" (Kingdon 1984) would be an appropriate conceptualization of the process. Such a transfer stream could be seen as a part of the policy stream, but given how political transfer appeared across the cases, it might well be connected to both the policy and the politics streams.

In the case of Morocco, mounting pressure due to increasing demand, insufficient installed electricity generation capacity and rising fossil fuel prices in 2008 made policy decisions necessary. A national energy strategy that grants an important place to renewable energy was announced the same year, and ambitions increased further with the announcement of the Moroccan Solar Plan in 2009. Moroccan pro-renewables advocates strategically used this window of opportunity to join forces with German transfer agents and introduced relevant information on energy scenario options into the process. In South Africa, years of stalemate in renewable energy policy, combined with increasing electricity security of supply issues and the prospect of hosting the COP17 conference created a similar context between 2008 and 2011. In particular, the unilateral move by the South African regulator to announce feed-in tariffs prompted further policy action. The resulting REIPPPP, although limited in size, can be considered a successful first step into the direction of an energy transition in South Africa. In California, the initial lack of satisfying results from the RPS opened an opportunity for evidence about the Energiewende to enter the debate in the mid-2000s.

Although windows of opportunity are important, they need not necessarily be unforeseeable events or crises. The scheduled updating of California's net metering policy is one such example of a scheduled reform in which transferred knowledge was used. More generally, windows of opportunity can also be linked to global trends, which in turn can be triggered and accelerated by pioneering policies and diffusion. The global move towards more sustainable, climate-friendly energy systems was part of the background on which policy decisions in all three cases were taken, but did not appear as their root cause.

In summary, transfer outcomes in all three cases need to be analyzed taking the policy layer into account. Transfer outcomes at the intermediary level of renewable energy promotion instruments were constrained by the highly redistributive and system-changing nature of the German Energiewende, which was at odds with regulatory traditions, policy-makers' preferences and powerful groups in all three cases.

Table 23: Summary of Hypothesis III by case study

Case	Evaluation of Hypothesis III
Morocco	Strong, recognized impact of transferred knowledge regarding renewables in Moroccan energy strategy and its translation into wind and solar programs; Involvement of German transfer agents in the policy-making process helped shift balances in favor of a pro-renewables approach; Targeted empowerment of policy entrepreneurs within the network through training, evidence; In particular, strong impact on technology choice (CSP), in this particular instance combined with monetary incentives and the activities of other donors; No transfer of feed-in tariffs / policy instruments due to perceived lack of fit with national structures and priorities, level of economic development, as well as opposition from incumbents.
South Africa	Strong impact of the German example and of measures taken by German transfer agents (studies, tours), in collaboration with local pro-renewables advocates, leading to the proposal of feed-in tariffs; Reflection of transferred knowledge in eventual renewables program limited due to dominant forces in policy network adopting different positions (but not related to a negative view on German example); Influence on implementation and adjustments to REIPPPP (introduction of small-scale component, monitoring) to ensure stabilization of the move toward renewable energy. Transferred knowledge contributed to agenda-setting for biogas and co-generation, translating into programs that are in turn supported by active leadership efforts (institution-building, capacity-building); Successful attenuation of opposition to renewables from incumbent, through advice from within utility and studies developed in cooperation.
California	Transfer outcomes strongly dependent on policy layer concerned; Opposition to transferred policy (feed-in tariffs) by CPUC, who holds decision-making competence and by incumbents because of a perceived lack of regulatory fit and disruptiveness of proposed policy -> feed-in tariff example strongly modified and adapted to local context at the state level; Transfer with direct outcomes in triggering policy action and design at the technical level (permitting, smart meters); Meta-level of transfer (shared understanding of leadership) creates additional legitimacy for ambitious renewable energy targets.

9.1.6 *Effective leadership: A question of objectives*

Evaluating the effectiveness of German leadership through transfer eventually depends on the objectives Germany set out to reach in each case. This was a less straightforward question to answer than expected. The identification of Germany's leadership agenda is complicated by the absence of an explicit international Energiewende strategy and by the presence of players with different preferences in German outreach activities. Nevertheless, common patterns of leadership and transfer clearly emerge from the case studies.

Across the cases, Germany's efforts were seen as primarily targeted at contributing to global climate protection and were generally perceived as being carried out in a partnership perspective. But in Morocco and South Africa, there also was uncertainty with regard to Germany's agenda and some disbelief that it could really be altruistic (especially in light of the experience with other donors). Reasons linked to German economic interests were of secondary importance, in German outreach as well as in the perception of recipient country stakeholders and for companies interviewed.

In terms of the content of communicated policies, the relatively clear focus on feed-in tariffs that dominated German outreach until the end of the 2000s, and played out in all three cases, has shifted. Pushback from important groups in recipient countries, the multiplication of country experiences with alternative instruments, and the evolution of policy instruments in the Energiewende are reasons for Germany's evolving leadership. Interview partners involved in the definition of external Energiewende measures and in partner countries (D1, D7, D11, M1 M13, S45) confirmed the shift away from an approach focused primarily on classic feed-in tariffs. Regardless of the underlying driver, the general objective of facilitating renewable energy uptake seems best suited as a basis for the assessment of the effectiveness of German leadership through transfer.

In each of the cases, several examples can be found to illustrate how the Energiewende as an example or German outreach spurred renewable energy policy. In Morocco, interviewees agreed that the new energy strategy with the role it grants to renewables, and in particular the Moroccan Solar Plan, were decisively impacted by evidence provided by German transfer agents and their cooperation with Moroccan stakeholders. Maintaining the issue of decentralized renewable energy generation on the agenda is another indication of effective leadership by transfer. In the case of South Africa, the regulator's decision to adopt feed-in tariffs is directly linked to lesson drawing from the German experience. German advisors and consultants now facilitate the REIPPPP's implementation, thereby reducing opposition from the incumbent utility Eskom and increasing the likelihood of continuing success of the program. Issues such as biogas and cogeneration were pushed to the agenda by German advisors, and consultants facilitated the development of a net-energy metering position by municipalities.

In California, transfer outcomes differed, depending on the addressees of transferred policies and the issue at stake. Negative lesson drawing and a strong perception of a lack of regulatory fit meant that German-style feed-in tariffs were refused by the CPUC and transfer did not translate into a feed-in tariff program of comparable scope and philosophy. Regulations requiring specific technical equipment for solar installations (smart inverters) and the simplification of permitting procedures were examples of direct transfer from Germany. A strong role for the Energiewende as a motivator, source of positive competition and of Germany as a "partner in leadership" could be observed at a meta-level of transfer.

The value of leadership, and German Energiewende leadership in particular, was very much recognized across the cases and independently of stakeholders' view on the transferability of particular elements of the Energiewende or of the entire model. Since the promotion of renewables can clearly be identified as the overarching goal of German leadership, the effectiveness of leadership by transfer in all three cases can be confirmed despite selective transfer.

9.1.7 How much "Energiewende" is there in transfer?

One of the analytical challenges of tracing transfer was deciding to what extent observed transfer is "Energiewende transfer". Since Morocco, South Africa, and California were deliberately selected as cases where no effective implementation of the German model of feed-in tariffs had taken place, a full "export" of the Energiewende was not expected. But comprehensive copy pasting is also unlikely in other cases, given the complexity, historical anchoring, and contextual factors shaping the Energiewende. So the question remains of where to draw a line between *Energiewende* transfer and other sources of transfer or even "normal policy-making". In other words, how does one identify Energiewende transfer when one sees it?

This book set out with a dialectic approach to German leadership, with the model (unilateral leadership) on the one hand and outreach to recipients (active leadership) on the other hand. This distinction was useful for analytical purposes, but a clear line between these two variants of leadership cannot be drawn in practice. Only when the two dimensions are considered together can Energiewende leadership and its effectiveness be understood.

The potential of the Energiewende example *alone* in *triggering* policy action without any interference from active leadership was limited to particular instances such as requiring smart inverters in California. Even in the example of the California Energy Commission (CEC) bringing German-style feed-in tariffs into the debate, the trigger for action was still a perceived policy failure of existing regulations in California rather than *only* the German Energiewende example. Indirectly, unilateral leadership can trigger action by contributing to the emergence of "global trends" and of critical masses of countries moving in one general policy direction, e.g., toward decarbonization or renewables (see Finnemore 1993 on norm "band-wagoning"). Without leaders and the multi-faceted contribution they can make to transfer and diffusion, such global trends are unthinkable. Although its impact in the global trend toward renewables is impossible to isolate, Germany has certainly facilitated it through its contribution to decreasing technology prices, the creation of numerous dedicated fora for policy dialogue, and its role in partner countries like those considered in this book.

A main finding across the cases was that different variants of leadership through transfer are closely linked and that active outreach could not be imagined without unilateral leadership and the Energiewende as a basis. Firstly, much of the financial means available for active leadership (GIZ and KfW programs, but also funding for the Transatlantic Climate Bridge or special climate funds made available for the work of German political foundations) are a consequence of Germany's own ambition in climate and energy policy and of the international aim that has accompanied the implementation of the Energiewende from the start. Secondly, the Energiewende provides a unique source of legitimacy to German transfer agents, which clearly distinguishes them from other actors. The recognition of German leadership, political will, and the particular nature of its Energiewende, were not put into question by criticism concerning specific aspects of the project. Independently of whether elements were directly transferred from the Energiewende to the recipient context, this legitimacy allowed German transfer agents to speak with more authority and credibility than other external agents, and led to an assumption of expertise in the field that set German activities apart. Thirdly, German transfer agents abroad generally drew from personal experience in the Energiewende in Germany. Agents interviewed agreed that the German example was a "natural" choice to draw from, both because of its salience and "objective" pertinence (e.g., effectiveness in increasing the share of renewables, long-term targets) and because of the sheer amount of information, industry connections and expertise available as a consequence of the Energiewende. With an increase in the number of available alternative models of renewable energy policy – including from countries where Germany is active and has contributed to their design – the pool of expertise increases and the role of the Energiewende as the reference model for active leadership might become diluted. This is all the more likely as Germany's own policy model is changing and, despite its pioneering status, has become part of a more global transition effort.

Even though knowledge communicated by German transfer agents for "mediated transfer" often concerned general policy orientations for more renewables without direct reference to Germany, it should be stressed that these transfer agents operate abroad precisely because of the Energiewende and because of Germany's aim of inspiring policies which, if not mimicked, at least go in the same direction as its own policy model.

9.2 Theory: Developing Transfer Research Further

9.2.1 What about classic determinants for transfer?

As can be seen from the discussion of findings above, strict boundaries between "classic" mechanisms of transfer, such as learning, lesson-drawing, emulation, competition-driven or conditionality-driven transfer cannot be distinguished in a clear-cut manner based on the results in this book. For example, Germany's prominence as a sender – a commonly assumed driver for emulation – was outstanding, but did not lead to a mimicking of its policy model. Likewise, learning – a supposedly purely demand-driven transfer process – was in fact facilitated by active leadership efforts, which impacted how stakeholders updated their knowledge. From the findings, it also appears impossible to reduce transfer-related individual decisions to one specific logic of action, as in recipients who would only act based on a "logic of appropriateness" and favor transfer for norm-related reasons or stakeholders who would be exclusively driven by a desire to improve policies or to maximize personal gains. The book' results therefore suggest that the transfer mechanisms Dolowitz and Marsh (2000) present as a widely used continuum should be employed as a guideline rather than as a checklist for transfer. Calls by authors such as Dobbin et al. (2007) for transfer researchers to systematically test one transfer mechanism against another, e.g., competition versus demand-driven learning, are thus very unlikely to translate into research that can capture the complexity, not only of transfer, but of political decision-making and human action in general. The strong presence of mechanisms-based thinking in transfer and diffusion research (Heinze 2011) might well be due to the common treatment of transfer and diffusion as phenomena that are different from "normal" policy-making and which therefore obey by different, more automatic rules. As the political use of Energiewende-related transferred knowledge shows, transfer is integrated as an asset and as an instrument into the policy-making process. This can be true regardless of whether "only" ideas are at stake or whether policies come with incentives provided by senders.

Classic mechanisms of transfer appeared useful in structuring thinking in this book, but the complexity of transfer as part of political decision-making makes processes that obey by one particular logic of action unlikely. The development of crosscutting categories of transfer thus appears necessary.

9.2.2 Mediated learning and lesson drawing

In addition to the categorization of transfer mechanisms according to their underlying logic of action, Dolowitz and Marsh's (2000) transfer continuum ranges from purely demand-driven learning to fully sender-driven transfer through conditionality and coercion. Transfer processes studied in Morocco and South Africa in particular make a strong case for the overcoming of the dichotomy between demand-driven and sender-driven transfer, or between "push" and "pull"-driven transfer (Majone 1991: 103). The updating of knowledge that is decisively facilitated by the sender could be described as "mediated learning" and was particularly important in Morocco and South Africa. What this means concretely is that senders, by tailoring information about preferred policies to the needs of local stakeholders, both in terms of content and in terms of strategy, can facilitate learning, especially when resources are scarce. This does not mean that information is forced upon recipients, but that it is framed by the sender and made more accessible. Sender-related transfer agents in Morocco and South Africa engaged with policy networks to understand needs for information and evidence and produced knowledge in cooperation with receivers. Information about the Energiewende in particular and sustainable energy/climate protection in general was "translated" by sending agents to the local context, where it was then received, treated and used. The absence of this type of sender-facilitated learning in classifications of transfer processes is likely linked to the lack of transfer studies involving developing countries and/or active senders. But the phenomenon was not only limited to Morocco and South Africa. In California as well, German transfer agents communicated information, and learning was often not through first-hand knowledge derived directly from primary sources and data stakeholders researched themselves.

Mediated learning and lesson drawing are not identical to learning without intermediaries because information about the sender's model goes through an additional filter. This can lead to an underestimation of actual transfer when only policy outcomes are considered: transferred knowledge can be modified in several steps from the original policy model to the sending agent, and then to communication to a recipient stakeholder, and to their use of information. Deciding up until what point transfer is "transfer" and when information about a model becomes too diluted in general knowledge is all the more relevant when mediated learning occurs.

9.2.3 Empowerment as a transfer mechanism

Closely linked to "mediated learning", another non-classic transfer mechanism – "empowerment" – was identified in the cases studies. Empowerment refers to the fact that information linked to the Energiewende and/or provided by transfer agents improved the negotiating position and argument base of *particular* stakeholders in the recipient policy network. As a transfer mechanism, empowerment is strictly incompatible with a view of recipients as monoliths. Different members of the policy network may see different interest in using elements provided by German transfer agents or the Energiewende (Sabatier & Weible 2007: 194). As a transfer mechanism, empowerment is closely linked to the politics of transfer and the instrumentalization of transferred knowledge discussed below. It is also similar to the "boomerang pattern" identified by Keck and Sikkink (1999), where members of local advocacy networks connect with transnational networks to reinforce their claims and increase the resonance and legitimacy of their position. This mechanism is expected to be particularly relevant in political systems that are closed to outside influence and challengers, but empowerment was observed in all cases and was also very strong with regard to environmental advocates in California.

Empowerment can be used strategically by senders, who can equip key stakeholders with information tailored to their needs in policy discussions. It can also occur without the sender's active involvement, when, for example, certain features of the policy model are more in line with a stakeholder's negotiating position and can therefore be used strategically. Sender-driven empowerment was observed in particular in Morocco and South Africa. Pro-renewables stakeholders there were equipped with information that closed decisive epistemic gaps and improved the arguments these stakeholders wanted to make. One case in point is the study on the job creation potential of CSP in South Africa, which an advocate had asked GIZ to carry out, and which could then be used to support claims in policy debates with the additional legitimacy that German support provided (GIZ et al. 2013). Although the active and passive empowerment of pro-transfer stakeholders is an important mechanism observed, the effectiveness of leadership through transfer also depends on whether the concerns of opponents/incumbents can be mitigated, as discussed below. It needs to be underlined again that in the three cases, a full implementation of transferred knowledge would have led to a fundamental transformation not only of the energy system, but also the ownership structure of electricity assets and potentially the institutional landscape. Resistance to transfer hence equates resistance to policy change that was perceived as threatening. Taking the qualities of transfer items (as is already done in much of transfer research) and the policy field into account hence appears necessary.

9.2.4 Conditionality, emulation, and competition: Overestimated mechanisms?

One of the most unexpected findings for Morocco and South Africa was that conditionality did virtually not appear to play a role as driver for policy transfer. Although Germany contributed decisively to financing energy projects in both countries, conditionality cannot be considered a dominant mechanism for the transfer of policies based on interview results. The case of the Sere wind farm and the Upington CSP project in South Africa, which were required by KfW and other donors in exchange for support to two large coal-fired power plants, fall in the category of conditionality, but did not lead to the transfer of policies. In Morocco, the availability of funding for CSP as a technology might have facilitated the decision to engage in this technological path within the Moroccan energy strategy, but interviewees almost unanimously refused the idea that the availability of funds was causal in the decision to increase the role of renewables in the country's energy mix. The question then remains why conditionality was arguably not an important factor in either case. Two main lines of argumentation can be brought forward and might be considered in future applications. One is Germany's approach as a sender and the other are the characteristics of energy as a policy field.

Even though Germany's agenda and motivation to engage in sustainable energy cooperation were subject to questioning, most interviewees saw Germany as an "honest broker" and partner-oriented sender, an approach that makes conditionality less likely. Contrary to an early expectation, German companies did not play a visible role in policy transfer in any of the cases. Interviewees from German companies stressed that a stable and growing market and clarity about the direction regulation would take were far more important than policy instruments that are similar to the ones used in their home market Germany. Also, capacities to engage in lobbying in export markets were seen as very limited within these companies, independently of their size.

Other partners, in particular France in the case of Morocco, were perceived as far more inclined to attach conditions to the provision of aid. Germany's outreach is however conditional insofar as it only supports non-nuclear energy and has drastically increased the requirements for any support to coal projects (Bundesregierung 2015). Even prior to that, Germany's strategy was very focused on renewables and efficiency. The second potential explanation for the relative unimportance of conditionality is inherent in the policy field studied. Major global donors pursue different energy strategies domestically, which is reflected in their cooperation efforts. This results in a variety of choices in terms of cooperation partners available to recipients, reducing the likelihood of renewables policies being adopted out of hard conditionality. For transfer theory, this means that the relevance of conditionality as a hypothesis is likely to vary from sender to sender and depends on the policy area studied. Nevertheless, the likely reluctance of interviewees to acknowledge the extent to which their decisions were influenced by conditionality has to be taken into account, and the views of respondents outside institutions (NGOs, civil society) should therefore be considered as well for hints at conditionality.

With a view to their comparable degree of ambition in climate protection, technological leadership and global orientation, it was expected that regulatory competition could play a role for transfer between California and Germany. However, the instances of competition observed were purely reputational in nature. Advocates in California pointed to Germany as an example with a view to appeal to California's sense of leadership and spur action. The rationale that California had to improve its policies or transfer policies from Germany with a view to assert economic stakes in the global renewable energy industry was not confirmed in interviews.

The likelihood of pure emulation, i.e., the full imitation of a policy based on a positive impression of the country, was already viewed critically before starting empirical research. The selection of negative cases regarding a central dimension of the sender's policy also made it highly unlikely to identify instances of strictly symbolic imitation or emulation. These mechanisms did not appear in either of the cases, despite a positive view of Germany as a country and the Energiewende as a model. A particular expectation formulated by Shipan and Volden (2008) with regard to emulation can in any case be strongly put into question. The authors expect different mechanisms to be at play depending on whether a country or a policy is what triggers the interest of stakeholders in the recipient context: "In learning, policymakers focus on the policy itself—how was it adopted, was it effective, what were its political consequences? In contrast, imitation involves a focus on the other government—what did that government do and how can we appear to be the same?" (Shipan & Volden 2008: 842). Case study results very strongly suggest that Germany as a senders and their models are closely intertwined and that the source of a policy matters (Saint-Martin 2010). The consideration of "only" a policy, independently of where it comes from, is therefore unlikely and a hypothesis to be critically assessed in future studies.

9.2.5 A strong case for bringing the sender in

One of the central lessons learned from results in this book is the relevance of bringing the sender in to transfer analysis. The dominant approach to transfer studies has been to focus on transfer from the moment transferred knowledge enters the recipient case, with little regard to whether a model's origin is of relevance (Biedenkopf 2012, Saint-Martin 2010) and whether senders actively influence transfer. The role of pioneering countries as active senders has received particularly little attention, as opposed to their well-studied role in international and European negotiations (Saul & Seidel 2011, Schreurs & Tiberghien 2007, Wurzel & Connelly 2011) and the determinants that make countries frontrunners (Jänicke 2005). Senders who have not *implemented* a policy model themselves dominate the landscape in transfer studies. These include transnational networks and their secretariats, international institutions, epistemic communities, NGOs, and the EU. While these actors are undoubtedly relevant, pioneering countries as senders are different. Policy instruments, as becomes clear from a historical institutionalist view on transfer, come with their particular history and in that history, countries, which pioneered a certain instrument, play an important role. How pioneers frame their own policies, whether they put them in a more global context and want to attract followers and whether they back up this aim with resources and a presence in recipient states affects the prospects and outcomes of transfer, as the case studies have shown. A potential explanation of why pioneers as senders have been so largely absent from transfer studies is that many policies do not have an immediate global implication and senders might be relatively indifferent as to whether other countries adopt similar policies. The situation is of course very different with policies that yield global externalities such as those contributing to climate protection. Whether pioneers' motivation to "export" (or not) domestic policies generally affects the likelihood of transfer, and in which policy fields active senders are more likely to operate, are interesting questions for further research.

A classic expectation from the transfer and diffusion literature is for ideas to be more likely to travel when sender and recipient are similar. This hypothesis could well be due to transfer studies often focusing on similar states. Results presented suggest that a reconceptualization of what "similarity" between sender and recipient means is necessary. In the policy field studied here, renewable energy, the dichotomy between "developing" and "developed" countries is much less pronounced than it possibly would be in other fields. Almost all countries are "developing" when it comes to energy transitions and the challenges of pursuing a shift to renewables might in fact be greater where they replace existing fossil-fuel capacities. Also, important actors such as incumbent utilities exist in developed and developing countries and are more likely than not to oppose systemic change. Rather than looking at GDP or language, another type of similarity between sender and receiver was found to facilitate transfer: the sense of belonging to a group of leaders and of being at the forefront of a global movement toward renewable energy (Steinbacher 2015). This finding applies in particular to California and Morocco, with the latter being a surprising case of similarity in its leadership aim. In South Africa, similarities with Germany such as the desire to create jobs through green technologies were mentioned despite a number of identified differences. This points to a need for a more diverse set of characteristics to be taken into account for the fit between senders and receivers as a determinant for transfer.

Finally, the role of similar policy objectives for transfer needs to be assessed on a case-by-case basis. Rather than impeding transfer, differing policy objectives might lead to selective transfer and strong adaptation of a foreign model to the local context (Mossberger & Wolman 2003, Steinbacher 2015). The multi-faceted nature of the Energiewende, which is seen as pursuing several objectives at once overall facilitated transfer since it appealed to a broader range of stakeholders for different reasons.

9.2.6 Politics of transfer: Harnessing ideas

Across the cases, it has become clear that transfer is a highly political endeavor, in three main regards. First, stakeholders look at the same piece of information in dramatically different ways, depending on whether the information supports or challenges prior held beliefs and interests. Also, whether transferred policies are likely to lead to gains or losses for their personal position or groups they represent is an essential determinant for transfer. There is no such thing as "a" transferred model since each instrument and piece of evidence can be interpreted in different ways, and "instrument choice can become the choice of a political goal itself" (Steuwer 2013: 74). This is true both across different recipient cases and within single cases of transfer. For example, whereas some California stakeholders found the decentralization of electricity generation achieved in the Energiewende to be a main advantage, utilities saw it as the major disadvantage. This means that in-depth case studies are needed to assess how a policy model is received and who uses which elements, how, and to what end. Literature on the political use of knowledge (e.g., Daviter 2015, Robertson 1991) and on the politics of policy instrument selection (Hood 2007: 136, Lascoumes & Le Galès 2007) can provide useful insights into these questions.

The Energiewende was strategically used by different stakeholders to advance their own agenda, as a source of negative and of positive lessons. This "instrumentalization" or harnessing of transferred knowledge led stakeholders to invoke Germany as an example of how things had to be done, or as a warning of what was to happen if certain policy elements were transferred. A strategic use of information was also observed with regard to the aim of advancing one's role and importance in the policy network. By bringing in new information from abroad, stakeholders in California, Morocco and South Africa were able to draw attention to themselves and their causes, independently of whether their positions were eventually reflected in policy decisions. Senders can enable such instrumentalization by supporting certain stakeholders with information or capacities to advance policy transfer ("empowerment", see above).

A highly interesting variant of instrumentalization occurred both in California and South Africa, where representatives of one institution (CEC and NERSA) used information about the German experience to bring their institutions to the forefront in turf wars. In California, the CEC was seen as using the issue of feed-in tariffs to steer a discussion on the track record of the RPS and the future design of renewable energy policies in the state. The CEC's policy recommendations in the Integrated Energy Policy Report to the CPUC (CEC 2007) are a clear sign of using foreign evidence to push an institutional agenda. In the case of South Africa, as mentioned above, evidence from the German feed-in tariff was used in the design of a similar policy at the national level. This program, the REFIT, was unilaterally announced by NERSA, whose leadership had been exasperated by the lack of policy action on renewables. The announcement created an urgent need to act for other departments, the DoE and the National Treasury, to regain control of energy policy-making. These findings strongly support the argument that "conflict over the distribution of policy competences can be a catalyst of knowledge use and affect the policy process in fundamental ways", a notion that has, however, largely "remained uncharted territory" so far (Daviter 2015: 495).

9.2.7 Internalized transfer: Dynamic filters

A common criticism of the transfer literature is that it does not make its case clear as to how it is different from "classic" policy-making. Drawing such a distinction might in fact be unnecessary and even misleading. The line between "internal" and "external" influence factors on policy decisions indeed becomes blurred when transfer is used strategically. If members of the policy network integrate transferred knowledge into their strategies, transfer becomes an integral part of policy-making. Decision-making processes that *exclusively* concern information about foreign models and where such knowledge is not mixed with prior experience or domestic considerations are difficult to imagine. Another argument in favor of looking at transfer as a potential ingredient to policy-making rather than a process of a different nature is the fact that active senders and even transferred policies can change the environment for decisions in the recipient case. For example, if senders attach incentives to certain policies but not others, the relative attractiveness of different options changes. Also, where senders enhance capacities or even contribute to the creation of institutions, the realm of feasible options and the structure of policy networks can be affected. It then becomes impossible to completely distinguish between "exogenous" and "endogenous" determinants for decision-making. But while an integration of transfer into local policy making occurred, this does not mean that the source of knowledge can be ignored; the origin of knowledge, and support by the sender were highly relevant in the three cases studied.

Transfer can lead to change at many levels: policy networks may be impacted in terms of their structure, membership and the knowledge and information that are available; individual agents may harness or feel threatened by knowledge from abroad; and new actors as transfer agents may enter the stage. Changes in the sender's model and in recipients' domestic politics and structures impact transfer as well, underlining how important the temporal dimension of transfer is (Dussauge-Laguna 2012). This also means that understanding the impact of the time lag between the implementation of a policy by a sender and an element's transfer is a highly relevant point to consider.

9.2.8 *Transfer dependent on policy layers*

The discussion of findings with regard to Hypothesis III shows an interesting phenomenon across the cases. While transfer met fierce resistance at the level of specific policy instruments – feed-in tariffs – it was very much possible at the level of other "policy layers". Policy elements transfer more or less easily depending on their characteristics, and there can also be differentiation among transfer items belonging to the same model. Knill et al. (2010: 417) propose a differentiation among three dimensions, akin to Hall's orders of change (Hall 1993) and the classification used by Howlett (2009) for different "policy orders". Although the categories are used by Knill et al. (2010) to assess convergence, they can be applied to more general transfer studies with complex transfer outcomes as well. The authors suggest a first layer that concerns the mere existence of any policy in a given area. Transfer could then impact fundamental policy change, contributing to the design of a first program in a policy field, as was the case in Morocco and South Africa. The second layer concerns policy instruments. In this category, resistance was reflected in transfer outcomes, although knowledge had been considered and used, pointing to the existence of structural filters between consideration and utilization, as well as between utilization and adoption. The third layer concerns the calibration of policy instruments, for example the level of a standard or the adjustment of a procedure. Concrete lesson drawing and transfer occurred at this level in all cases. Following the results presented in this book, it appears worthwhile to add a fourth policy layer as a meta-level of transfer that is similar to the diffusion of norms. Knowledge about the "trailblazing" efforts of another country validated the move into a similar direction, in particular in California, but also in Morocco.

The findings call for more qualitative studies of apparent non-transfer to uncover where in the process transferred knowledge hits filters, is harnessed negatively or conflicts with path dependencies and obstacles too big to overcome, as has been aimed for in this book. For senders, it is important to understand these obstacles. In the cases analyzed in this book, alternative instruments were adopted that are also conducive to Germany's goal of increasing the share of renewables and interviewees appreciated the relevance of negative lessons to devise stronger regulatory frameworks, but this can not always be expected to be the case in selective transfer.

9.3 Methods Discussion: How to Observe Transfer

9.3.1 In-depth case studies, process tracing, and interviews

In its very early stages, a research design for this book based on a large-n survey coupled with exploratory interviews and a stronger emphasis on document analysis was considered to assess the spread of regulatory instruments linked to the Energiewende. Apart from the fact that elements of the Energiewende are challenging to trace when they do not directly result in similar policy output, even observed convergence would not have permitted insights into causal mechanisms in the same way as in-depth case studies have. Even more important than the choice of a small-n research design was the focus on semi-structured interviews for data collection. Much of the information gathered in personal conversations would have been impossible to collect through public sources. Many of the processes and debates that were relevant for transfer happened behind closed doors and in small meetings, but other aspects of transfer also made it largely invisible from public records. Even in the case of California, where minutes from meetings at the public utilities commission are published, transfer was not consistently mentioned publicly and several examples of transfer agents "disguising" information about German experiences in other examples such as Gainesville or Ontario were reported. Interview partners were not uncomfortable with speaking about transfer in interviews and were happy to explain how they used information about Germany, but some did not see an added value of pointing to their use of the example publicly. The reason mentioned for this was mostly that decision-makers had to be convinced by using examples they could more easily relate to, i.e., from North America. More general hypotheses on why transfer was often dealt with outside the public eye could be a perceived risk of being seen as lacking own ideas and following a distant country out of a deficit in creativity. Also, the public or voters might not immediately see the benefit of learning from an example like the Energiewende that, at first sight, appears rather unique and challenging to transfer to the domestic context.

Some interviewees in California and South Africa were concerned about being seen as "taking direction" from another country. In California, this was due to a sense of leadership based on the state's own model, while in South Africa the colonial past and a view of South Africa being in a unique geographic and historic position were arguments mentioned as to why transfer was not systematically discussed outside the policy network. This does not apply to all interviewees, but was a finding that confirmed the need for interviewing and granting interviewees anonymity.

Another methodological decision taken was to carry out interviews in a semi-structured manner, where general topics to be addressed were determined before field research and questionnaires were then tailored to each interviewee's role, position, experience, and involvement in particular policy processes and transfer activities. In retrospect, this choice had advantages and disadvantages. On the negative side, the lack of standardization in questions made it impossible to rigorously report *quantitative* interview outcomes in percentages or number of interviewees who stated a certain point or agreed to a certain view. Interviewees referred to abstract concepts (leadership, learning etc.) in different ways and each statement also needs to be understood in the context of interview partners' responsibilities, institutional affiliation, and more or less direct involvement in policy processes and cooperation with Germany. Given the limited availability of secondary and primary literature, not only on transfer but also regarding renewable energy policy-making in all three cases, and particularly in Morocco and South Africa, semi-structured interviews could yield the depth and width of information required to trace policy processes (Collier 2011). Since it was unclear at the start of research what transfer processes would be explained and what events could serve to structure the cases, open questions were required. Semi-structured interviews offered the opportunity to integrate what was available in prior information from different sources and earlier interviews and add new and sometimes diametrically opposed views. Process tracing is a methodological framework of choice in many studies of transfer, but a mix of classic approaches to studying transfer and methods more related to diffusion or other political science fields appears promising for future research, as outlined below (Lee & Strang 2006: 886).

9.3.2 Alternative approaches

The idea of conceptualizing the policy-making environment in recipient cases in the form of policy networks emerged in the course of field research. Using more formalized social network analysis for transfer studies is a promising avenue for further research, as shown by Garrett and Jansa (2015) in their network-analysis study of the role of interest groups in diffusion. In this book, the connections between different transfer agents and members of the policy network were assessed qualitatively and yielded interesting insights into who transferred information to whom and which transfer channels were predominantly used. But by gathering standardized data on the number and intensity of interactions between members of the policy network and transfer agents, the weight and nodes of relations between members of the network could be represented in a formalized social network analysis (Borgatti et al. 2009). Given the challenges of data collection on unrecorded, sometimes also coincidental meetings, this type of social network analysis would need to be integrated with participant observation and is more likely to be conclusive in cases where comprehensive data is publicly available, and as a complement to case studies.

Another promising type of method to employ in transfer studies is media analysis. Since the public opinion on a given foreign policy model, a type of instrument, or a particular sender is likely to influence the attractiveness of transferred data in the eyes of decision-makers, media analysis could be used to trace the presence of references to such models in general media in recipient states. Again, this approach is more likely to be beneficial in cases with relatively open political systems. Media analysis would also need to be combined with other methods because it will necessarily yield fewer results for very technical regulations, in early stages of transfer, when the model might only be known to a small number of members of a transfer network, and when the source "label" is removed from transferred knowledge.

A main insight from this book is the potential added value of combining standardized and non-standardized questions in interviews. A few standardized questions, e.g., regarding the perception of different senders and most common transfer channels used, could have provided a useful interesting comparative complement to open interview questions. The experience gained in this book with a combination of semi-structured questionnaires and more formalized ranking exercises in Morocco and South Africa showed that a sequence where standardized questions follow the open ones and where some time is kept in reserve for points that may arise during the standardized questions appears most useful. An integration of methods used in diffusion and in transfer studies appears possible and desirable and can be realized by mixing methods (Bennett 1997: 215). Such integration, possibly in a larger research program, would include a large-n study of diffusion patterns combined with a more in-depth study of some cases using qualitative methods, as done by Busch and Jörgens (2010). Whereas this approach holds the promise of greater generalizability, the risk of selection biases by omitting cases of non-transfer or selective transfer should be mitigated.

9.4 Policy Lessons for Effective Leadership by Transfer

A main motivation for the empirical focus in this book was Germany's aim of being an active sender in order to find followers to its Energiewende. Beyond an academic interest, a better understanding of the role of senders in policy transfer is of policy relevance. The following sections explore key elements of effective sending and areas for improvement for the German approach (see also Li 2016, Messner & Morgan 2013, Morgan & Weischer 2013, Quitzow et al. 2016, Steinbacher & Pahle 2015, Tänzler & Wolters 2014).

9.4.1 Unilateral leadership as an asset

Throughout the cases, Germany's pioneering into the "age of renewables" by implementing the Energiewende domestically was a formidable backdrop for its active leadership efforts. The legitimacy and credibility of outreach activities were enhanced by the domestic implementation of ambitious renewable energy policies.

For unilateral leadership to constitute a powerful basis for active leadership, requirements of legitimacy and credibility have to be met (Karlsson et al. 2012, Steinbacher & Pahle 2015, Torney 2015: 32). If climate protection is the main objective of German leadership, as the analysis of political statements and parliamentary debates suggests, the domestic model needs to be targeted at the achievement of this objective and yield visible results in this direction. The Energiewende's "climate paradox" – a temporary increase in emissions in 2013 and 2014 (Graichen & Redl 2014) – was widely noticed by interview partners and could affect how convincing Germany's model is in the long run. The importance of the Energiewende's short-term climate record in Germany for the prospects of transfer should however not be overstated either, given the diversity of objectives that were identified as motivating renewable energy policies in follower states. What appears decisive is the perception of Germany as a country with ambitious goals and strong commitment, even if implementation is imperfect. Clear commitment to a collective objective is thus likely to be more important for the credibility and legitimacy of outreach than a flawless short-term track record of frontrunners.

Unmediated and "untranslated" information on the Energiewende alone is however rarely sufficient to trigger policy transfer given limitations in terms of resources and capacities. Germany's leadership efforts acknowledge this, in particular with regard to developing countries. Through a range of initiatives and programs, Energiewende experience is translated into the local context. As mentioned above, mediated learning is particularly relevant where fewer capacities are available. But the importance of communicating more on the Energiewende was also seen in California, where interviewees critiqued the lack of easily accessible, official information. In particular, "digested" information, beyond raw data and legislative texts, was sought for, also to debunk myths related to the Energiewende. Stakeholders were eager to learn more about questions such as how end user prices for electricity in Germany are exactly composed and how they evolved, what the Energiewende's impact on the quality of electricity supply in Germany is, and how electricity exports evolved.

All this information is already available online, but scattered across different websites and provided by different sources. A one-stop-shop for official information on the Energiewende, in addition to the increasing and useful information offer provided by think tanks such as Agora Energiewende or services such as the Clean Energy Wire, still appears necessary. The announcement of a tender for a €3.5 million project on communicating the Energiewende internationally in spring 2016 by BMWi[91] is a clear sign that the need for more communication, in addition to the numerous existing initiatives, has been recognized.

9.4.2 *Knowing leaders' and followers' objectives*

As the ranking exercises carried out in Morocco and South Africa revealed, the reasons for engaging in the promotion of renewable energy or any other policy may differ widely among potential recipient states and between recipients and the sender. Even though the objectives of greatest importance to Morocco and

91 Tender announcement "Unterstützungsleistungen zur Kommunikation der deutschen Energiewende im internationalen Kontext" [Support for the communication of the German Energiewende in the international context], Bundsamt für Außenwirtschaft und Ausfuhrkontrolle, 25 February 2016. No longer available.

South Africa – security of supply and socio-economic objectives, in particular job creation – are of relevance in Germany as well, the degree of priority and urgency is not the same. On the other hand, objectives at the center of Germany's Energiewende, phasing out/replacing nuclear power through renewables and climate protection, were of secondary importance (climate) or not important (nuclear). What this means for leaders willing to promote policies abroad is that recipient objectives need to be understood and communication and other leadership activities be tailored to them. As pointed out by Underdal (1992: 2), supply of information by the leader has to ultimately match demand (see also Torney 2015: 201). The longstanding working relationships Germany has established with many potential followers are a sound basis to understand what political drivers are behind the (non-) adoption of policies. Energy partnerships and any other form of continuous, institutionalized dialogue can be used to adapt policy promotion locally. These capacities were not only found to be a decisive factor in this book, but their lack also partly explained the limited effectiveness of active EU climate leadership toward China and India described by Torney (2015: 199): "External capacity matters because of the importance of understanding the positions and underlying domestic politics of third countries. This requires institutional capacity and resources, particularly on the ground in third countries".

Across the cases, questions remained as to what objectives Germany pursues with its international efforts in the field of sustainable energy. While most interview partners concluded Germany's goals were more altruistic than self-driven, interview partners in Morocco and South Africa would not have seen a problem with a stronger focus on mutual economic benefits – as long as the agenda was communicated clearly. The lack of clarity among stakeholders in recipient states as to why Germany is so active in the field of renewable energy is not only a potential issue for the effectiveness of leadership, but is also ethically questionable.

Germany's diverse agenda for leadership abroad mirrors uncertainty regarding the hierarchy of political objectives for the Energiewende at home (Joas et al. 2016). Whether the Energiewende is mostly a climate policy or more targeted at nuclear phase-out, and how important industrial policy aspects, local pollution or democratization of electricity supply are, is subject to debate. Turf wars and competition between projects sponsored by different governmental departments were nevertheless hardly observed in Morocco, South Africa, and California, and if anything the possibility of pointing to different dimensions of the Energiewende appeared as an advantage. As an interview partner with a long career in energy policy cooperation stressed, being able to emphasize different aspects of the Energiewende in front of different audiences is actually a main advantage for German Energiewende leadership (D6).

The question also remains as to whether it is possible to speak of an actual external Energiewende strategy that Germany pursues. Findings suggest that such a strategy emerges implicitly, in a bottom-up manner, as the sum of projects, initiatives, and the work of individual transfer agents, rather than as a consistent top-down plan as to how the Energiewende should be promoted abroad. One reason why the approach appears effective in the absence of a unified framework is the strong identification of transfer agents with the Energiewende and strong political consensus around the general direction of German energy policy. This leads to a consistent narrative around the Energiewende in leadership activities and a *de facto* strategy in the cases studied.

9.4.3 Empowering challengers and incumbents

With senders' resources being limited, decisions need to be made not only regarding which countries to target, but also what groups to concentrate efforts on in recipient states. Potential addressees can be located within official institutions or be challengers outside decision-making processes, such as representatives of citizen groups, the general public (as in public diplomacy) or NGOs. Even within systems, the question remains of whether actors in favor of the sender's policies, in the German case pro-renewables advocates, should be supported or, less intuitively, opponents to the transferred ideas should be targeted. Results from this book suggest that a two-way strategy is most effective. While pro-renewables actors (within and outside institutions) were empowered and therefore could bring new arguments to policy-making, the impact of transferred knowledge could fully unfold only where incumbents were brought on board through targeted information and capacity building. A better understanding of issues at stake and training to enable utility staff to manage changes in the electricity system were particularly decisive in South Africa, both at the level of official development cooperation and individual German transfer agents. In Morocco, early exposure of utility staff to renewables through the rural electrification program mitigated concerns about the grid impact of renewables and facilitated a utility-led renewables program where solar capacity is added at the end of electricity lines. In California, utility representatives were in contact with counterparts in Germany, which not only led to negative lesson drawing due to the losses utilities there were experiencing, but also led to proactively embracing change through the adjustment of requirements for technical equipment. Promoting dialogue between similar groups of actors, e.g., municipal utilities or grid operators, is thus a promising avenue for leadership as energy transitions are scaled up.

9.4.4 Is there still a need for renewable energy policy transfer?

As of 2015, more than 160 countries globally had adopted targets for renewable energy (REN21 Secretariat 2015: 18) and prices for renewable energy technologies have decreased tremendously in recent years (IRENA 2015b: 27). The question is therefore raised whether there will be any need for ongoing German Energiewende leadership or other leadership in renewable energy in the future. Interviewees' desire to continue and intensify cooperation with Germany and willingness to learn from it as the country moves toward ever-higher shares of renewable electricity clearly point in this direction. The focus of Energiewende leadership and the way it is exerted are however in transition.

Until the end of the 2000s, feed-in tariffs were at the center of Germany's policy promotion efforts in terms of instruments. The regulatory toolbox has since been extended, due to alternative routes taken in recipient countries. Germany has been a learning leader, transmitting information about these new models to other partner countries (Lovinfosse et al. 2013, Tietjen et al. 2015). In parallel, Germany's own Energiewende model is undergoing fundamental change with the introduction of auctions (BMWi 2015a) and soft program caps (Agora Energiewende 2015b). How this shift in the domestic example affects messages that are communicated and transfer is an interesting question for future research.

As results from interviews in California show, even when policies are already advanced, stakeholders can see great benefits in being able to transfer experiences and share a common sense of leadership. The focus of Germany's leadership in some of its partner countries will increasingly shift from the promotion of renewables as niche technologies to questions of integrating large shares of renewables. But for countries starting to introduce renewables, Germany's initial Energiewende experience will also remain valid. Germany's presence around the globe, in particular through GIZ and KfW, offers opportunities for cross-transfer among different partner countries and reverse transfer back to Germany.

In the global energy transition that is underway, Germany's experience as an early mover and the specificities of its model – the simultaneous pursuit of decarbonization, nuclear phase-out and reduction of energy consumption, the democratization of electricity generation and the overall degree of ambition – will continue to make it a relevant source of lessons. Whether these lessons are effectively considered and used will also depend on Germany's continued aim to back up its domestic model with outreach to partners worldwide.

10 Conclusion

10.1 Summary

This book set out to explore what determines the effectiveness of Germany's leadership in promoting renewable energy policy transfer to Morocco, South Africa, and California. Chapter 1 introduced the research aim and positioned the book at the intersection of policy transfer and leadership literature.

Chapter 2 presented these strands of literature, as well as literature on policy networks, in order to place policy transfer in the context of decision-making in recipient states. Research gaps were identified with regard to the role of pioneering countries as senders in policy transfer, selection biases in terms of cases studied, and a lack of integration among different strands of theory regarding leadership, transfer, and policy making in the recipient context.

Based on expectations derived from theory, Chapter 3 presented the analytical framework employed in this book. Three hypotheses were developed to cover the transfer process from a model's source to its consideration, use, and potential reflection in policy outcomes, which are the three categories of transfer explored in the case studies. The analytical framework puts a particular emphasis on how senders can influence determinants for transfer in these three categories.

Chapter 4 presented the methods used to explore transfer processes in depth in Morocco, South Africa, and California. The selection of these cases is justified by a triple aim of exploring a novel empirical field, studying neglected cases of selective transfer, and overcoming the lack of comparative studies of developing and industrialized countries in policy transfer research. The chapter explains the use of semi-structured interviews carried out in the three recipient states and in Germany. These were coded and triangulated with findings from the qualitative analysis of primary and secondary sources with a view to trace debates and decision-making processes.

Chapter 5 explored the unilateral and active dimensions of German Energiewende leadership. The Energiewende is presented as a longstanding, bottom-up effort to transform an energy system, away from nuclear and fossil energy and toward high shares of renewables in the energy mix. The transition's interna-

K. Steinbacher, *Exporting the Energiewende*, Energiepolitik und Klimaschutz.
Energy Policy and Climate Protection, https://doi.org/10.1007/978-3-658-22496-7_10

tional dimension is outlined in terms of leadership aims and the diverse forms of active outreach to partner countries Germany has established. The chapter concludes with a discussion of Germany's leadership aim and actions.

Chapters 6, 7, and 8 constitute the empirical core of the book. In Chapter 6, policy transfer and German leadership in Morocco were explored. Based on close working relationships between German transfer agents and Moroccan stakeholders, the provision of crucial evidence when a window of opportunity for policy action opened had a decisive impact on the orientation of the Moroccan energy strategy toward renewables. In South Africa, lesson drawing from Germany and support to key pro-renewables stakeholders in the policy network led to the adoption of a feed-in tariff proposal which spurred the adoption of South Africa's first effective renewable energy program. Its implementation is now facilitated by German transfer agents and through transferred knowledge. In California, the Energiewende example was used intensively and in a highly political manner in various policy debates. Even though evidence from the German example on the level of policy instruments hit barriers of regulatory traditions and opposition from key stakeholders, transfer is reflected in outcomes at the level of technical regulations and as a validating factor for general policy orientations.

Chapter 9 discussed findings from the case studies in a comparative manner. Several overarching conclusions can be drawn. First, the absence of convergence between Germany and the three recipients in terms of feed-in tariffs as policy instruments cannot lead to the conclusion that these are cases of "non-transfer". The Energiewende and/or its principles were intensively used in policy making, and transfer was valued as a facilitator and catalyst for policy developments in the direction of stronger renewable energy policy. Secondly, transfer appears as a highly political process across the cases. This is certainly related to the particularly disruptive nature of the Energiewende and the consequences a full transfer of the model would have on the structure of energy systems in recipient states. The preferences of powerful incumbents, but also local policy objectives and adherence to regulatory traditions and concerns over the fit with these, shaped transfer outcomes. Transfer from Germany acted as an empowerment to challengers of current energy systems in all cases.

Another common finding is the tight link between Germany's Energiewende example and active leadership measures. The Energiewende provided a strong basis for outreach in recipients and was recognized as a source of legitimacy, regardless of stakeholders' positions concerning transferability. The effectiveness of leadership in the absence of a coherent or formalized external Energiewende strategy is surprising at first. However, the absence of such a strategy also implies a high degree of flexibility for German transfer agents to translate transferred knowledge to the recipient context.

In summary, findings call for a systematic integration of pioneers as senders into the policy transfer literature. They also emphasize the need to consider transfer as a dynamic process that happens within policy-making processes rather than in a vacuum. It can not only change the knowledge available to members of policy networks, but was also found across the case studies to affect structures and the relative strength of different groups. Finally, the identification of transfer in the absence of convergence is an argument in favor of more studies of cases of selective and negative transfer.

10.2 Avenues for Further Research

Transfer and diffusion research has exciting days ahead. As policies and the issues they address constantly evolve and communication channels multiply, there is a growing need to understand how countries coordinate in a decentralized manner, how knowledge about programs that work (or not) travel, and what the effects of such transfers on the likelihood and quality of policy action are. The following lines of further research emerge from the findings and the lessons learned throughout the research process.

10.2.1 Transferring transitions

The field of environmental policies is well covered in transfer and diffusion studies, but rather from the angle of single regulatory instruments and standards. A broader view on how *transitions* diffuse, and how different broader developments toward sustainability are connected, appears as a relevant new empirical research field. The study of the diffusion of energy or sustainability transitions is likely to be more similar to studies of democratization or social changes than the spread of single instruments. A larger research program looking at how the diffusion and transfer of particular policy instruments can trigger and facilitate transitions or sustain them is a research gap at the intersection of transition studies and policy transfer and diffusion (Marquardt et al. 2015). The role of pioneers in promoting transitions should be included in such studies to understand how they can contribute to speeding up the spread of policies and facilitate their implementation. As outlined above, there is likely to be a difference between transfer agents who draw from the experience of implementing a policy instrument domestically (nation states, regional and local units of government) and intermediaries such as networks and international organizations – a hypothesis that requires further testing.

Another interesting aspect with regard to the transfer of transitions is whether and how the narrative surrounding particular policy instruments affects the likelihood and shape of transfer. Does a broader "transition" narrative or the perspective of a policy contributing to a global objective like climate protection increase the likelihood of transfer? How can senders contribute to the construction of narratives and how are these used to frame transferred policies? In answering these questions, insights from the policy instrument selection literature, especially concerning policies' images and discourse surrounding instrument selection (Böcher 2012: 17, Linder & Peters 1989), will likely be relevant.

Of particular importance to the field of "transition transfer" is the question of the success or effectiveness of transferred policies. Dolowitz and Marsh (2000) introduce the notion of "failed" and "successful" transfer to distinguish between transferred policies that unfold a desired effect when implemented in a follower constituency and those who are so ill-adapted, ill-transferred or poorly implemented that they fail in reaching policy objectives. Again, it is important to clearly establish whose objectives are to be reached to establish whether transfer was effective or not. If the transfer of energy transitions is studied, a contribution to climate protection efforts should at least be made more likely through what is transferred, even if other objectives are of greater importance to recipients or if results are not immediately visible. Assessing the environmental effectiveness of policies is therefore a necessary extension of transfer research in this policy field and requires enlarging the scope beyond agenda-setting and policy formulation, toward implementation.

10.2.2 Policy transfer beyond policy formation

This research project's focus was on the role of (sender-promoted) transfer in agenda-setting and policy formulation given the relatively recent nature of most of the policy-making processes studied in the three cases. There is a strong argument for looking at transfer in other steps of the policy cycle as well. The necessity to trace transfer up to policy implementation is linked to a concern for transferred policies potentially being ineffective in a foreign environment they were not designed for. Although selective transfer, and the adaptation of transferred knowledge into the local context are likely to reduce this risk, capacities to perform the adaptation of transferred policies are likely to vary from case to case. Research here could explore whether there are differences in the implementation of transferred policies versus those developed exclusively in one setting (although one might argue that policies are hardly ever created *ex nihilo*). Possible hypotheses are that agents on the ground might only implement transferred policies reluctantly if their transfer was linked to conditionality or is seen as ill-adapted to the local context. On the other hand, the implementation of transferred policies could be easier given that they were tried and tested elsewhere and practical knowledge about implementation can be shared.

Transfer has been linked to the policy-making step of evaluation in that it can serve as a way to "prospectively evaluate" the likely impact of a policy by looking at another country's experience with it (Mossberger & Wolman 2003). The evaluation *of* transferred policies is a rather understudied question (Dolowitz & Marsh 2000). Long-range studies of transferred policies that go from agenda setting to evaluation can shed light on whether the transferred model is for example constantly compared to the performance of the "original" or whether it is possibly benchmarked against new international experiences.

10.2.3 Technology and policy transfer

The transfer of renewable energy policy and its promotion by Germany and other senders is not an end in and of itself. The policies discussed in this book aim at the uptake, deployment and scaling-up of renewable energy technologies, hence at technology transfer and diffusion. Numerous parallels exist between expectations from the technology diffusion and the policy diffusion and transfer literatures (see section 2.2.4). These parallels were largely confirmed. Structural path dependencies such as existing infrastructure are directly reflected in path dependencies and inertia within policy networks, where incumbents may oppose the uptake of new policies, including transferred ones, and of foreign policy models potentially challenging their business model. A combined look at policy and technology transfer studies is adopted in the lead markets literature (Beise & Rennings 2005, Jänicke 2014: 38-39), but it is not the rule in transfer studies. More studies in the many different policy areas that have a physical, technological component, such as healthcare, telecommunications, energy or transport, are needed to understand the sequence of policy and technology transfer, the link between flows of information and commercial exchange, and whether the same agents are involved in both aspects of transfer. One of the findings, the surprising absence of companies in policy transfer, should be critically tested in studies that explicitly take into account technology transfer and policy transfer at the same time.

10.2.4 Senders in transfer

A main aim of this book was to explore the role of Germany as an active sender in transfer. Across the cases, where a policy comes from and how senders promote it was of crucial importance. Many additional questions surrounding senders in transfer arose in the course of research, both linked to the specific example of Germany and regarding senders more generally. With regard to Germany and its sending approach, a shift from a focus on promoting its own policy model toward a more "model-neutral" transfer strategy could be observed. In future transfer studies, it would be highly interesting to assess how the shift in strategy affects the content and outcome of transfer over the long term. One hypothesis could be that a more neutral strategy could increase the number of entry points into a recipient policy network. On the other hand, the findings presented in this book show that only a model that is domestically implemented confers additional credibility to sending agents and provides lessons others can draw from.

A second sender-related avenue for further research is the genesis of "policy export" strategies. The internal factors shaping Germany's Energiewende outreach were discussed in this book, but a detailed, possibly narrative-based examination of how preferences regarding Energiewende outreach developed within Germany or other countries would be a desirable complement. This book took a close look at Germany, which meant "zooming in" to better understand the role of senders through in-depth process tracing. Such a study could also form the basis for comparative research with other senders, and with inter-temporal and multi-level transfer within the recipient state. Comparing different leaders' approaches to energy policy promotion is an interesting focus for future research endeavors. Particularly salient questions for further research in this area are whether these senders have consistent lines of argumentation and preferences across different cases and an equally strong desire to gain followers for their own approach to energy policy (e.g., France and nuclear, etc.). Comparative sender studies can also be embedded in bigger frameworks, such as different member states' approaches in the framework of the EU's external energy strategy.

Exploring the multi-level dimension of transfer is a promising field for further research. Following the idea of "multi-level reinforcement" (Jänicke 2014, Schreurs & Tiberghien 2007), all levels of transfer with regard to one sender could be integrated. The example of transfer from Germany to Californian cities was one step in this direction, but integrating the links between one sender and one or several recipients at the local, regional, national and international level in one research endeavor promises highly interesting insights and an even more complete view of transfer processes. A question here could be whether strategies and transferred content vary according to the sending and receiving level. The use by Germany of several "arenas" at different levels to promote renewable energy, from international institutions to local government networks, could again be a very relevant empirical case. Finally, an additional line of potential further research concerns "learning leaders" and reverse transfer. In the case of Germany, experience on the South African auction scheme model, which Germany had helped implement, has to a limited extent been considered in Germany (Bundesregierung 2016b) and lesson drawing has been bidirectional in the case of California. Studying how common reverse transfer is – especially in transfer between developing and developed countries – could help further break down silos in transfer research.

10.3 Concluding Remarks

The architecture for global climate governance set out by the 2015 Paris Agreement grants an important role to national policy initiatives. This outcome is in conflict with a longstanding belief that global climate protection efforts need to be based on top-down coordination and that unilateral action by countries involves a risk of free-riding and is therefore likely to be ineffective. The findings presented in this book challenge this view and provide arguments for the relevance of unilateral and active leadership. The Energiewende model and its principles, often translated into the recipient context by German transfer agents, spark policy debates and can bring a new dynamic into policy making processes. This is all the more relevant in a policy field characterized by path dependencies and concentrated incumbent interests, in developing and industrialized countries alike.

Germany's move toward a more sustainable energy system was overwhelmingly recognized by stakeholders interviewed as a valuable source of evidence and inspiration, regardless of whether the entire model was seen as transferable. Germany's "daring to dare" an energy transition by far outweighed concerns over challenges it encounters as a first-mover. This insight puts into question a dominant theme in Energiewende discourse in Germany, namely that it would first need to "succeed" before others could consider the Energiewende example or taken lessons from it. Quite to the contrary, it is the very process of implementing an ambitious transition – with its achievements and challenges, its controversial discussions and negotiated compromises – that is most relevant for those willing to engage in energy transitions elsewhere around the globe.

Where a policy comes from, and who promotes it, matters. This puts a responsibility on senders – countries, regions, cities – whose policies can provide positive and negative lessons for the development of frameworks in other parts of the world. The findings presented in this book give reason for hope regarding the prospects of a polycentric climate regime in which the exchange of knowledge, learning from and with pioneers, experimentation, and partnerships can be means of moving toward a common objective. Leaders willing to share experiences, reach out to followers at eye-level, and adapt strategies to the objectives of their partners are decisive for the success of this approach.

References

Adam, C. (2015). Simulating policy diffusion through learning: Reducing the risk of false positive conclusions. *Journal of Theoretical Politics:* 0951629815581461. https://doi.org/10.1177/0951629815581461

Adam, S., & Kriesi, H. (2007). The Network Approach. In P. A. Sabatier (Ed.), *Theories of the policy process* (pp. 129–154). Boulder, Colo: Westview Press.

Adler, E., & Haas, P. M. (1992). Conclusion: epistemic communities, world order, and the creation of a reflective research program. *International Organization, 46*(01): 367–390. https://doi.org/10.1017/S0020818300001533

AFD. (2012). *Prêt d'un montant maximum de 100 M EUR ou son équivalent en USD pour le financement de la première centrale solaire à Ouarzazate*. Agence Francaise de Dévelоppment. Retrieved from http://www.afd.fr/base-projets/downloadDocument.action?idDocument=1244

Afrikaverein der Deutschen Wirtschaft. (2016). 10 Years of German-African Energy Business Promotion. Retrieved from http://www.energyafrica.de

Agentur für Erneuerbare Energien. (2014a). *Die deutsche Energiewende in der internationalen Presse.* (Renews Kompakt). Berlin. Retrieved from http://www.unendlich-viel-energie.de/mediathek/hintergrundpapiere/die-deutsche-energiewende-in-der-internationalen-presse

Agentur für Erneuerbare Energien. (2014b). *Großteil der Erneuerbaren Energien kommt aus Bürgerhand.* (Renews Kompakt). Berlin. Retrieved from http://www.unendlich-viel-energie.de/media/file/284.AEE_RenewsKompakt_Buergerenergie.pdf

Agentur für Erneuerbare Energien. (2015). *Akzeptanz für Erneuerbare weiterhin hoch.* (Renews Kompakt). Retrieved from https://www.unendlich-viel-energie.de/media/file/416.AEE_RenewsKompakt_Akzeptanzumfrage2015.pdf

Agentur für Erneuerbare Energien. (2016). Erfolgsgeschichte EEG - das Erneuerbare-Energien-Gesetz. Retrieved from https://www.unendlich-viel-energie.de/themen/politik/erneuerbare-energien-gesetz-eeg/erfolgsgeschichte-eeg-das-erneuerbare-energien-gesetz

Agora Energiewende. (n.d.). About us. Retrieved from https://www.agora-energiewende.de/en/about-us/

Agora Energiewende. (2015a). *Insights from Germany's Energiewende: State of affairs, trends and challenges*. Berlin. Retrieved from http://www.agora-energiewende.de/fileadmin/Projekte/2015/Understanding_the_EW/Key_Insights_Energy_Transition_EN_Stand_7.10.2015_web.pdf

K. Steinbacher, *Exporting the Energiewende*, Energiepolitik und Klimaschutz. Energy Policy and Climate Protection, https://doi.org/10.1007/978-3-658-22496-7

Agora Energiewende. (2015b). *Understanding the Energiewende: FAQ on the ongoing transition of the German power system*. Berlin. Retrieved from http://www.agora-energiewende.de/fileadmin/Projekte/2015/Understanding_the_EW/Agora_Understanding_the_Energiewende.pdf

Agora Energiewende. (2016a). *The Energiewende in the Power Sector: State of Affairs*. Berlin. Retrieved from http://www.agora-energiewende.de/fileadmin/Projekte/2016/Jahresauswertung_2016/Agora_Jahresauswertung_2015_Slides_web_EN.pdf

Agora Energiewende. (2016b). *Eleven Principles for a Consensus on Coal: Concept for a stepwise decarbonisation of the German power sector (Short Version)*. Berlin. Retrieved from http://www.agora-energiewende.de/fileadmin/Projekte/2015/Kohlekonsens/Agora_Kohlekonsens_KF_EN_WEB.pdf

AHK. (n.d.). About AHK. Retrieved from http://www.ahk.de/en/about-ahk/ahk-organization/

AidData. (n.d.). Open Data for International Development. Retrieved from http://aiddata.org

Aklin, M., & Urpelainen, J. (2014). The Global Spread of Environmental Ministries: Domestic–International Interactions. *International Studies Quarterly, 58*(4): 764–780. https://doi.org/10.1111/isqu.12119

Altmaier, P. (27 August 2012a). Wir brauchen einen Klub der Energiewendestaaten. *Financial Times Deutschland*. Retrieved from http://www.ftd.de/politik/deutschland/gastkommentar-von-peter-altmaier-wir-brauchen-einen-klub-der-energiewendestaaten/70081887.html

Altmaier, P. (30 October 2012b). "Die Energiewende war unsere Mondlandung". *Die Welt Online*. Retrieved from http://www.welt.de/politik/deutschland/article110381084/Die-Energiewende-war-unsere-Mondlandung.html

Altmaier, P. (2012c). Energiewende ist das größte wirtschaftspolitische Projekt seit dem Wiederaufbau Deutschlands: Rede zum Bundesministerium für Umwelt, Naturschutz und Reaktorsicherheit 2.) Erste Beratung Bundesregierung hier: Einzelplan 16 (Umwelt, Naturschutz und Reaktorsicherheit) - Drs 17/10200 -. Retrieved from https://www.cducsu.de/themen/energie/energiewende-ist-das-groesste-wirtschaftspolitische-projekt-seit-dem-wiederaufbau-deutschlands

Altmaier, P. (2013). Die Energiewende ist die größte umwelt- und wirtschaftspolitische Herausforderung zu Beginn des 21. Jahrhunderts. In Peter Altmaier & Johannes Varwick (Eds.), *Politische Bildung: Vol. 46. Energiewende* (pp. 7–25). Schwalbach/Ts: Wochenschau-Verlag.

Altmaier, P., & Batho, D. (2013a). *Gemeinsame Erklärung zwischen dem Bundesminister für Umwelt, Naturschutz und Reaktorsicherheit der Bundesrepublik Deutschland und der Ministerin für Umwelt, nachhaltige Entwicklung und Energie der Französischen Republik über die Zusammenarbeit im Bereich erneuerbarer Energien und die Schaffung eines Deutsch-französischen Büros für Erneuerbare Energien im Rahmen der Energiewende*. 7 February 2013. Paris. Retrieved from http://www.developpement-durable.gouv.fr/IMG/pdf/2013-02-07_Decl-_P-_ALTMAIER-D-_BATHO_-allemand-.pdf

Altmaier, P., & Batho, D. (14 May 2013b). Vereint für die Energiewende. *Tagesspiegel*. Retrieved from https://www.bundesregierung.de/ContentArchiv/DE/Archiv17/Namensbeitrag/2013/05/2013-05-14-altmaier-batho-tagesspiegel.html

Andresen, S., & Agrawala, S. (2002). Leaders, pushers and laggards in the making of the climate regime. *Global Environmental Change, 12*(1): 41–51. https://doi.org/10.1016/S0959-3780(01)00023-1

Andriani, B., Lignieres, P., Barges, M., Bennis, A., & Mokhtari, G. (2013). *L'énergie au Royaume du Maroc: Stratégie énergétique et développements récents*. Retrieved from http://www.linklaters.com/pdfs/mkt/paris/6143_Paris_Office_Morocco_Energy_Newsletter_FRENCH_FINAL.pdf

Appunn, K. (2016). EEG reform 2016 – switching to auctions for renewables. Clean Energy Wire. Retrieved from https://www.cleanenergywire.org/factsheets/eeg-reform-2016-switching-auctions-renewables

Arts, B., Knill, C., & Holzinger, K. (2008). *Environmental policy convergence in Europe: The impact of international institutions and trade*. Cambridge, UK, New York: Cambridge University Press.

Asare, B. E., & Studlar, D. T. (2009). Lesson-drawing and public policy: secondhand smoking restrictions in Scotland and England. *Policy Studies, 30*(3): 365–382. https://doi.org/ 10.1080/01442870902863935

Auswärtiges Amt. (2014). Energieexperten und -expertinnen zu Gast im Auswärtigen Amt. Retrieved from http://www.auswaertiges-amt.de/DE/Aussenpolitik/GlobaleFragen/Energie/141007_Themenreise_Energiewende.html

Auswärtiges Amt. (2015a). *The Energiewende: Secure, sustainable and affordable energy for the 21st century*. Retrieved from https://www.auswaertiges-amt.de/cae/servelet/contentblob/718164/publicationFile/210297/The_Energiewende.pdf

Auswärtiges Amt. (2015b). Marokko: Innenpolitik. Retrieved from http://www.auswaertiges-amt.de/DE/Aussenpolitik/Laender/Laenderinfos/Marokko/Innenpolitik_node.html

Auswärtiges Amt. (2015c). *Organisationsplan des Auswärtigen Amtes*. Berlin, Bonn: Auswärtiges Amt.

Auswärtiges Amt. (2015d). South Africa. Retrieved from http://www.auswaertiges-amt.de/sid_10A9DE436640F21259220DE5FD2396EC/EN/Aussenpolitik/Laender/Laenderinfos/01-Nodes/Suedafrika_node.html#doc354424bodyText2

Auswärtiges Amt. (2015e). *Who is Who of the Energiewende in Germany: Contact Partners in Politics, Industry and Society*. Berlin, Bonn. Retrieved from https://www.auswaertiges-amt.de/cae/servlet/contentblob/701026/publicationFile/209413/EnergiewendeWhoisWho.pdf

Auswärtiges Amt. (2016a). *Die deutsche Energiewende*. Berlin: Auswärtiges Amt.

Auswärtiges Amt. (2016b). *Rede von Außenminister Steinmeier bei der Eröffnung der Energiewende-Ausstellung in Peking*. Retrieved from http://www.auswaertiges-amt.de/DE/Infoservice/Presse/Reden/2016/160408_BM_Energiewendeausstellung_Peking.html?nn=382708

Bahadur, A., Ashley, B., Hilliger, C., Hallowes, D., Sanders, D., Forslund, D. … Aroun, W. (2011). *One Million Climate Jobs: A just transition to a low carbon economy to combat unemployment and climate change.* Retrieved from http://greeneconomynet.ca/wp-content/uploads/sites/43/2014/12/ClimateJobsBooklet2011-2_South-Africa.pdf

Baker, L. (2016). *Post-apartheid electricity policy and the emergence of South Africa's renewable energy sector. (WIDER Working Paper, N° 2016/15).* Retrieved from https://www.wider.unu.edu/sites/default/files/WP2016-15.pdf

Baker, L., Newell, P., & Phillips, J. (2014). The Political Economy of Energy Transitions: The Case of South Africa. *New Political Economy, 19*(6): 1–28. https://doi.org/10.1080/13563467.2013.849674

Barabasch, A., Huang, S., & Lawson, R. (2009). Planned policy transfer: the impact of the German model on Chinese vocational education. *Compare: A Journal of Comparative and International Education, 39*(1): 5–20. https://doi.org/10.1080/03057920802265566

Baumgartner, F. R., & Jones, B. D. (1991). Agenda Dynamics and Policy Subsystems. *Journal of Politics, 53*(4): 1044-1074.

Baumgartner, F. R., & Jones, B. D. (1993). *Agendas and instability in American politics.* Chicago: University of Chicago Press.

Baybeck, B., Berry, W. D., & Siegel, D. A. (2011). A Strategic Theory of Policy Diffusion via Intergovernmental Competition. *Journal of Politics, 73*(1): 232–247. https://doi.org/10.1017/S0022381610000988

Bayer, B., Schäuble, D., Langenheld, A., & Jenner, S. (2014). *Demand response: what can we learn from California?* (IASS Working Paper, February 2014). Retrieved from http://www.iass-potsdam.de/sites/default/files/files/working_paper_demand-response-california.pdf

BDEW. (2016). *Strompreisanalyse Januar 2016.* Retrieved from https://www.bdew.de/internet.nsf/id/DC9ABD3F2D97604DC1257F42002E5075/$file/160122%20BDEW%20zum%20Strompreis%20der%20Haushalte%20Anhang.pdf

Bechberger, M. (2000). *Das Erneuerbare-Energien-Gesetz (EEG): Eine Analyse des Politikformulierungsprozesses.* (FFU-Report, 00-06). Retrieved from http://www.polsoz.fu-berlin.de/polwiss/forschung/systeme/ffu/publikationen/2000/bechberger_mischa_2000/rep_00-06.PDF

Bechberger, M., Körner, S., & Reiche, D. (2003). *Erfolgsbedingungen von Instrumenten zur Förderung Erneuerbarer Energien im Strommarkt.* (FFU-Report, 01-2003). Retrieved from http://www.polsoz.fu-berlin.de/polwiss/forschung/systeme/ffu/publikationen/2003/bechberger_mischa_koerner_stefan_reiche_danyel_2003/rep_2003rep_2003-01.pdf

Beise, M. (2003). *Lead Markets: Drivers of the Global Diffusion of Innovations.* (Discussion Papers Research Institute for Economics & Business Administration, Kobe University, N°141). Retrieved from http://www.rieb.kobe-u.ac.jp/academic/ra/dp/English/dp141.pdf

Beise, M. (2004). Lead markets: country-specific drivers of the global diffusion of innovations. *Research Policy, 33*(6–7): 997–1018. https://doi.org/10.1016/j.respol.2004.03.003

Beise, M., & Gemünden, H. (2004). Lead Markets: A New Framework for the International Diffusion of Innovation. *Management International Review, 44*(3): 83–102.

Beise, M.,& Rennings, K. (2003), *Lead Markets of Environmental Innovations: A Framework for Innovation and Environmental Economics*. (ZEW Discussion Paper No. 03-01). Mannheim: ZEW. Retrieved from ftp://ftp.zew.de/pub/zew-docs/dp/dp0301.pdf

Beise, M., & Rennings, K. (2005). Lead markets and regulation: a framework for analyzing the international diffusion of environmental innovations. *Ecological Economics, 52*(1): 5–17. https://doi.org/10.1016/j.ecolecon.2004.06.007

Ben Hayoun, M. (3 February 2016). Programme solaire de l'ONEE Noor-Argana portera sur 200 MW pour plus de 250 millions d'euros. *Le Matin*. Retrieved from http://www.lemag.ma/Exclusif-Programme-solaire-de-l-ONEE-Noor-Argana-portera-sur-200-MW-pour-plus-de-250-millions-d-euros_a95009.html

Ben-Meir, Y. (2015). Human Development in the Arab Spring: Morocco's Efforts to Shape Its Global Future. *Mediterranean Quarterly, 26*(3): 67–93.

Bennett, C. J. (1991). How States Utilize Foreign Evidence. *Journal of Public Policy, 11*(01): 31–54. https://doi.org/10.1017/S0143814X0000492X

Bennett, C. J. (1997). Understanding Ripple Effects: The Cross–National Adoption of Policy Instruments for Bureaucratic Accountability. *Governance, 10*(3): 213–233. https://doi.org/10.1111/0952-1895.401997040

Bennett, C. J., & Howlett, M. (1992). The lessons of learning: Reconciling theories of policy learning and policy change. *Policy Sciences, 25*(3): 275–294.

Bennouna, A., Zejli, D., & Benchrifa, R. (2007). Les énergies renouvelables - Pour un développement durable et global. *Revue des Energies Renouvelables*, (CER'07): 1–8.

Benson, D., & Jordan, A. (2011). What Have We Learned from Policy Transfer Research? Dolowitz and Marsh Revisited. *Political Studies Review, 9*(3): 366–378. https://doi.org/10.1111/j.1478-9302.2011.00240.x

Berliner Energieagentur. (2012). Mongolische Energiebehörde und BEA vereinbaren Zusammenarbeit. Retrieved from http://www.berliner-e-agentur.de/news/mongolische-energiebehoerde-und-bea-vereinbaren-zusammenarbeit

Berman, D. M. (1996). *Who owns the sun?: People, politics, and the struggle for a solar economy*. White River Junction, Vt.: Chelsea Green Pub. Co

Bernhardt, J. (2010). Packed Room Learns About the Economic Benefits of Feed-In Tariffs. Clean Coalition. Retrieved from http://www.clean-coalition.org/press-releases/packed-room-learns-about-the-economic-benefits-of-feed-in-tariffs/

Berry, F. S., & Berry, W. D. (1990). State Lottery Adoptions as Policy Innovations: An Event History Analysis. *The American Political Science Review, 84*(2): 395–415. https://doi.org/10.2307/1963526

Berry, F. S., & Berry, W. D. (1999). Innovation and Diffusion Models in Policy Research. In P. A. Sabatier (Ed.), *Theoretical lenses on public policy. Theories of the policy process* (pp. 223–260). Boulder, Colo.: Westview Press.

Berry, J. M. (2002). Validity and Reliability Issues In Elite Interviewing. *PS: Political Science and Politics, 35*(4): 679–682.

Betsill, M. M., & Corell, E. (2001). NGO Influence in International Environmental Negotiations: A Framework for Analysis. *Global Environmental Politics, 1*(4): 65–85. https://doi.org/10.1162/152638001317146372

Beucker, S., Clausen, J., Fichter, K., Jacob, K., & Bär, H., (2014). *Angebote und Bedarfe von Technologien und Dienstleistungen für Klimaschutz und Klimaanpassung*. Commissioned by BMWi. Berlin: Borderstep Institute.

Biedenkopf, K. (2011). *Policy Recycling? The External Effects of EU Environmental Legislation on the United States.* (Doctoral thesis). Brussels: Vrije Universiteit Brussels.

Biedenkopf, K. (2012). Environmental Leadership Through the Diffusion of Environmental Pioneering Policy. In D. R. Gallagher (Ed.), *Environmental leadership. A reference handbook. Vol. 1* (pp. 105–112). Los Angeles, London, New Delhi, Singapore: SAGE Publications.

Bischof-Niemz, T. (2015). *How to stimulate the South African rooftop PV market without putting municipalities' financial stability at risk: A Net Feed-in Tariff proposal*. June 2015. Pretoria. Retrieved from http://www.whatnext.org/resources/What-Next-Seminars/2015_03_30-04-01-(Delhi-meeting)/For-participants/Presentations-from-sessions/Net-FIT-Proposal-v14-short.pdf

Bleich, E., & Pekkanen, R. (2013). How to Report Interview Data. In L. Mosley (Ed.), *Interview research in political science* (pp. 84–106). Ithaca: Cornell University Press.

Blum, S., & Schubert, K. (Eds.). (2009). *Politikfeldanalyse*. Wiesbaden: VS Verlag für Sozialwissenschaften.

Blyth, M., Helgadottir, O., & Kring, W. (2016). Ideas and Historical Institutionalism. In Fioretos, O., Falleti, T.G., & Sheingate, A. *The Oxford Handbook of Historical Institutionalism.* (pp. 142-162). Oxford: Oxford University Press.

BMU. (2003). 100.000 Dächer-Solarstrom-Programm kurz vor dem Ziel. Press release 108/03. Retrieved from http://www.bmub.bund.de/presse/pressemitteilungen/pm/artikel/100000-daecher-solarstrom-programm-kurz-vor-dem-ziel/

BMU. (23 March 2004). *Von Ökosteuer bis Emissionshandel: Trittin bekräftigt Instrumentenmix im Klimaschutz*. Press release 078/04. Retrieved from http://www.bmub.bund.de/presse/pressemitteilungen/pm/artikel/von-oekosteuer-bis-emissionshandel/

BMU. (2005). *Joint Declaration between the Ministry for Industry, Tourism and Trade of the Kingdom of Spain and the Ministry for the Environment, Nature Conservation and Nuclear Safety of the Federal Republic of Germany on cooperation on the development and promotion of a feed-in system to increase the use of renewable energy sources in the production of electricity*. Retrieved from http://www.bmub.bund.de/fileadmin/bmu-import/files/english/pdf/application/pdf/feed_in_des_en.pdf

BMU. (2007). *EEG-Das Erneuerbare Energien Gesetz: Die Erfolgsgeschichte nachhaltiger Politik für den Standort Deutschland.* Berlin: Bundesministerium für Umwelt, Naturschutz und Reaktorsicherheit.

BMU. (2010). Petersberger Klimadialog I: "Building Momentum for Mexico". Retrieved from http://www.bmub.bund.de/themen/klima-energie/klimaschutz/internationale-klimapolitik/petersberger-klimadialog/petersberger-klimadialog-i/

BMU. (2013). *Club der Energiewende-Staaten: Kommuniqué.* Abu Dhabi. Retrieved from http://www.bmu.de/fileadmin/Daten_BMU/Download_PDF/Energiewende/club_communique_deutsch_final_bf.pdf

BMUB. (2014a). *Aktionsprogramm Klimaschutz 2020.* Retrieved from http://www.bmub.bund.de/fileadmin/Daten_BMU/Download_PDF/Aktionsprogramm_Klimaschutz/aktionsprogramm_klimaschutz_2020_broschuere.pdf

BMUB. (2014b). Rede von Dr. Barbara Hendricks in der 57. Sitzung des Deutschen Bundestages. 9 October 2014. Retrieved from http://www.bmub.bund.de/presse/reden/detailansicht/artikel/rede-von-dr-barbara-hendricks-in-der-57-sitzung-des-deutschen-bundestages/

BMUB. (2015). *Funding the Future: Review of Activities of the International Climate Initiative from 2008 to 2014.* Retrieved from http://www.international-climate-initiative.com/fileadmin/Dokumente/2015/160125_IKI_Bilanzbericht_engl.pdf

BMWi. (2012a). *Gemeinsame Absichtserklärung über die Errichtung einer Energiepartnerschaft zwischen dem Königreich Marokko und der Bundesrepublik Deutschland.* Berlin. Retrieved from http://www.bmwi.de/BMWi/Redaktion/PDF/G/gemeinsame-erklaerung-deutschland-marokko-energiepartnerschaft ,property=pdf,bereich=bmwi2012,sprache=de,rwb=true.pdf

BMWi. (2012b). *Gemeinsame Absichtserklärung über eine Energiepartnerschaft zwischen der Bundesrepublik Deutschland und der Tunesischen Republik.* Tunis. Retrieved from http://www.bmwi.de/BMWi/Redaktion/PDF/G/gemeinsame-absichtserklaerung-energiepatrnerschaft-deutschland-tunesien,property=pdf,bereich=bmwi2012,sprache=de,rwb=true.pdf

BMWi. (2013). *Joint Declaration of Intent between the Government of the Federal Republic of Germany and the Federal Government of the Republic of South Africa on the Establishment of an Energy Partnership.* Cape Town. Retrieved from http://www.bmwi.de/BMWi/Redaktion/PDF/A/absichtserklaerung-energiepartnerschaft-deutschland-suedafrika,property=pdf,bereich=bmwi2012,sprache=de,rwb=true.pdf

BMWi. (2014a). Gabriel: Energiewende soll Nachahmer finden. (BMWi Newsletter Energiewende, 17.02.2014). Retrieved from http://www.bmwi-energiewende.de/EWD/Redaktion/Newsletter/2014/03/Meldung/nachahmer.html

BMWi. (2014b). Gabriel: Keine Hermesdeckungen mehr für Nuklearanlagen im Ausland. Retrieved from http://www.bmwi.de/DE/Presse/pressemitteilungen,did=642020.html

BMWi. (2015a). *Ausschreibungen für die Förderung von Erneuerbare-Energien-Anlagen: Eckpunktepapier*. Retrieved from https://www.bmwi.de/BMWi/Redaktion/PDF/Publikationen/ausschreibungen-foerderung-erneuerbare-energien-anlage,property=pdf,bereich=bmwi2012,sprache=de,rwb=true.pdf

BMWi. (2015b). *Die Energie der Zukunft: Vierter Monitoring-Bericht zur Energiewende*. Berlin: Bundesministerium für Wirtschaft und Energie. Retrieved from https://www.bmwi.de/BMWi/Redaktion/PDF/V/vierter-monitoring-bericht-energie-der-zukunft,property=pdf,bereich=bmwi2012,sprache=de,rwb=true.pdf

BMWi. (2015c). *Das Erneuerbare-Energien-Gesetz 2014: Die wichtigsten Fakten zur Reform des EEG*. Retrieved from http://www.bmwi.de/BMWi/Redaktion/PDF/Publikationen/das-erneuerbare-energien-gesetz-2014,property=pdf,bereich=bmwi2012,sprache=de,rwb=true.pdf

BMWi. (2015d). *Overview of legislation governing Germany's energy supply system: Key strategies, acts, directives, and regulations / ordinances*. Retrieved from http://www.bmwi.de/English/Redaktion/Pdf/gesetzeskarte,property=pdf,bereich=bmwi2012,sprache=en,rwb=true.pdf

BMWi. (2015e). *Studien des BMWi zum Thema Beschäftigungswirkung im Energiesektor*. Retrieved from http://www.bmwi.de/BMWi/Redaktion/PDF/S-T/studien-des-bmwi-zum-thema-beschaeftigungswirkung-im-energiesektor,property=pdf,bereich=bmwi2012,sprache=de,rwb=true.pdf

BMWi. (2015f). IEA-Ministertreffen vor Klimakonferenz in Paris: Globale Energiewende zentral zur Eindämmung des Klimawandels. Retrieved from http://www.bmwi.de/DE/Presse/pressemitteilungen,did=738904.html

BMWi. (2015g). *Joint Declaration for Regional Cooperation on Security of Electricity Supply in the Framework of the Internal Energy Market*. Retrieved from http://www.bmwi.de/BMWi/Redaktion/PDF/J-L/joint-declaration-for-regional-cooperation-on-security-of-electricity-supply-in-the-framework-of-the-internal-energy-market,property=pdf,bereich=bmwi2012,sprache=de,rwb=true.pdf

BMWi. (2016a). The Energy Transition. Retrieved from http://www.bmwi.de/EN/Topics/Energy/energy-transition.html

BMWi. (2016b). *The energy transition: key projects of the 18th legislative term*. Retrieved from http://www.bmwi.de/English/Redaktion/Pdf/10-punkte-energie-agenda-fortschreibung,property=pdf,bereich=bmwi2012,sprache=en,rwb=true.pdf

BMWi. (2016c). Internationale Energiepolitik. Retrieved from http://www.bmwi.de/DE/Themen/Energie/Europaische-und-internationale-Energiepolitik/internationale-energiepolitik,did=551754.html

BMWi. (2016d). *Organisationsplan*. Retrieved from http://www.bmwi.de/BMWi/Redaktion/PDF/M-O/organisationsplan-bmwi,property=pdf,bereich=bmwi2012,sprache=de,rwb=true.pdf

BMWi. (2016e). Die Struktur der Exportinitiative. Retrieved from http://www.export-erneuerbare.de/EEE/Navigation/DE/Ueber_uns/Akteure/akteure.html

BMWi. (2016f). Zuschläge in der vierten Ausschreibungsrunde für Photovoltaik-Freiflächenanlagen erteilt. Retrieved from http://www.bmwi.de/DE/Themen/energie, did=762846.html

BMWi. (2016g). *Zeitreihen zur Entwicklung der erneuerbaren Energien in Deutschland 1990-2015. Unter Verwendung von Daten der Arbeitsgruppe Erneuerbare Energien-Statistik (AGEE-Stat).* Retrieved from http://www.erneuerbare-energien.de/EE/Redaktion/DE/Downloads/zeitreihen-zur-entwicklung-der-erneuerbaren-energien-in-deutschland-1990-2015.pdf?__blob=publicationFile&v=6

BMZ. (n.d.-a). Morocco: Situation and Cooperation. Retrieved from http://www.bmz.de/en/what_we_do/countries_regions/naher_osten_mittelmeer/marokko/zusammenarbeit.html

BMZ. (n.d.-b). Südafrika: Situation und Zusammenarbeit. Retrieved from http://www.bmz.de/de/laender_regionen/subsahara/suedafrika/zusammenarbeit/

BMZ. (2010). Follow-up process to the International Conference for Renewable Energies – Renewables 2004. Retrieved from http://www.bmz.de/en/what_we_do/issues /energie/international_energy_policy/renewables/index.html?follow=adword

BMZ. (2013). *Maroc-Allemagne: 50 Ans de coopération.* Retrieved from http://www.rabat.diplo.de/contentblob/4060832/Daten/3726657/BroschuereBMZ131211.pdf

BMZ. (2014). *Nachhaltige Energie für Entwicklung: Die Deutsche Entwicklungszusammenarbeit im Energiesektor.* Retrieved from http://www.bmz.de/de/mediathek/publikationen/reihen/infobroschueren_flyer/infobroschueren/Materialie240_Informationsbroschuere_01_2014.pdf

BMZ. (2015). *Entwicklungspolitische Zusammenarbeit mit Globalen Entwicklungspartner: Gemeinsam Verantwortung wahrnehmen - globale Entwicklung gestalten.* (BMZ-Strategiepapier). Retrieved from https://www.bmz.de/de/mediathek/publikationen/reihen/strategiepapiere/Strategiepapier353_04_2015.pdf

BMZ. (2016a). Deutsche ODA-Leistungen. Retrieved from http://www.bmz.de/de/ministerium/zahlen_fakten/oda/leistungen/index.html

BMZ. (2016b). *Morocco: Green mosques.* Retrieved from https://www.giz.de/en/downloads/giz2016-en-Marokko-Green-mosques.pdf

Böcher, M. (2012). A theoretical framework for explaining the choice of instruments in environmental policy. *Forest Policy and Economics, 16.* https://doi.org/10.1016/j.forpol.2011.03.012

Bodansky, D. (1999). The Legitimacy of International Governance: A Coming Challenge for International Environmental Law? *The American Journal of International Law, 93*(3): 596–624. https://doi.org/10.2307/2555262

Bodansky, D. M., Hoedl, S. A., Metcalf, G. E., & Stavins, R. N. (2015). Facilitating linkage of climate policies through the Paris outcome. *Climate Policy:* 1–17. https://doi.org/10.1080/14693062.2015.1069175

Borgatti, S. P., Mehra, A., Brass, D. J., & Labianca, G. (2009). Network Analysis in the Social Sciences. *Science, 323*(5916): 892–895. https://doi.org/10.1126/science.1165821

Börzel, T. A. (1997). What's So Special About Policy Networks? An Exploration of the Concept and Its Usefulness in Studying European Governance. *European Integration Online Papers, 1*(16).

Börzel, T. A. (2002). Pace-Setting, Foot-Dragging, and Fence-Sitting: Member State Responses to Europeanization. *Journal of Common Market Studies, 40*(2): 193–214. https://doi.org/10.1111/1468-5965.00351

Börzel, T. A., & Risse, T. (2009). *The Transformative Power of Europe: The European Union and the Diffusion of Ideas.* (KFG Working Paper Series, N°1). Retrieved from http://www.polsoz.fu-berlin.de/en/v/transformeurope/publications/working_paper/WP_01_Juni_Boerzel_Risse.pdf

Börzel, T., & Risse, T. (2012). From Europeanisation to Diffusion: Introduction. *West European Politics, 35*(1): 1–19. https://doi.org/10.1080/01402382.2012.631310

Botschaft des Königreichs Marokko. (2010). Newsletter der Botschaft der Königreichs Marokko. (Newsletter N°1-2010). Retrieved from http://www.mkfd.de/sites/default/files/Botschaft_Newsletter_%2001_2010.pdf

Brady, H. E., & Collier, D. (Eds.). (2004). *Rethinking social inquiry: Diverse tools, shared standards.* Lanham, MD: Rowman & Littlefield.

Brandt, U. S. (2004). Unilateral actions, the case of international environmental problems. *Resource and Energy Economics, 26*(4): 373–391. https://doi.org/10.1016/j.reseneeco.2004.03.001

Braun, D., Gilardi, F., Füglister, K., & Luyet, S. (2007). Ex Pluribus Unum: Integrating the Different Strands of Policy Diffusion Theory. In K. Holzinger, H. Jörgens, & C. Knill (Eds.), *Transfer, Diffusion und Konvergenz von Politiken* (pp. 39-55). Wiesbaden: VS Verlag für Sozialwissenschaften.

Bridges, D. (2014). The ethics and politics of the international transfer of educational policy and practice. *Ethics and Education, 9*(1): 84–96. https://doi.org/10.1080/17449642.2014.890274

Bridoux, J. (2014). *Democracy promotion: A critical introduction.* Milton Park, Abingdon, Oxon: Routledge.

Brooks, S. (2007). When Does Diffusion Matter?: Explaining the Spread of Structural Pension Reforms Across Nations. *Journal of Politics, 69*(3): 701–715.

Brooks, S. (2004). International Financial Institutions and the Diffusion of Foreign Models for Social Security Reform in Latin America. In K. G. Weyland (Ed.), *Learning from foreign models in Latin American policy reform* (pp. 53–80). Washington, DC: Woodrow Wilson Center Press.

Brown, E. G. Jr. (2010). *Clean Energy Jobs Plan.* Retrieved from https://www.gov.ca.gov/docs/Clean_Energy_Plan.pdf

BSW. (2008). *Gesetz für den Vorrang Erneuerbarer Energien (EEG). Positionspapier des Bundesverbandes Solarwirtschaft e.V.* Bundesverband Solarwirtschaft. Retrieved from https://www.solarwirtschaft.de/fileadmin/content_files/eeg_position_bswsolar.pdf

Bulmer, S. (2007). *Policy transfer in European Union governance: regulating the utilities*. Milton Park, Abingdon, Oxon: Routledge.

Bulmer, S., & Padgett, S. (2005). Policy Transfer in the European Union: An Institutionalist Perspective. *Brit. J. Polit. Sci., 35*(1): 103–126. https://doi.org/10.1017/S0007123405000050

BUND. (2014). EEG-Novelle: harter Rückschlag für die Bürgerenergiewende. Bund für Umwelt und Naturschutz - Friends of the Earth Germany. Retrieved from http://www.bund.net/themen_und_projekte/klima_und_energie/energiewende/energie politik/eeg_reform/bewertung_eeg_novelle/

BUND. (2015). *Stellungnahme zum Eckpunktepapier "Ausschreibungen für die Förderung von Erneuerbare-Energien-Anlagen"*. Bund für Umwelt und Naturschutz - Friends of the Earth Germany. Retrieved from http://www.bmwi.de/BMWi/Redaktion/PDF/Stellungnahmen/Stellungnahmen-Eckpunktepapier-EE-Foerderung/20151001-bund-bund-umwelt-naturschutz,property=pdf,bereich=bmwi2012,sprache=de,rwb=true.pdf

BMBF. (2015). *Statusbericht: Südafrika (Kooperationen mit Deutschland)*. Bundesministerium für Bildung und Forschung. Retrieved from http://www.kooperation-international.de/buf/suedafrika/statusbericht-kooperationen-mit-deutschland.html

Bundesregierung. (n.d.). Beschlüsse des Bundeskabinetts zur Energiewende vom 6. Juni 2011. Retrieved from https://www.bundesregierung.de/ContentArchiv/DE/Archiv17/Artikel/2012/06/2012-06-04-artikel-hintergrund-energiewende-gesetzespaket.html

Bundesregierung. (1981). *Finanzierung des Baus von sogenannten Solardörfern und anderen solarthermischen Anlagen im Ausland: Antwort der Bundesregierung auf die Kleine Anfrage der Abgeordneten Lenzer, Pfeifer, Dr. Probst, Gerstein, Dr. Bugl, Engelsberger, Eymer (Lübeck), Dr. Hubrig, Maaß, Neuhaus, Prangenberg, Weirich, Dr. Riesenhuber, Dr. Stavenhagen, Spilker, Dr. Laufs, Dr. Kunz (Weiden), Bühler (Bruchsal) und der Fraktion der CDU/CSU — Drucksache 9/193.* (Bundestagsdrucksache, 9/224). Retrieved from http://dipbt.bundestag.de/doc/btd/09/002/0900224.pdf

Bundesregierung. (1982). *Beitrag der deutschen Entwicklungshilfe zur Lösung der Energieprobleme der Dritten Welt: Antwort der Bundesregierung auf die Kleine Anfrage der Abgeordneten Dr. Hornhues, Dr. Köhler (Wolfsburg), Dr. Riesenhuber, Dr. Pinger, Boroffka, Dr. Bugl, Frau Fischer, Gerstein, Höffkes, Dr. Hüsch, Dr.-Ing. Kansy, Dr. Kunz (Weiden), Lamers, Dr. Laufs, Lenzer, Magin, Müller (Wadern), Dr. Pohlmeier, Prangenberg, Repnik, Schmöle, Dr. Freiherr Spies von Büllesheim, Spilker, Dr. Stavenhagen, Bahner, Graf von Waldburg-Zeit, Herkenrath und der Fraktion der CDU/CSU — Drucksache 9/1867.* (Bundestagsdrucksache, 9/1980). Retrieved from http://dipbt.bundestag.de/doc/btd/09/019/0901980.pdf

Bundesregierung. (1988). Antwort der Bundesregierung auf die Große Anfrage zur Förderung und Nutzung "Erneuerbarer Energiequellen“ in der Bundesrepublik Deutschland. (Bundestagsdrucksache, 11/2684). Retrieved from http://dipbt.bundestag.de/doc/btd/11/026/1102684.pdf

Bundesregierung. (1996). *Unterstützung der Photovoltaik durch die Bundesregierung: Antwort der Bundesregierung auf die Große Anfrage der Abgeordneten Michaele Hustedt, Dr. Uschi Eid, Simone Probst, weiterer Abgeordneter und der Fraktion BÜNDNIS 90/DIE GRÜNEN — Drucksache 13/5230.* (Bundestagsdrucksache, 13/6393). Retrieved from http://dipbt.bundestag.de/doc/btd/13/063/1306393.pdf

Bundesregierung. (2009a). *Zur Energieaußenpolitik der Bundesregierung: Antwort der Bundesregierung auf die Große Anfrage der Abgeordneten Jürgen Trittin, Winfried Nachtwei, Volker Beck (Köln), weiterer Abgeordneter und der Fraktion BÜNDNIS 90/DIE GRÜNEN – Drucksache 16/10386.* (Bundestagsdrucksache, 16/13276). Retrieved from http://dipbt.bundestag.de/dip21/btd/16/132/1613276.pdf

Bundesregierung. (2009b). *Rolle der Bundesregierung und der Union für das Mittelmeer bei der Nutzung erneuerbarer Energien im Rahmen des Energieprojekts Solarplan/DESERTEC: Antwort der Bundesregierung auf die Kleine Anfrage der Abgeordneten Michael Kauch, Marina Schuster, Angelika Brunkhorst, weiterer Abgeordneter und der Fraktion der FDP – Drucksache 16/12163.* (Bundestagsdrucksache, 16/12363). Retrieved from http://dipbt.bundestag.de/doc/btd/16/123/1612363.pdf

Bundesregierung. (2010a). *Energiekonzept für eine umweltschonende, zuverlässige und bezahlbare Energieversorgung.* Retrieved from http://www.bundesregierung.de/ContentArchiv/DE/Archiv17/_Anlagen/2012/02/energiekonzept-final.pdf?__blob=publicationFile&v=5%20

Bundesregierung. (2010b). *Antwort der Bundesregierung auf die Kleine Anfrage der Abgeordneten Sevim Dagdelen, Christine Buchholz, Annette Groth, weiterer Abgeordneter und der Fraktion DIE LINKE. – Drucksache 17/1329 – Zum völkerrechtlichen Status der Westsahara und Projekten zur Förderung erneuerbarer Energien in Marokko und der Westsahara.* (Bundestagsdrucksache, 17/1521). Retrieved from http://dip21.bundestag.de/dip21/btd/17/015/1701521.pdf

Bundesregierung. (30 May 2011). *Mitschrift der Pressekonferenz zum Energiekonzept der Bundesregierung mit Bundeskanzlerin Merkel, BM Rösler, BM Röttgen und BM Ramsauer.* Retrieved from http://www.bundesregierung.de/Content/DE/Mitschrift/Pressekonferenzen/2011/05/2011-05-30-pk-bk-bm-energiekonzept.html

Bundesregierung. (2012). Globalisierung gestalten - Partnerschaften ausbauen - Verantwortung teilen: Konzept der Bundesregierung. Retrieved from http://www.auswaertiges-amt.de/sid_1BFFEA9AAAB6C7A0D47625BFD0628863/DE/Aussenpolitik/Schwerpunkte_Aussenpolitik_node.html

Bundesregierung. (2013a). *Antwort der Bundesregierung auf die Kleine Anfrage der Abgeordneten, Dr. Hermann E. Ott, Bärbel Höhn, Hans-Josef Fell, weiterer Abgeordneter und der Fraktion BÜNDNIS 90/DIE GRÜNEN.* (Bundestagsdrucksache, 17/14315). Retrieved from http://dip21.bundestag.de/dip21/btd/17/143/1714315.pdf

Bundesregierung. (2013b). *Stand und Bewertung der Exportinitiative Erneuerbare Energien für die Jahre 2010 und 2011.* (Bundestagsdrucksache, 17/12772). Retrieved from http://dip21.bundestag.de/dip21/btd/17/127/1712772.pdf

Bundesregierung. (2015). *Bericht der Bundesregierung zur internationalen Kohlefinanzierung für den Wirtschaftsausschuss des Deutschen Bundestages.* Retrieved from http://www.bmwi.de/BMWi/Redaktion/PDF/B/bericht-der-bundesregierung-zur-internationalen-kohlefinanzierung-fuer-den-wirtschaftsausschuss-des-deutschen-bundestages,property=pdf,bereich=bmwi2012,sprache=de,rwb=true.pdf

Bundesregierung. (2016a). *Antwort der Bundesregierung auf die Kleine Anfrage der Abgeordneten Eva Bulling-Schröter, Caren Lay, Kerstin Kassner, weiterer Abgeordneter und der Fraktion DIE LINKE. – Drucksache 18/7057: Genese sowie Kosten und Wirkung des Klimaschutzbeitrags der Stromwirtschaft bis zum Jahr 2020.* (Bundestagsdrucksache, 18/7321). Retrieved from http://dip21.bundestag.de/dip21 /btd/18/073/1807321.pdf

Bundesregierung. (2016b). *Unterrichtung durch die Bundesregierung: Ausschreibungsbericht nach § 99 des Erneuerbare-Energien-Gesetzes.* (Bundestagsdrucksache, 18/7287). Retrieved from http://dip21.bundestag.de/dip21/btd/18/072/1807287.pdf

Bundestag. (1989). *Stenographischer Bericht 128. Sitzung.* (Plenarprotokoll, 11/128). Retrieved from http://dip21.bundestag.de/dip21/btp/11/11128.pdf

Bundestag. (1999). *Stenographischer Bericht 79. Sitzung: 16 December 1999.* (Plenarprotokoll, 14/79). Retrieved from http://dipbt.bundestag.de/doc/btp/14/14079.pdf #P.7272

Bundestag. (2000). *Gesetz für den Vorrang Erneuerbarer Energien (Erneuerbare-Energien-Gesetz – EEG) sowie zur Änderung des Energiewirtschaftsgesetzes und des Mineralölsteuergesetzes.* BGBl. I 2000 S. 305.

Bundestag. (2004). *Konsolidierte Fassung der Begründung zu dem Gesetz für den Vorrang Erneuerbarer Energien (Erneuerbare-Energien-Gesetz – EEG).* BGBl. 2004 I S. 1918.

Bundestag. (2010). *Entwicklung der Solarpartnerschaft in der Mittelmeerunion: Antwort der Bundesregierung auf die Kleine Anfrage der Fraktion der SPD-Drucksache 17/2506.* (Bundestagsdrucksache, 17/2676). Retrieved from http://dip21.bundestag.de/dip21/btd/17/026/1702676.pdf

Bundestag. (2013a). *Schriftliche Fragen mit den in der Woche vom 2. September 2013 eingegangenen Antworten der Bundesregierung.* (Drucksache, 17/14712). Retrieved from http://dipbt.bundestag.de/dip21/btd/17/147/1714712.pdf

Bundestag. (2013b). *Stenografischer Bericht. 3. Sitzung*: 28 November 2013. (Plenarprotokoll, 18/3). Retrieved from http://dip21.bundestag.de/dip21/btp/18/18003.pdf

Bundestag. (2014a). *Stenografischer Bericht 38. Sitzung: 4 June 2014.* (Plenarprotokoll, 18/38). Retrieved from http://dipbt.bundestag.de/doc/btp/18/18038.pdf

Bundestag. (2014b). *Stenografischer Bericht 63. Sitzung: 06 November 2014.* (Plenarprotokoll, 18/63). Retrieved from http://dipbt.bundestag.de/doc/btp/18/18063.pdf

Bundestag. (2014c). *Stenografischer Bericht. 66. Sitzung: 13 November 2014.* (Plenarprotokoll, 18/66). Retrieved from http://dip21.bundestag.de/dip21/btp/18/18066.pdf

Bundestag. (2014d). *Stenografischer Bericht 73. Sitzung: 4 December 2014.* (Plenarprotokoll, 18/73). Retrieved from http://dipbt.bundestag.de/doc/btp/18/18073.pdf

Bundestag. (2014e). *Stenografischer Bericht. 77. Sitzung: 19 December 2014.* (Plenarprotokoll, 18/77). Retrieved from http://dipbt.bundestag.de/doc/btp/18/18077.pdf

Bundestag. (2015a). Einzelplan 05 - Auswärtiges Amt. In *Gesetz über die Feststellung des Bundeshaushaltsplans für das Haushaltsjahr 2016 (Haushaltsgesetz 2016).* BGBl I S. 2378.

Bundestag. (2015b). Einzelplan 16 - Bundesministerium für Umwelt, Naturschutz, Bau und Reaktorsicherheit. In *Gesetz über die Feststellung des Bundeshaushaltsplans für das Haushaltsjahr 2016 (Haushaltsgesetz 2016).* BGBl I S. 2378.

Bundestag. (2015c). Einzeplan 09 - Bundesministerium für Wirtschaft und Energie. In *Gesetz über die Feststellung des Bundeshaushaltsplans für das Haushaltsjahr 2016 (Haushaltsgesetz 2016).* BGBl I S. 2378.

Bundestag. (2015d). *Stenografisches Protokoll 97. Sitzung: 26 March 2015.* (Plenarprotokoll, 18/97). Retrieved from http://dipbt.bundestag.de/doc/btp/18/18097.pdf

Bundesverband Erneuerbare Energien e.V. (2015). *BEE-Stellungnahme zum BMWi-Eckpunktepapier für Ausschreibungen für die Förderung von Erneuerbare-Energien-Anlagen vom 31.07.2015.* Retrieved from http://www.bee-ev.de/fileadmin/Publikationen/Positionspapiere_Stellungnahmen/20151001_BEE-Stellungnahme_zum_Eckpunktepapier_zu_Ausschreibungen_fin.pdf

Bündnis 90/Die Grünen. (2016). Energiewende: EEG 2016 wird zur Vollbremsung. Retrieved from https://www.gruene-bundestag.de/themen/energie/eeg-2016-wird-zur-vollbremsung.html

Bündnis 90/Die Grünen Bayern. (27 March 2015). *15 Jahre Erneuerbare-Energien-Gesetz: Das erfolgreichste Klimaschutzgesetz der Welt auf der Kippe.* Retrieved from http://www.hans-josef-fell.de/content/index.php/dokumente/weitere-themenbereiche/881-pressekonferenz-15-jahre-eeg/file

Bündnis Bürgerenergie e.V. (2015). *Stellungnahme des Bündnis Bürgerenergie zum Eckpunkte Papier "Ausschreibungen für die Förderung von Erneuerbaren Energien".* Retrieved from https://www.buendnis-buergerenergie.de/fileadmin/user_upload/Ausschreibungen_Stellungnahme_BBEn_300915_final.pdf

Busch, P.-O. (2003). *Die Diffusion von Einspeisevergütungen und Quotenmodellen: Konkurrenz der Modelle in Europa.* (FFU-Report 03-2003). Retrieved from http://www.polsoz.fu-berlin.de/polwiss/forschung/systeme/ffu/publikationen/2003/busch_per-olof_2003/rep-2003-03.pdf

Busch, P.-O., & Jörgens, H. (2005a). *Globale Ausbreitungsmuster umweltpolitischer Innovationen.* (FFU-Report, 02-2005). Retrieved from http://www.polsoz.fu-berlin.de/polwiss/forschung/systeme/ffu/publikationen/2005/busch_per_olof_joergens_helge_2005/rep_2005_02.pdf

Busch, P.-O., & Jörgens, H. (2005b). International patterns of environmental policy change and convergence. *European Environment, 15*(2): 80–101. https://doi.org/10.1002/eet.374

Busch, P.-O., & Jörgens, H. (2005c). The international sources of policy convergence: explaining the spread of environmental policy innovations. *Journal of European Public Policy, 12*(5): 860–884.

Busch, P.-O., & Jörgens, H. (2010). *Governance by diffusion. International environmental policy coordination in the era of globalization* (Doctoral thesis). Berlin: Freie Universität.

Busch, P.-O., & Jörgens, H. (2012). Governance by diffusion: exploring a new mechanism of international policy-coordination. In J. Meadowcroft, O. Langhelle, & A. Rudd (Eds.), *Governance, democracy and sustainable development. Moving beyond the impasse* (pp. 221–248). Cheltenham: Edward Elgar.

Busch, P.-O., Jörgens, H., & Tews, K. (2005). The Global Diffusion of Regulatory Instruments: The Making of a New International Environmental Regime. *The ANNALS of the American Academy of Political and Social Science, vol. 598*(1): 146–167.

California ISO. (2015). Leading the way to 50%. Agenda. 2015 Stakeholder Symposium. October 22 & 23, 2015. Retrieved from https://www.caiso.com/Documents/2015StakeholderSymposiumAgendaFinal.pdf

California State Legislature. (2009). *Senate Bill 32. Negrete McLeod. Renewable electric generation facilities.* Retrieved from http://leginfo.legislature.ca.gov/faces/billTextClient.xhtml?bill_id=200920100SB32

Calitz, J., Mushwana, C., & Bischof-Niemz, T. (2015). *Financial benefits of renewables in South Africa in 2015: Actual diesel- and coal-fuel savings and avoided "unserved energy" from the first operational 1.8 GW of wind and PV projects in a constrained South African power system.* CSIR Energy Centre. Retrieved from http://www.csir.co.za/media_releases/docs/Financial%20benefits%20of%20Wind%20and%20PV%202015.pdf

Carafa, L. (25 June 2010). *When The Birds Fly Together. Analysing Conditions of Rule Extension in the Context of EU-Turkey Energy Cooperation.* (ECPR Standing Group on the EU: Fifth Pan-European Conference on EU Politics, 25 June 2010). Retrieved from http://www.jhubc.it/ecpr-porto/virtualpaperroom/168.pdf

Carafa, L. (2011). *The Mediterranean Solar Plan through the Prism of External Governance.* (EUSA Twelfth Biennial International Conference, March 3-5 2011). Boston. Retrieved from http://euce.org/eusa/2011/papers/1k_carafa.pdf

Carafa, L. (2015). Policy and Markets in the MENA: The Nexus between Governance and Renewable Energy Finance. *International Conference on Concentrating Solar Power and Chemical Energy Systems, SolarPACES 2014, 69:* 1696-1703. https://doi.org/10.1016/j.egypro.2015.03.132

Carlyle, R. (4 October 2013). Should Other Nations Follow Germany's Lead On Promoting Solar Power?. *Forbes*. Retrieved from http://www.forbes.com/sites/quora/2013/10/04/should-other-nations-follow-germanys-lead-on-promoting-solar-power/#78e1f9c03866

Carroll, P., & Common, R. (Eds.). (2013). *Policy transfer and learning in public policy and management: International contexts, content and development*. Milton Park, Abingdon, Oxon: Routledge.

Carstensen, M. B., & Schmidt, V. A. (2016). Power through, over and in ideas: conceptualizing ideational power in discursive institutionalism. *Journal of European Public Policy, 23*(3): 318–337. https://doi.org/10.1080/13501763.2015.1115534

CEC. (n.d.). Nuclear Energy & Nuclear Issues: Nuclear Energy in California. California Energy Commission. Retrieved from http://www.energy.ca.gov/nuclear/

CEC. (2007). *Integrated Energy Policy Report*. Sacramento: California Energy Commission.

CEC. (2015). *Renewable Energy – Overview*. California Energy Commission. Retrieved from http://www.energy.ca.gov/renewables/tracking_progress/documents/renewable.pdf

CEC. (2016). Renewable Portfolio Standard. California Energy Commission. Retrieved from http://www.energy.ca.gov/portfolio/

Chandler, J. (2009). Trendy solutions: Why do states adopt Sustainable Energy Portfolio Standards? *Energy Policy, 37*(8): 3274–3281. https://doi.org/10.1016/j.enpol.2009.04.032

Charmaz, K. (2006). *Constructing grounded theory: A practical guide through qualitative analysis*. London, Thousand Oaks, CA: SAGE Publications.

Charmaz, K. (2014). *Constructing grounded theory* (2nd ed). London, Thousand Oaks, Calif: Sage.

Checkel, J. (1997). International Norms and Domestic Politics: Bridging the Rationalist--Constructivist Divide. *European Journal of International Relations, 3*(4): 473–495. https://doi.org/10.1177/1354066197003004003

Checkel, J. T. (2008). Process Tracing. In A. Klotz & D. Prakash (Eds.), *Qualitative Methods in International Relations. A Pluralist Guide* (pp. 114–127). Basingstoke, New York: Palgrave Macmillan.

Clark, P. (08 October 2014). German energy market a disaster, says EDF chief. *Financial Times Online*. Retrieved from http://www.ft.com/intl/cms/s/0/9cbf4d8c-4f0d-11e4-9c88-00144feab7de.html#axzz42bwFs2tw

CLEW. (n.d.). Clean Energy Wire. Retrieved from https://www.cleanenergywire.org

Cludius, J., Hermann, H., Matthes, F. C., & Graichen, V. (2014). The merit order effect of wind and photovoltaic electricity generation in Germany 2008–2016: Estimation and distributional implications. *Energy economics, 44:* 302–313. https://doi.org/10.1016/j.eneco.2014.04.020

Cohen, M. D., March, J. G., & Olsen, J. P. (1972). A garbage can model of organizational choice. *Administrative science quarterly (ASQ), 17*(1): 1–25.

Cole, D. H. (2015). Advantages of a polycentric approach to climate change policy. *Nature Climate Change, 5*(2): 114–118. https://doi.org/ dx.doi.org/10.1038/nclimate2490

Coleman, W. D., & Perl, A. (1999). Internationalized Policy Environments and Policy Network Analysis. *Political Studies, 47*(4): 691–709. https://doi.org/10.1111/1467-9248.00225

Collier, D. (2011). Understanding Process Tracing. *Political Science & Politics, 44*(04): 823–830. https://doi.org/10.1017/S1049096511001429

Collier, P. (2000). Conditionality, dependence and coordination: Three current debates in aid policy. In C. L. Gilbert & D. Vines (Eds.), *The World Bank. Structure and policies* (pp. 299–324). Cambridge, New York: Cambridge University Press.

Common, R. (2013). When policy diffusion does not lead to policy transfer: Explaining resistance to international learning in public management reform. In P. Carroll & R. Common (Eds.), *Policy transfer and learning in public policy and management. International contexts, content and development* (pp. 13–29). Milton Park, Abingdon, Oxon: Routledge.

Corbin, J., & Strauss, A. (1990). Grounded theory research: Procedures, canons, and evaluative criteria. *Qualitative Sociology, 13*(1): 3–21. https://doi.org/10.1007/BF00988593

Cortell, A. P., & Davis Jr, J. W. (2000). Understanding the Domestic Impact of International Norms: A Research Agenda. *International Studies Review, 2*(1): 65–87. https://doi.org/10.1111/1521-9488.00184

Couture, T., Cory, K., Kreycik, C., & Williams, E. (2010). *A Policymaker's Guide to Feed-in Tariff Policy Design.* (NREL Technical Report, NREL/TP-6A2-44849). Retrieved from http://www.nrel.gov/docs/fy10osti/44849.pdf

Cox, R. H., & Dekanozishvili, M. (2015) German Efforts to Shape European Renewable Energy Policy. In J. Tosun, S. Biesenbender, & K. Schulze (Eds.), *Lecture Notes in Energy: Vol. 28. Energy policy making in the EU. Building the agenda* (pp. 167–184). London: Springer-Verlag.

CPUC. (2009). An international panel of experts discuss feed-in tariffs: August 27, 2009. Retrieved from http://www.cpuc.ca.gov/General.aspx?id=6822

CPUC. (2015a). *Beyond 33% Renewables: Grid Integration Policy for a Low-Carbon Future: A CPUC Staff White Paper.* Retrieved from http://www.cpuc.ca.gov/uploadedFiles/CPUC_Website/Content/Utilities_and_Industries/Energy/Reports_and_White_Papers/Beyond33PercentRenewables_GridIntegrationPolicy_Final.pdf

CPUC. (2015b). Rule 21 Smart Inverter Working Group Technical Reference Materials. Retrieved from http://www.energy.ca.gov/electricity_analysis/rule21/

CPUC. (2016a). About the California Solar Initiative. Retrieved from http://www.cpuc.ca.gov/General.aspx?id=6047

CPUC. (2016b). *Order Instituting Rulemaking to Develop a Successor to Existing Net Energy Metering Tariffs Pursuant to Public Utilities Code Section 2827.1, and to Address Other Issues Related to Net Energy Metering*. Decision 16-01-044 January 28, 2016. Retrieved from http://docs.cpuc.ca.gov/PublishedDocs/Published/G000/M158/K181/158181678.pdf

Crow, D. (2012). Policy Diffusion and Innovation: Media and Experts in Colorado Recreational Water Rights. *Journal of Natural Resources Policy Research, 4*(1): 27–41. https://doi.org/10.1080/19390459.2012.642635

Cull, N. (2010). Public diplomacy: Seven lessons for its future from its past. *Place Branding and Public Diplomacy, 6*(1): 11–17.

Curren, J., Makhele, L., Jakubowski, A., Goldblatt, M., Langiss, O., Basteck, T., & Schiffner, A. (2009). *Regional Regulatory Action Plan for the Western Cape*. Commissioned by the Western Cape Department of Environmental Affairs and. Retrieved from https://www.westerncape.gov.za/other/2010/5/regional_regulatory_action_plan_final_report.pdf

Daguerre, A. (2004). Importing Workfare: Policy Transfer of Social and Labour Market Policies from the USA to Britain under New Labour. *Social Policy & Administration, 38*(1), 41-56.

Dalmasso, E. (2012). Surfing the Democratic Tsunami in Morocco: Apolitical Society and the Reconfiguration of a Sustainable Authoritarian Regime. *Mediterranean Politics, 17*(2): 217–232. https://doi.org/10.1080/13629395.2012.694045

Damro, C. (2010). *Market Power Europe: EU Externalisation of Market-Related Policies.* Mercury. (Mercury E-Paper, 5). Retrieved from http://www.san.ed.ac.uk/_data/assets/pdf_file/0004/85288/E-paper_no5_r2010_Market_Power_Europe.pdf

Daviter, F. (2015). The political use of knowledge in the policy process. *Integrating Knowledge and Practice to Advance Human Dignity, 48*(4): 491–505. https://doi.org/10.1007/s11077-015-9232-y

DB Climate Change Advisors (Ed.). (2011). *GET FiT Plus: De- Risking Clean Energy Business Models in a Developing Country Context.* Retrieved from https://www.db.com/cr/de/docs/GET_FiT_Plus.pdf

Deegan, H. (2011). *Politics South Africa* (2nd ed). Harlow, Essex, New York: Longman/Pearson.

Dehmer, D. (25 September 2007). Die Klimakanzlerin. *Der Tagesspiegel*. Retrieved from http://www.tagesspiegel.de/politik/die-klimakanzlerin/1051226.html

Dehmer, D. (2013). The German Energiewende: The First Year. *The Electricity Journal, 26*(1): 71–78. https://doi.org/10.1016/j.tej.2012.12.001

de Jong, M. (2009). Rose's '10 steps': why process messiness, history and culture are not vague and banal. *Policy and politics, 37*(1): 145–150.

Della Porta, D., & Keating, M. (2008). How many approaches in the social sciences? An epistemological introduction. In D. Della Porta & M. Keating (Eds.), *Approaches and methodologies in the social sciences. A pluralist perspective* (pp. 19–39). Cambridge New York: Cambridge University Press.

dena. (2016). International Cooperation. Retrieved from http://www.dena.de/en/international.html

Département de l'Environnement. (2009). *Plan national de lutte contre le réchauffement climatique*. Retrieved from http://climateobserver.org/wp-content/uploads/2014/09/Morocco_PNRC-2009.pdf

Department of Economic Development. (2010). *The New Growth Path: The Framework*. Retrieved from http://www.gov.za/sites/www.gov.za/files/NGP%20Framework%20for%20public%20release%20FINAL_1.pdf

Department of Minerals and Energy. (1998). *White Paper on the Energy Policy of the Republic of South Africa*. Pretoria. Retrieved from http://www.energy.gov.za/files/policies/whitepaper_energypolicy_1998.pdf

Department of Minerals and Energy. (2003). *White Paper on Renewable Energy*. Retrieved from https://unfccc.int/files/meetings/seminar/application/pdf/sem_sup1_south_africa.pdf

DeShazo, J. R., & Matulka, R. (2009). *Best Practices for Implementing a Feed-in Tariff Program*. UCLA Luskin Center for Innovation Report. Retrieved from http://innovation.luskin.ucla.edu/sites/default/files/Best%20Practices%20for%20Implementing%20a%20Feed%20in%20Tariff%20Program.pdf

DeShazo, J. R., & Matulka, R. (2010a). *Bringing Solar Energy to Los Angeles: An Assessment of the Feasibility and Impacts of an In-basin Solar Feed-in Tariff Program*. UCLA Luskin Center for Innovation Report. Retrieved from http://innovation.luskin.ucla.edu/sites/default/files/Bringing%20Solar%20to%20Los%20Angeles.pdf

DeShazo, J. R., & Matulka, R. (2010b). *Designing an Effective Feed-in Tariff for Greater Los Angeles*. LA Business Council Study in partnership with UCLA Luskin Center for Innovation. Retrieved from http://innovation.luskin.ucla.edu/sites/default/files/35-Designing_an_Effective_Feed-in_Tariff.pdf

Dey, I. (2007). Grounding Categories. In A. Bryant & K. Charmaz (Eds.), *The SAGE Handbook of Grounded Theory* (pp. 168–189). Los Angeles, London, New Delhi, Singapore: SAGE Publications.

Diekmann, J., Kemfert, C., Neuhoff, K., Schill, W.-P., & Traber, T. (2012). Erneuerbare Energien: Quotenmodell keine Alternative zum EEG. *DIW Wochenbericht*, (45/2012): 15–20.

DIgSILENT. (2009). *Grid Integration of Wind Energy in the Western Cape: Final Report*. Commissioned by GTZ, D:EA&DP and Eskom. Retrieved from https://www.westerncape.gov.za/other/2010/5/grid_study_gtz-deadp-eskom_final_report.pdf

Dimaggio, P., & Powell, W. (1983). The iron cage revisited: institutional isomorphism and collective rationality in organizational fields. *American Sociological Review, 48*(2): 147–160.

Direction de l'Observation et de la Programmation. (n.d.). Chiffres clés du secteur de l'énergie année 2011. Retrieved from http://www.mem.gov.ma/ChiffresCles/Energie/ChiffreEnergie%20annuelhtm.html

Direction de l'Observation et de la Programmation. (2014). *Statistiques Enérgétiques: Avril 2014*. MEMEE. Retrieved from http://www.mem.gov.ma/SiteAssets/PdfCHCle1/energie/mensuel/2014/chifcles04-14.pdf

DIW. (15 December 2010). Dringender Appell zur Rettung des Erneuerbare-Energien-Gesetzes seitens deutscher Energiewissenschaftler. Press release. Retrieved from http://www.diw.de/de/diw_01.c.364745.de/themen_nachrichten/dringender_appell_zur_rettung_des_erneuerbare_energien_gesetzes_seitens_deutscher_energiewissenschaftler.html

DoE. (2013). *Integrated Resource Plan for Electricity (IRP) 2010-2030*. Department of Energy. Retrieved from http://www.doe-irp.co.za/content/IRP2010_updatea.pdf

DoE. (2015). *State of Renewable Energy in South Africa*. Department of Energy. Retrieved from http://www.gov.za/sites/www.gov.za/files/State%20of%20Renewable%20Energy%20in%20South%20Africa_s.pdf

Dobbin, F., Simmons, B., & Garrett, G. (2007). The Global Diffusion of Public Policies: Social Construction, Coercion, Competition, or Learning? *Annual Review of Sociology, 33:* 449–472. https://doi.org/10.1146/annurev.soc.33.090106.142507

Dolowitz, D., & Marsh, D. (1996). Who learns what from whom: A review of the policy transfer literature. *Political Studies, 44*(2): 343–357.

Dolowitz, D. P., & Marsh, D. (2000). Learning from Abroad: The Role of Policy Transfer in Contemporary Policy-Making. *Governance, 13*(1): 5–24.

Dolowitz, D. P., & Marsh, D. (2012). The Future of Policy Transfer Research. *Political Studies Review, 10*(3): 339–345. https://doi.org/10.1111/j.1478-9302.2012.00274.x

Drezner, D. W. (2001). Globalization and Policy Convergence. *International Studies Review, 3*(1): 53–78.

Drezner, D. W. (2008). *All politics is global: Explaining international regulatory regimes*. Princeton, N.J, Woodstock: Princeton University Press.

DSIRE. (2015). Renewable Market Adjusting Tariff (ReMAT). Database of State Incentives for Renewables & Efficiency. Retrieved from http://programs.dsireusa.org/system/program/detail/5665

Duit, A., Feindt, P., & Meadowcroft, J. (2016). Greening Leviathan: the rise of the environmental state? *Environmental Politics, 25*(1): 1–23. https://doi.org/10.1080/09644016.2015.1085218

Dunlop, C. A. (2009). Policy transfer as learning: capturing variation in what decision-makers learn from epistemic communities. *Policy Studies, 30*(3): 289–311. https://doi.org/ 10.1080/01442870902863869

Dussauge-Laguna, M. I. (2012). The neglected dimension: bringing time back into cross-national policy transfer studies. *Policy Studies, 33*(6): 567–585.

Dussauge-Laguna, M. I. (2013). Contested policy transfer: when Chile's 'Programa de Mejoramiento de la Gestión' travelled to Mexico. In P. Carroll & R. Common (Eds.), *Policy transfer and learning in public policy and management. International contexts, content and development* (pp. 167–187). Milton Park, Abingdon, Oxon: Routledge.

Eastern Cape Sustainable Energy. (n.d.). Connecting for a green energy future. Retrieved from http://greenenergy-ec.co.za

Eberhard, A. (2013). *South Africa's electricity crisis: How did we get here? And how do we put things right?* (CDE Round Table, 10). Retrieved from http://www.cde.org.za/wp-content/uploads/2013/02/South%20Africas%20Electricity%20Crisis%20Full%20Report.pdf

Eberhard, A. (2014). *South Africa's Renewable Energy IPP Procurement Program: Success Factors and Lessons: Presented to BMZ and KfW, Berlin. 1 December 2014.* Retrieved from http://www.gsb.uct.ac.za/files/BerlinREIPPPpresentation1Dec14.pdf

Eberhard, A., Kolker, J., & Leigland, J. (2014). *South Africa's Renewable Energy IPP Procurement Program: Success Factors and Lessons.* Public Private Infrastructure Advisory Group (PPIAF). World Bank Group. Retrieved from http://www.gsb.uct.ac.za/files/PPIAFReport.pdf

Eckermann, F. (2016). *Die Umweltwirtschaft in Deutschland 2015: Entwicklung, Struktur und internationale Wettbewerbsfähigkeit.* Umweltbundesamt. Retrieved from https://www.umweltbundesamt.de/sites/default/files/medien/376/publikationen/dieumweltwirtschaft_in_deutschland_2015.pdf

Eckersley, R. (2011). *Does climate leadership matter?* Australian Political Studies Association Annual Conference 26-28 September. (Australian Political Studies Association Conference Proceedings). Acton. Retrieved from https://web.archive.org/web/20120201000000*/http://law.anu.edu.au/COAST/events/APSA/papers/89.pdf

Eckersley, R. (2016). National identities, international roles, and the legitimation of climate leadership: Germany and Norway compared. *Environmental Politics, 25*(1): 180–201.

Ecologic. (n.d.). Visitors Program. Retrieved from http://ecologic.eu/de/visitors-program

Economic Freedom Fighters. (06 October 2015). *EFF urges the South African government to strongly consider Germany's advice on nuclear energy.* Press release. Retrieved from http://effighters.org.za/eff-urges-the-south-african-government-to-strongly-consider-germanys-advice-on-nuclear-energy/

Eddy, M. (10 December 2015). California and Germany, Opposites With a Common Energy Goal. *The New York Times.* Retrieved from http://www.nytimes.com/2015/12/11/world/europe/california-and-germany-opposites-with-a-common-energy-goal.html?_r=0

Edenhofer, O., Flachsland, C., Jakob, M., & Lessmann, K. (2013). *The Atmosphere as a Global Commons – Challenges for International Cooperation and Governance.* (Discussion Paper 2013-58). Cambridge, Mass.: Harvard Project on International Climate Agreements.

Edkins, Max: Marquard, Andrew, & Winkler, H. (2010). *Assessing the effectiveness of national solar and wind energy policies in South Africa.* Final Report for the United Nations Environment Programme Research Project "Enhancing information for renewable energy technology deployment in Brazil, China and South Africa". Energy Research Centre, University of Cape Town. Retrieved from http://www.erc.uct.ac.za/sites/default/files/image_tool/images/119/Papers-2010/10Edkinesetal-Solar_and_wind_policies.pdf

Edwards, S. (1997). Trade liberalization reforms and the World Bank. *American Economic Review, 87*(2): 43–48.

El Aidi, A. (2013). *Maroc et Union européenne: perspetives pour un partenariat énergétique*. Rabat: Konrad Adenauer Stiftung. Retrieved from http://www.kas.de/wf/doc/kas_35380-1522-3-30.pdf?130912152857

Elkins, Z., & Simmons, B. (2005). On Waves, Clusters, and Diffusion: A Conceptual Framework. *The ANNALS of the American Academy of Political and Social Science, 598*(1): 33–51. https://doi.org/10.1177/0002716204272516

Ellerman, D. (2000). Knowledge-based development assistance. *Knowledge, technology and policy, 12*(4): 17-43.

Elliott, D. E. (2013). Why the United States does not have a renewable energy policy. *Environmental Law Reporter, 43*(2): 10095–10101.

Energieagentur.NRW. (n.d.). Die Außenwirtschaftsaktivitäten der EnergieAgentur.NRW. Retrieved from http://www.energieagentur.nrw/international/ueber-uns3

Enkhardt, S. (9 May 2016). Agora Energiewende: 95 Prozent Anteil von Photovoltaik, Windkraft & Co. *pv magazine*. Retrieved from http://www.pv-magazine.de/nachrichten/details/beitrag/agora-energiewende--95-prozent-anteil-von-photovoltaik-windkraft--co_100022960/

Enzili, M., Nayysa, A., Affani, F., & Simonis, P. (1998). *Wind Energy in Morocco Potential - State of the Art - Perspectives.* (DEWI Magazin, N°12). Retrieved from http://www.dewi.de/dewi/fileadmin/pdf/publications/Magazin_12/07.pdf

Essop, T. (2008). *Budget Speech by Tasneem Essop, Minister of Environment, Planning, and Economic Development.* Retrieved from https://www.westerncape.gov.za/other/2008/8/budget_speech_2008_environment__and_dev_planning_vote_9.pdf

Ethik-Kommission Sichere Energieversorgung. (2011). *Deutschlands Energiewende – Ein Gemeinschaftswerk für die Zukunft.* Berlin. Retrieved from http://www.bundesregierung.de/ContentArchiv/DE/Archiv17/_Anlagen/2011/07/2011-07-28-abschlussbericht-ethikkommission.pdf?__blob=publicationFile&v=4

European Commission. (2005). *The support of electricity from renewable energy sources.* (Communication from the European Commission, COM(2005) 627).

EUROSOLAR. (2012). *Scheinheilige Debatte um das Erneuerbare-Energien-Gesetz EEG*. Press release. Bonn. Retrieved from http://www.eurosolar.de/de/index.php/pressemitteilungen-2012-archivmenupressemitteil-379/1649-scheinheilige-debatte-um-das-erneuerbare-energien-gesetz-eeg

EUROSOLAR, & WCRE (Eds.). (2009 (2001)). *The Long Road to IRENA: From the Idea to the Foundation of the International Renewable Energy Agency*. Bochum: Ponte Press Verlags-GmbH.

Evans, M. (2004a). Introduction: Is Policy Transfer Rational Policy-making? In M. Evans (Ed.), *Policy transfer in global perspective* (pp. 1–9). Aldershot, Hants, England; Burlington, VT: Ashgate.

Evans, M. (Ed.). (2004b). *Policy transfer in global perspective*. Aldershot, Hants, England; Burlington, VT: Ashgate.

Evans, M. (2004c). Understanding Policy Transfer. In M. Evans (Ed.), *Policy transfer in global perspective* (pp. 10–42). Aldershot, Hants, England; Burlington, VT: Ashgate.

Evans, M. (2009a). New directions in the study of policy transfer. *Policy Studies, 30*(3): 237–241. https://doi.org/10.1080/01442870902863810

Evans, M. (2009b). Policy transfer in critical perspective. *Policy Studies, 30*(3): 243–268.

Evans, M. (2010a). In conclusion: Parting shots. In M. Evans (Ed.), *New directions in the study of policy transfer* (pp. 157–162). London: Routledge.

Evans, M. (2010b). Policy transfer in critical perspective. In M. Evans (Ed.), *New directions in the study of policy transfer* (pp. 6–31). London: Routledge.

Evans, M., & Davies, J. (1999). Understanding Policy Transfer: A Multi-Level, Multi-Disciplinary Perspective. *Public administration, 77*(2): 361–385.

Fabeck, W. von. (2006). Die kostendeckende Vergütung - Eine Idee geht um die Welt. Solarenergie Förderverein Deutschland E.V. 12 December 2006. Retrieved from http://www.sfv.de/artikel/die_kostendeckende_verguetung_-_eine_idee_geht_um_die_welt.htm

Falkner, R. (2007). The political economy of 'normative power' Europe: EU environmental leadership in international biotechnology regulation. *Journal of European Public Policy, 14*(4): 507–526. https://doi.org/10.1080/13501760701314326

Fankhauser, S., Gennaioli, C., & Collins, M. (2015). Do international factors influence the passage of climate change legislation? *Climate Policy*: 1–14. https://doi.org/10.1080/14693062.2014.1000814

Faull, L. (07 October 2011). Battle for South Africa's R1-trillion nuclear contract. *Mail & Guardian Online*. Retrieved from http://mg.co.za/article/2011-10-07-r1trillion-nuclear-tender-bidding-war

Fell, H.-J. (2014). Bewertung der EEG-Novelle: Allgemeine Rahmenbedingungen. *pv magazine*. Retrieved from http://www.pv-magazine.de/nachrichten/details/beitrag/bewertung-der-eeg-novelle--allgemeine-rahmenbedingungen_100014873/?L=1%255C&cHash=5e5ce63354ec8aaa8c1652d0723f45ad

Fell, H.-J. & Ehring, G. (2014). “Gabriel bremst den Ausbau erneuerbarer Energien”. Interviewer: Ehring, Georg. Deutschlandfunk. Retrieved from http://www.deutschlandfunk.de/eeg-novelle-gabriel-bremst-den-ausbau-erneuerbarer-energien.697.de.html?dram:article_id=293367

Fine, B., and Z. Rustomjee (1996). *The Political Economy of South Africa: From Minerals-Energy-Complex to Industrialisation.* London: C. Hurst & Co. Ltd.

Finkel, T., Koch, C., & Roloff, N. (2013). *Endbericht: Evaluierung der Exportinitiative Energieeffizienz: Studie im Auftrag des Bundesministeriums für Wirtschaft und Technologie*. Commissioned by BMWi. Retrieved from https://www.efficiency-from-germany.info/ENEFF/Redaktion/DE/Downloads/Publikationen/Zur_Exportinitiative/evaluierungsbericht_como.pdf?__blob=publicationFile&v

Finnemore, M. (1993). International organizations as teachers of norms: The United Nations' Educational, Scientific, and Cultural Organization and science policy. *International Organization, 47*(4): 565–597.

Finnemore, M., & Sikkink, K. (1998). International Norm Dynamics and Political Change. *International Organization, 52*(4): 887–917.

Fischer, F. (2003). *Reframing public policy: Discursive politics and deliberative practices*. Oxford, New York: Oxford University Press.

Fischer, S., & Geden, O. (2011). *Die deutsche Energiewende europäisch denken.* (SWP Aktuell, N°47). Retrieved from https://www.swp-berlin.org/fileadmin/contents/products/aktuell/2011A47_fis_gdn_ks.pdf

Flachsland, C., Lessmann, K., & Edenhofer, O. (2012). Climate Policy in a Decentralised World. In O. Edenhofer (Ed.), *Climate change, justice and sustainability. Linking climate and development policy* (pp. 257–268). Dordrecht: Springer.

Fourie, D., Kritzinger-van Niekerk, L., & Nel, M. (2015). An overview of the renewable energy independent power producers procurement programme (REIPPPP). *Energize RE, 3*(June 2015): 9–12.

Frankfurt School-UNEP Centre, & BNEF. (2016). *Global trends in renewable energy investment.* Frankfurt School of Finance & Management. Frankfurt am Main. Retrieved from http://fs-unep-centre.org/sites/default/files/publications/globaltrendsinrenewableenergyinvestment2016lowres_0.pdf

Friedman, T. L. (06 May 2015). Germany, the Green Superpower. *The New York Times.* Retrieved from http://www.nytimes.com/2015/05/06/opinion/thomas-friedman-germany-the-green-superpower.html?_r=0

Friedrich Ebert Stiftung. (n.d.). FES International. Retrieved from https://www.fes.de/de/standorte/fes-international/

Friedrich Naumann Stiftung. (n.d.). Standorte International. Retrieved from https://www.freiheit.org/content/standorte-international

Fritsche, U., & Schmidt, K. (2008). *Schwerpunktanalyse Regenerative Energien für die Region Nordafrika/Naher Osten (MENA) mit Ergänzungen zur Energieeffizienz.* Darmstadt, Berlin: BMZ and Öko-Institut e.V.

Fukuda-Parr, S., Lopes, C., & Malik, K. (2002). *Capacity for development: New solutions to old problems*. Sterling, Va.: Earthscan Publications.

Gabriel, S. (2008). *Bundestagsrede von Bundesminister Sigmar Gabriel zur Verabschiedung des Erneuerbare-EnergienGesetzes (EEG) und des Erneuerbare-Energien-Wärmegesetzes (EEWärmeG)*. BMU. Retrieved from http://www.bmub.bund.de/fileadmin/bmu-import/files/pdfs/allgemein/application/pdf/rede_gabriel_verabschg_eeg_080605.pdf

Gabriel, S. (08 May 2014). *Erste Beratung des von der Bundesregierung eingebrachten Entwurfs eines Gesetzes zur grundlegenden Reform des Erneuerbare-Energien-Gesetzes und zur Änderung weiterer Bestimmungen des Energiewirtschaftsrechts*. Speech in the German Bundestag. (Bundestagsdrucksache, 18/33: 2697-2700). Retrieved from http://dip21.bundestag.de/dip21/btp/18/18033.pdf

Galiteva, A., & Moss, D. (2014). *Germany California Learning and Collaboration Tour: Toward an Integrated Renewable Energy System*. Renewables 100 Institute. Retrieved from http://www.renewables100.org/fileadmin/templates/renewables/media/pdf/Germany-CA_Learning__Collaboration_Tour2014.pdf

Gallagher, D. R. (Ed.). (2012). *Environmental leadership: A reference handbook*. Vol. 1. Los Angeles, London, New Delhi, Singapore: SAGE Publications.

Gallagher, K. S. (2014). *The globalization of clean energy technology: Lessons from China. Urban and industrial environments*. Cambridge, Mass.: MIT Press.

Garrett, K. N., & Jansa, J. M. (2015). Interest Group Influence in Policy Diffusion Networks. *State Politics & Policy Quarterly, 15*(3): 387–417. https://doi.org/10.1177/1532440015592776

Gawel, E., Strunz, S., & Lehmann, P. (2016). *Support Policies for Renewables: Instrument choice and instrument change from a Public Choice perspective*. (WIDER Working Paper, 2016/6). Retrieved from https://www.wider.unu.edu/sites/default/files/wp2016-6.pdf

Gehrke, B., & Schasse, U. (February 2013). *Position Deutschlands im Außenhandel mit Gütern zur Nutzung erneuerbarer Energien und zur Steigerung der Energieeffizienz*. Niedersächsisches Institut für Wirtschaftsforschung. (Studien zum deutschen Innovationssystem, 9-2013). Retrieved from http://www.niw.de/uploads/pdf/publikationen/StuDIS_2013_09_Position_Deutschland_Aussenhandel_Energien.pdf

Gehrke, B., & Schasse, U. (2015). *Die Umweltschutzwirtschaft in Deutschland: Produktion, Umsatz und Außenhandel*. Commissioned by Umweltbundesamt. Retrieved from https://www.umweltbundesamt.de/sites/default/files/medien/378/publikationen/uib_04_2015_umweltschutzwirtschaft_in_deutschland.pdf

Genovese, F., Kern, F., & Martin, C. (2015). *Policy Osmosis: Rethinking diffusion processes when policies have substitutes*. Research Group Comparative Politics, Interdependence and Globalization, Christian-Albrechts-Universität zu Kiel. (Working Papers, 16 February 2015). Retrieved from http://www.cpig.uni-kiel.de/de/research/working-papers/documents/policy-osmosis-rethinking-diffusion-processes-when-policies-have-alternatives

George, A. L., & Bennett, A. (2005). *Case studies and theory development in the social sciences*. Cambridge, Mass: MIT Press.

Gerding, H. (2011). *German financing of energy projects in the region: The role of Renewables and CCS.* KfW Development Bank. Retrieved from http://siteresources.worldbank.org/INTENERGY2/Resources/4114191-1311604578704/Harald_Gerding_kfw.pdf

German Embassy Rabat. (n.d.). Relations économiques bilatérales avec l'Allemagne. Retrieved from http://www.rabat.diplo.de/Vertretung/rabat/fr/05/Bilaterale__wirtschaftsbeziehungen.html

German Missions in the United States. (2015). Transatlantic Climate Bridge. Retrieved from http://www.germany.info/climatebridge

German Missions in South Africa, Lesotho, and Swaziland. (n.d.). Germany: a green nation. Retrieved from http://www.southafrica.diplo.de/Vertretung/suedafrika/en/08__Science__Environment/Environment__Climate/Green__Germany/Green_Germany.html

German Missions in South Africa, Lesotho, and Swaziland. (2014). South African parliamentarians visit Germany. Retrieved from http://www.southafrica.diplo.de/Vertretung/suedafrika/en/_pr/__Embassy/2014/4th__Q/12-Parliamentarians-Study-Visit.html

German Missions in South Africa, Lesotho, and Swaziland. (2015). Journalist Workshop on "Green Topics": Tools and techniques of investigative journalism. Retrieved from http://www.southafrica.diplo.de/Vertretung/suedafrika/en/__pr/__Embassy/2015/1st__Q/03-JournoWorshop.html

Germanwatch. (n.d.). *Die zehn größten CO2-emittierenden Länder nach Anteil an den weltweiten CO2-Emissionen im Jahr 2015.* (Statista - Das Statistik-Portal). Retrieved from http://de.statista.com/statistik/daten/studie/179260/umfrage/die-zehn-groessten-c02-emittenten-weltweit/.

Gerring, J. (2004). What Is a Case Study and What Is It Good for? *American Political Science Review*, *98*(2 2): 341–354.

Gerring, J. (2007). Is There a (Viable) Crucial-Case Method? *Comparative Political Studies, 40*(3): 231–253.

Gerring, J. (2008). Case Selection for Case-Study Analysis: Qualitative and Quantitative Techniques. In J. M. Box-Steffensmeier (Ed.), *The Oxford Handbook of Political Methodology* (pp. 645–682). Oxford: Oxford University Press.

Gilardi, F. (2012). Methods for the analysis of policy interdependence. Retrieved from http://www.fabriziogilardi.org/resources/papers/policy-interdep-v2.pdf

Gilardi, F. (2013a). *Four theses on policy diffusion research in political science.* Paper prepared for the workshop "How to study diffusion: Theories, methods, and research designs". Max Planck Institute for the Study of Religious and Ethnic Diversity. Göttingen, 25-26 October 2013.

Gilardi, F. (2013b). Transnational diffusion: Norms, ideas, and policies. In W. Carlsnaes, T. Risse-Kappen, & B. A. Simmons (Eds.), *Handbook of international relations* (pp. 453–478). London, Thousand Oaks, CA: SAGE Publications.

Gilardi, F. (2014). Methods for the Analysis of Policy Interdependence. In I. Engeli & C. R. Allison (Eds.), *Comparative Policy Studies: Conceptual and Methodological Challenges* (pp. 185-204). London: Palgrave Macmillan UK.

Gilardi, F., & Foglister, K. (2008). Empirical Modeling of Policy Diffusion in Federal States: The Dyadic Approach. *Swiss Political Science Review, 14*(3): 413–450.

Giljova, S. (2015). *National Biogas Platform: Feedback.* Presentation at National Biogas Conference, 5-6 March 2015, IDC. Retrieved from http://www.energy.gov.za/files/biogas/2015-Biogas-Conference/Day-1/National-Biogas-Platform-Feedback.pdf

Gipe, P. (2015). Debunking Myths about Germany's EEG. http://www.wind-works.org/cms/index.php?id=368,

GIZ. (n.d.-a). Questions and answers about GIZ. Retrieved from https://www.giz.de/en/press/9785.html

GIZ. (n.d.-b). Project Data: Morocco. Retrieved from https://www.giz.de/projektdaten/index.action?request_locale=en_EN#?region=3&countries=MA

GIZ. (2012). Kooperation beim Klimaschutz. Retrieved from https://www.giz.de/de/mediathek/6159.html

GIZ. (2013). *Energy connects.* Eschborn. Retrieved from http://www.giz.de/fachexpertise/downloads/Giz2013-en-Energy-Connects.pdf

GIZ, Department of Trade and Industry, & SASTELA. (2013). *Assessment of the localisation, industrialisation and job creation potential of CSP infrastructure projects in South Africa: A 2030 vision for CSP.* Retrieved from http://www.record.org.za/resources/downloads/item/giz-satela-dti-csp-localisation-potential-study

GIZ. (2014a). Energy. Retrieved from http://www.giz.de/expertise/html/2023.html

GIZ. (2014b). Le Projet DKTI. Retrieved from http://dkti-maroc.org

GIZ. (2014c). *Les projets de la GIZ au Maroc.* Retrieved from http://www.giz.de/en/downloads/giz2014_fr_liste_projets_Maroc.pdf

GIZ. (2015a). *Germany in the Eyes of the World: Key findings of the second GIZ survey (2015).* Retrieved from https://www.giz.de/de/downloads/giz2015-en-germany-in-the-eyes-of-the-world-2015.pdf

GIZ. (2015b). *GIZ in South Africa: Programmes and Projects.* Retrieved from https://www.giz.de/en/downloads/giz2015-en-programmes-projects-south-africa.pdf

GIZ. (2016). Projektdaten. Retrieved from https://www.giz.de/projektdaten/index.action

GIZ, BWE, VDMA, & BMWi. (2015). *Programm: 10. Windenergie- und Entwicklungsdialog.* Berlin. Retrieved from https://www.wind-energie.de/sites/default/files/events/10-windenergie-und-entwicklungsdialog-2015/weed-2015-programm.pdf

Glaser, B. G., & Strauss, A. L. (1967). *The discovery of grounded theory: Strategies for qualitative research. Grounded theory*. Chicago: Aldine Pub. Co.

Glaser, B. G., & Strauss, A. L. (1999). *The discovery of grounded theory: Strategies for qualitative research.* New York: Aldine de Gruyter.

Global Environment Facility. (2015). *Equity Fund for the Small Projects Independent Power Producer Procurement Programme*. Retrieved from http://www.thegef.org/gef/sites/thegef.org/files/gef_prj_docs/GEFProjectDocuments/Climate%20Change/South%20Africa%20-%20(9085)%20-%20Equity%20Fund%20for%20the%20Small%20Projects%20Independent%20Pow/Equity_Fund_for_Small_IPPs_-_PIF_-_DBSA_revised.pdf

Gourevitch, P. (1978). The second image reversed: the international sources of domestic politics. *Int Org, 32*(4): 881–912. https://doi.org/10.1017/S002081830003201X

Govender, S. (26.01.2016). South African opposition seeks to harness discontent over Zuma. *Deutsche Welle*. Retrieved from http://dw.com/p/1HkL1

Governor's Press Office. (2015). Governor Brown, International Leaders Form Historic Partnership to Fight Climate Change. Press release. Retrieved from http://under2mou.org/?page_id=447

Grace, R. (2008). *California Feed-in Tariff Design & Policy Options*. California Energy Commission. Feed-in Tariff Workshop #2. October 1, 2008. Presentation. Retrieved from http://www.energy.ca.gov/portfolio/documents/2008-10-01_workshop/presentations/KEMA_Robert_Grace.pdf

Grace, R., Rickerson, W., Corfee, K., Porter, K., & Cleijne, H. (2008). *California Feed-In Tariff Design and Policy Options. California Energy Commission.* (CEC-300-2008-009F). Retrieved from http://www.energy.ca.gov/2008publications/CEC-300-2008-009/CEC-300-2008-009-F.PDF

Graichen, P. (2014). *10 Questions and Answers on the 2014 Reform of the German Renewable Energy Act*. Retrieved from http://www.agora-energiewende.de/fileadmin/Projekte/2013/EEG-20/Agora_Energiewende_Background_EEG_2014_08292014_web.pdf

Graichen, P., Kleiner, M., Litz, P., & Podewils, C. (2015). *Die Energiewende im Stromsektor: Stand der Dinge 2014: Rückblick auf die wesentlichen Entwicklungen sowie Ausblick auf 2015.* (Agora Energiewende, Analysis). Berlin.

Graichen, P., & Redl, C. (2014). *The German Energiewende and its Climate Paradox: An Analysis of Power Sector Trends for Renewables, Coal, Gas, Nuclear Power and CO2 Emissions, 2010-2030.* (Agora Energiewende, Analysis). Retrieved from http://www.agora-energiewende.org/fileadmin/downloads/publikationen/Analysen/Trends_im_deutschen_Stromsektor/Analysis_Energiewende_Paradox_web_EN.pdf

Greenpeace. (2007). *Mit dem EEG zum Ökostrom: Das Erneuerbare-Energien-Gesetz (EEG) sichert Strom aus Erneuerbaren Energien.* Hamburg. Retrieved from http://www.greenpeace.de/sites/www.greenpeace.de/files/fseeg2007_1.pdf

Grossback, L. J., Nicholson-Crotty, S., & Peterson, D. A. M. (2004). Ideology and Learning in Policy Diffusion. *American Politics Research, 32*(5): 521–545. https://doi.org/10.1177/1532673X04263801

GTZ. (2009). *TERNA Wind Energy Programme 1997-2009: Impact Report*. Retrieved from http://hinfo.humaninfo.ro/gsdl/genus/documents/s20880en/s20880en.pdf

GTZ. (2010). Erneuerbare Energien sind der Motor der Energiewende. Retrieved from http://www.netzhammerbreiholz.de/index.php/gtz-factsheets.html

GTZ, & DME. (2004). *Solar Cooking Compendium Vol. 4: The solar cooking toolkit: Conclusions from the South African Field Test for future solar cooking projects*. Retrieved from https://energypedia.info/images/b/b2/Solar_cooking_compendium_vol_4_toolkit.pdf

Gullberg, A. T., Ohlhorst, D., & Schreurs, M. (2014). Towards a low carbon energy future – Renewable energy cooperation between Germany and Norway. *Renewable Energy, 68:* 216–222.

Gunther, E. (2008). 1st Solar Symposium: Feed-in Tariff for California First. Retrieved from http://guntherportfolio.blogspot.de/2008/07/1st-solar-symposium-feed-in-tariff-for.html

Haas, P. M. (1992). Introduction: epistemic communities and international policy coordination. *International Organization, 46*(1): 1–35.

Haas, R., Resch, G., Panzer, C., Busch, S., Ragwitz, M., & Held, A. (2011). Efficiency and effectiveness of promotion systems for electricity generation from renewable energy sources - Lessons from EU countries. *Energy, 36*(4): 2186-2193. https://doi.org/10.1016/j.energy.2010.06.028

Hafidi, M. A. (29 October 2012). Le sotuien des Allemands n'est pas gratuit. Interview with Badr Ikken. *Le Soir Echos*. Retrieved from http://www.maroc-akhbar.com/nouvelle.php?id=3149

Hajer, M. A. (1993). Discourse Coalitions and the Institutionalisation of Practice: The Case of Acid Rain in Britain. In F. Fischer & J. Forester (Eds.), *The Argumentative turn in policy analysis and planning* (pp. 43–76). Durham, N.C: Duke University Press.

Hake, J.-F., Fischer, W., Venghaus, S., & Weckenbrock, C. (2015). The German Energiewende – History and status quo. *Sustainable Development of Energy, Water and Environment Systems, 92, Part 3:* 532–546. https://doi.org/10.1016/j.energy.2015.04.027

Hakelberg, L. (2014). Governance by Diffusion: Transnational Municipal Networks and the Spread of Local Climate Strategies in Europe. *Global Environmental Politics, 14*(1): 107–129. https://doi.org/10.1162/GLEP_a_00216

Hall, P. A. (1993). Policy paradigms, social learning, and the state: The case of economic policymaking in Britain. *Comparative Politics, 25*(3): 275–296.

Hanns Seidel Stiftung. (n.d.). International Cooperation. Retrieved from http://www.hss.de/english/international-relations/international-cooperation.html

Haselip, J., Nygaard, I., Hansen, U., & Ackom, E. (Eds.). (2011). *Technology Transfer Perspectives Series. Diffusion of renewable energy technologies: case studies of enabling frameworks in developing countries*. Roskilde: UNEP Risoe Centre.

Hawiger, M. (19 November 2008). Feed-in Tariffs—Right for Germany, Wrong for California. *San Francisco Chronicle*. Retrieved from http://www.sfgate.com/opininion/article/Feed-in-tariffs-right-for-Germany-wrong-for-3184612.php

Hay, C. (2006). Constructivist Institutionalism. In R.A.W., Rhodes, S. A. Binder, & B. A. Rockman (Eds.), *The Oxford handbook of political institutions* (pp. 55–74). Oxford, New York: Oxford University Press.

Hay, C., & Wincott, D. (1998). Structure, Agency and Historical Institutionalism. *Political Studies, 46*(5): 951–957. https://doi.org/10.1111/1467-9248.00177

Hayward, J. (2008). *Introduction: Inhibited Consensual Leadership within an Interdependent Confederal Europe.* In Hayward, J. *Leaderless Europe*. Oxford University Press: Oxford Scholarship Online.
https://doi.org/ 10.1093/acprof:oso/9780199535026.003.0001

Hecking, H., Kruse, J., Paschmann, M., Polisadov, A., & Wildgrube, T. (2016). *Ökonomische Effekte eines deutschen Kohleausstiegs auf den Strommarkt in Deutschland und der EU.* ewi Energy Research & Scenarios gGmbH. Retrieved from http://www.ewi.research-scenarios.de/cms/wp-content/uploads/2016/05/ewi_ers_oekonomische_effekte_deutscher_kohleausstieg.pdf

Heclo, H. (1974). *Modern social politics in Britain and Sweden: From relief to income maintenance.* New Haven, London: Yale University Press.

Heeter, J., Barbose, G. L., Bird, L., Weaver, S., Flores-Espinosa, F., Kuskova-Burns, K., & Wiser, R.(2014). *Survey of State-Level Cost and Benefit Estimates of Renewable Portfolio Standards*. National Renewable Energy Laboratory and Lawrence Berkeley National Laboratory. (NREL/TP-6A20-61042, LBNL-6589E). Retrieved from http://www.nrel.gov/docs/fy14osti/61042.pdf

Heichel, S., Pape, J., & Sommerer, T. (2005). Is there convergence in convergence research?: An overview of empirical studies on policy convergence. *Journal of European Public Policy, 12*(5): 817–840. https://doi.org/10.1080/13501760500161431

Heiden, N., & Strebel, F. (2012). What about non-diffusion? The effect of competitiveness in policy-comparative diffusion research. *Policy Sciences, 2012, 45*(4): 345–358.

Heinrich Böll Stiftung. (n.d.-a). Auslandsbüros. Retrieved from https://www.boell.de/de/stiftung/auslandsbueros

Heinrich Böll Stiftung. (n.d.-b). Energy Transition. The German Energiewende. Retrieved from http://energytransition.de

Heinze, T. (2011). *Mechanism-Based Thinking on Policy Diffusion. A Review of Current Approaches in Political Science.* (KFG Working Paper Series, 34). Retrieved from http://userpage.fu-berlin.de/kfgeu/kfgwp/wpseries/WorkingPaperKFG_34.pdf

Hekkert, M. P., Suurs, R. A. A., Negro, S. O., Kuhlmann, S., & Smits, R. E. H. M. (2007). Functions of innovation systems: A new approach for analysing technological change. *Technological Forecasting and Social Change, 74*(4): 413–432. https://doi.org/10.1016/j.techfore.2006.03.002

Héritier, A. (1996). The accommodation of diversity in European policy-making and its outcomes: Regulatory policy as a patchwork. *Journal of European Public Policy, 3*(2): 149–167. https://doi.org/10.1080/13501769608407026

Héritier, A. (2001). Differential Europe: New Opportunities and Restrictions for Policy-making in the Member State. In A. Windhoff-Héritier, D. Kerwer, C. Knill, D. Lemkuhl, M. Teutsch, & A.-C. Douillet (Eds.), *Differential Europe. The European Union impact on national policymaking* (pp. 1–22). Lanham, Md: Rowman & Little-field.

Héritier, A. (2008). Causal explanation. In D. Della Porta & M. Keating (Eds.), *Approaches and methodologies in the social sciences. A pluralist perspective* (pp. 61–79). Cambridge New York: Cambridge University Press.

Hierl, J. K. (2011). *Das integrierte Energie- und Klimaprogramm der großen Koalition: Das parlamentarische Verfahren zum Erneuerbare-Energien-Wärmegesetz und zum Gesetz zur Änderung der Förderung von Biokraftstoffen.* Frankfurt: Peter Lang.

Hirsch, T. (2015). *Learning from the "Energiewende": What developing countries expect from Germany.* Friedrich Ebert Stiftung. Berlin. Retrieved from http://library.fes.de/pdf-files/iez/11304.pdf

Hirschl, B. (2008). *Erneuerbare Energien-Politik: Eine Multi-Level Policy-Analyse mit Fokus auf den deutschen Strommarkt. VS Research.* Wiesbaden: VS Verlag für Sozialwissenschaften.

Hirschl, B. (2009). International renewable energy policy—between marginalization and initial approaches. *Energy Policy, 37*(11): 4407–4416. https://doi.org/10.1016/j.enpol.2009.05.059

Hoel, M. (1991). Global environmental problems: The effects of unilateral actions taken by one country. *Journal of Env. Economics and Management, 20*(1): 55–70. https://doi.org/10.1016/0095-0696(91)90023-C

Holden, C. (2009). Exporting public-private partnerships in healthcare: export strategy and policy transfer. *Policy Studies, 30*(3): 313–332. Retrieved from https://doi.org/10.1080/01442870902863885

Holzinger, K., & Knill, C. (2005a). Causes and conditions of cross-national policy convergence. *Journal of European Public Policy, 12*(5): 775–796. https://doi.org/10.1080/13501760500161357

Holzinger, K. & Knill, C. (2005b). Competition, Cooperation and Communication: A Theoretical Analysis of Different Sources of Environmental Policy Convergence and Their Interaction. (IHS Political Science Series, N°102). Retrieved from http://www.ihs.ac.at/publications/pol/pw_102.pdf

Holzinger, K., Jörgens, H., & Knill, C. (Eds.). (2007). *Transfer, Diffusion und Konvergenz von Politiken*: VS Verlag für Sozialwissenschaften.

Holzinger, K., Knill, C., & Sommerer, T. (2008). Environmental Policy Convergence: The Impact of International Harmonization, Transnational Communication, and Regulatory Competition. *International Organization*, (62): 553587.

Holzinger, K., Knill, C., & Sommerer, T. (2009). *Is There Convergence of National Environmental Policies?: An Analysis of Policy Outputs in 24 OECD Countries*. Konstanz: Universität Konstanz.

Holzinger, K., Knill, C., & Sommerer, T. (2010). Umweltpolitik zwischen Annäherung und Aufholjagd: Eine Analyse umweltpolitischer Konvergenz in 24 OECD-Ländern. *Zeitschrift für Umweltpolitik & Umweltrecht: ZfU 33*(1): 1–31.

Hood, C. (2007). Intellectual Obsolescence and Intellectual Makeovers: Reflections on the Tools of Government after Two Decades. *Governance, 20*(1): 127-144. https://doi.org/10.1111/j.1468-0491.2007.00347.x

Hoppmann, J., Huenteler, J., & Girod, B. (2014). Compulsive policy-making—The evolution of the German feed-in tariff system for solar photovoltaic power. *Research Policy, 43*(8): 1422–1441. https://doi.org/10.1016/j.respol.2014.01.014

Horowitz, M. (2007). *The diffusion of military power. Causes and consequences for international politics* (Doctoral thesis.). Cambridge, Mass.: Dept. of Government, Harvard University.

Howlett, M. (2009). Governance modes, policy regimes and operational plans: A multi-level nested model of policy instrument choice and policy design. *Policy Sciences, 42*(1): 73-89. https://doi.org/10.1007/s11077-009-9079-1

Howlett, M., & Joshi-Koop, S. (2011). Transnational learning, policy analytical capacity, and environmental policy convergence: Survey results from Canada. *Global Environmental Change, 21*(1): 85–92. https://doi.org/10.1016/j.gloenvcha.2010.10.002

Howlett, M., & Ramesh, M. (1993). Patterns of Policy Instrument Choice: Policy Styles, Policy Learning and the Privatization Experience. *Policy Studies Review, 12*(1/2): 3–24.

Howlett, M., & Ramesh, M. (2002). The policy effects of internationalization: A subsystem adjustment analysis of policy change. *Journal of Comparative Policy Analysis: Research and Practice, 4*(1): 31–50. https://doi.org/10.1080/13876980208412669

Howlett, M., Ramesh, M., & Perl, A. (2009). *Studying public policy: Policy cycles & policy subsystems* (3rd ed.). Ontario: Oxford University Press.

Howlett, M., & Rayner, J. (2008). Third Generation Policy Diffusion Studies and the Analysis of Policy Mixes: Two Steps Forward and One Step Back? *Journal of Comparative Policy Analysis: Research and Practice, 10*(4), 385-402

Huber, D. (2015). *Democracy promotion and foreign policy: Identity and interests in US, EU and non-Western democracies.* Basingstoke: Palgrave Macmillan.

Hugo, V. (1877). *The History of a crime*. New York: G. Munro.

IEA. (2014). *Morocco 2014: Energy Policies Beyond IEA Countries*. Paris: International Energy Agency. Retrieved from https://www.iea.org/publications/freepublications/publication/Morocco2014.pdf

IEA. (2015a). *Recent trends in world CO2 emissions from fuel combustion*. Retrieved from https://www.iea.org/media/news/2015/news/151104_webarticle_CO2_FINAL.pdf

IEA. (2015b). *World Energy Outlook 2015*. Paris: International Energy Agency.

IEA-RETD (2016), *RE Transition –Transitioning to Policy Frameworks for Cost-Competitive Renewables* .(Jacobs et al., IET –International Energy Transition GmbH). Utrecht: IEA Technology Collaboration Programme for Renewable Energy Technology Deployment (IEA-RETD).

IfaS & GTZ. (15.01.2010). *Etude sur les potentiels de biomasse pour la région Souss-Massa-Drâa et la province d'Essaouira*. Retrieved from http://www.giz.de/themen/en/26777.htm

Ikenberry, G. J., & Kupchan, C. A. (1990). Socialization and Hegemonic Power. *International Organization, 44*(3): 283–315. https://doi.org/10.2307/2706778

Brandt, W. (Ed.). (1980). *North-South, a programme for survival: Report of the Independent Commission on International Development Issues*. Cambridge, Mass: MIT Press.

International Climate Initiative. (2016). Projects. Retrieved from http://www.international-climate-initiative.com/en/nc/projects/projects/

International Feed-In Cooperation. (n.d.-a). Events. Retrieved from http://www.feed-in-cooperation.org/wDefault_7/content/events.php

International Feed-In Cooperation. (n.d.-b). Founding Declarations. Retrieved from http://www.feed-in-cooperation.org/wDefault_7/content/founding-declarations.php

IRENA. (2014). *REmap 2030: A Renewable Energy Roadmap*. Abu Dhabi: International Renewable Energy Agency. Retrieved from http://irena.org/remap/IRENA_REmap_Report_June_2014.pdf

IRENA. (2015a). IRENA Membership. Retrieved from http://www.irena.org/Menu/Index.aspx?mnu=Cat&PriMenuID=46&CatID=67

IRENA. (2015b). *Renewable Power Generation Costs in 2014*. Retrieved from http://www.irena.org/documentdownloads/publications/irena_re_power_costs_2014_report.pdf

IRENA. (2015c). *Renewable Energy Prospects: Germany*. (REmap 2030 analysis report). Abu Dhabi: International Renewable Energy Agency.

Jachtenfuchs, M. (1996). *International policy-making as a learning process?: The European Union and the greenhouse effect*. Aldershot: Avebury.

Jacob, K., Beise, M., Blazejczak Jürgen, Edler, D., Haum, R., Jänicke, M., . . . Rennings, K. (2005). *Lead Markets for Environmental Innovations. ZEW Economic Studies: Vol. 27*. Heidelberg: Physica.

Jacobs, D. (2012). *Renewable energy policy convergence in the EU: The evolution of feed-in tariffs in Germany, Spain and France*. Global environmental governance. Farnham: Ashgate.

Jacobs, D. (2014). Policy invention as evolutionary tinkering and codification: the emergence of feed-in tariffs for renewable electricity. *Environmental Politics, 23*(5): 755–773. https://doi.org/10.1080/09644016.2014.923627

Jacobs, D., & Mez, L. (2012). Zur internationalen Vorbildfunktion von StrEG und EEG. In T. Müller (Ed.), *20 Jahre Recht der Erneuerbaren Energien* (pp. 258–271). Baden-Baden: Nomos.

Jacobs, D., Peinl, H., Gotchev, B., Schäuble, D., Matschoss, P., Bayer, B…Goldammer, K. (2014). *Ausschreibungen für erneuerbare Energien in Deutschland – Ausgestaltungsoptionen für den Erhalt der Akteursvielfalt.* IASS. (IASS Working Paper, September 2014). https://doi.org/ doi.org/10.2312/iass.2014.015

Jäger, J. (2011). *Potentialstudie für photovoltaische Solarenergie für die Regionen Meknès-Tafilalet, Oriental und Souss-Massa-Drâa. Deutsche Gesellschaft für Internationale Zusammenarbeit (GIZ) GmbH.* Updated version November 2011. Retrieved from https://www.giz.de/de/downloads/giz2011-de-potentialstudie-marokko-solarenergie.pdf

Jahn, D. (1992). Nuclear power, energy policy and new politics in Sweden and Germany. *Environmental Politics, 1*(3): 383–417.

Jahn, D. (1997). Green Politics and Parties in Germany. *Political Quarterly, 68*(B), 174-182.

Jahn, D., & Korolczuk, S. (2012). German exceptionalism: the end of nuclear energy in Germany! *Environmental Politics, 21*(1): 159–164.

James, O., & Lodge, M. (2003). The Limitations of 'Policy Transfer' and 'Lesson Drawing' for Public Policy Research. *Political Studies Review, 1*(2): 179–193.

Jänicke, M. (1997). The political system's capacity for environmental policy. In H. Weidner & M. Jänicke (Eds.), *National environmental policies. A comparative study of capacity-building* (pp. 1–14). New York, Berlin: Springer.

Jänicke, M. (2005a). Trend-setters in environmental policy: the character and role of pioneer countries. *European Environment, 15*(2): 129–142. https://doi.org/10.1002/eet.375

Jänicke, M. (2011). German climate change policy: political and economic leadership. In R. Wurzel & J. Connelly (Eds.). *The European Union as a leader in international climate change politics* (pp. 129–146). London, New York: Routledge.

Jänicke, M. (2012a). *The Acceleration of Innovation in Climate Policy: Lessons from best practice.* (FFU-Report, 01-2011). Retrieved from http://www.policyinnovations.org/ideas/policy_library/data/01672/_res/id=sa_File1/Jänicke_2011_-_The_Acceleration_of_Innovationin_Climate_Policy_-_FFU-rep_01-2011.pdf

Jänicke, M. (2012b). Dynamic governance of clean-energy markets: how technical innovation could accelerate climate policies. *Journal of Cleaner Production, 22*(1): 50–59. https://doi.org/10.1016/j.jclepro.2011.09.006

Jänicke, M. (2013). *Accelerators of Global Energy Transition: Horizontal and Vertical Reinforcement in Multi-Level Climate Governance.* (IASS Working Paper). Potsdam. Retrieved from http://www.polsoz.fu-berlin.de/polwiss/forschung/systeme/ffu/files/working_paper_accelerators_of_global_energy_transition_4.pdf

Jänicke, M. (2014). Multi-Level Reinforcement in Climate Governance. In A. Brunnengräber & M. R. Di Nucci (Eds.), *Im Hürdenlauf zur Energiewende* (pp. 35–47). Wiesbaden: Springer Fachmedien.

Jänicke, M., & Jacob, K. (2004). Lead Markets for Environmental Innovations: A New Role for the Nation State. *Global Environmental Politics, 4*(1): 29–46. https://doi.org/10.1162/152638004773730202

Jänicke, M., & Jacob, K. (Eds.). (2007a). *Environmental governance in global perspective: New approaches to ecological and political modernisation* (2nd ed.). Berlin: Freie Universität Berlin, Environmental Policy Research Centre.

Jänicke, M., & Jacob, K. (2007b). Lead Markets for Environmental Innovations: A New Role for the Nation State. In M. Jänicke & K. Jacob (Eds.), *Environmental governance in global perspective. New approaches to ecological and political modernisation* (2nd ed., pp. 30–50). Berlin: Freie Universität Berlin, Environmental Policy Research Centre.

Jenkins-Smith, H. C., & Sabatier, P. A. (1993). The Dynamics of Policy-Oriented Learning. In P. A. Sabatier & H. C. Jenkins-Smith (Eds.), *Theoretical lenses on public policy. Policy change and learning. An advocacy coalition approach* (pp. 41–56). Boulder, Colo: Westview Press.

Joas, F., Pahle, M., & Flachsland, C. (2014). Die Ziele der Energiewende: Prioritäten: Eine Kartierung der Prioritäten. *ifo-Schnelldienst*, (09): 6–11.

Joas, F., Pahle, M., Flachsland, C., & Joas, A. (2016). Which goals are driving the Energiewende? Making sense of the German Energy Transformation. *Energy Policy, (95):* 42-51. https://doi.org/10.1016/j.enpol.2016.04.003

Joemat-Pattersson, T. (2015). *Expansion and Acceleration of the Independent Power Producer Procurement Programme*. Department of Energy. Pretoria. Retrieved from http://www.ipprenewables.co.za/#page/2179

Johnstone, B. (2011). *Switching to Solar: What We Can Learn from Germany's Success in Harnessing Clean Energy*. Kindle Edition. Amherst, New York: Prometheus Books.

Jones, B. D., & Baumgartner, F. R. (2012). From There to Here: Punctuated Equilibrium to the General Punctuation Thesis to a Theory of Government Information Processing. *The Policy Studies Journal, 40*(1): 1–20.

Jones, C. O. (1970). *An introduction to the study of public policy.* Belmont, Calif: Wadsworth Pub. Co.

Jordan, A., & Huitema, D. (2014). Innovations in climate policy: the politics of invention, diffusion, and evaluation. *Environmental Politics, 23*(5): 715–734. https://doi.org/10.1080/09644016.2014.923614

Jordana, J., Levi-Faur, D., & Fernandez i Marín, X. (2011). The Global Diffusion of Regulatory Agencies: Channels of Transfer and Stages of Diffusion. *Comparative Political Studies, 44*(10): 1343–1369. https://doi.org/10.1177/0010414011407466

Jung, S. (2015). *Vom Aachener Modell zum Erneuerbaren-Energien-Gesetz: Zur Historie und zu wesentlichen Regeln des heutigen EEG.* Presentation. Aachen, 18 May 2015. Retrieved from http://www.bdb-aachen.de/storages/BG/BDB-Aachen/user_upload/Seminare_2015/Skript_EEG_BDB-Aachen_Solarenergiefoerderverein.pdf

Kalman, A. (2007). Partner für Energie: Europa umwirbt die Maghreb-Staaten. Deutschlandfunk. Retrieved from http://www.deutschlandfunk.de/partner-fuer-energie.716.de.html?dram:article_id=90392

Karlsson, C., Hjerpe, M., Parker, C., & Linnér, B.-O. (2012). The Legitimacy of Leadership in International Climate Change Negotiations. *Ambio, 41*(Suppl 1): 46–55. https://doi.org/10.1007/s13280-011-0240-7

Karnitschnig, M. (26 August 2014). Germany's Expensive Gamble on Renewable Energy. *Wall Street Journal (Online).* Retrieved from http://www.wsj.com/articles/germanys-expensive-gamble-on-renewable-energy-1409106602

Karnitschnig, M. (25 November 2015). *Germany's green sticker shock: The laboratory for the planet's most progressive climate policies discovers hidden costs and unintended consequences.* Politico. Retrieved from http://www.politico.eu/article/the-good-green-german-gets-sticker-shock/

Keck, M., & Sikkink, K. (1999). Transnational Advocacy Networks in International and Regional Politics. *International Social Science Journal, 51*(1): 89–101.

Kelle, U. (2007). The Development of Categories: Different Approaches in Grounded Theory. In A. Bryant & K. Charmaz (Eds.), *The SAGE Handbook of Grounded Theory* (pp. 191–213). Los Angeles, London, New Delhi, Singapore: SAGE Publications.

Keller, W. (1996). Absorptive capacity: On the creation and acquisition of technology in development. *Journal of Development Economics 49*(1): 199–227. https://doi.org/10.1016/0304-3878(95)00060-7

KEMA. (2008). *Exploring Feed-in Tariffs for California.* (CEC-300-2008-003-D). Retrieved from http://www.energy.ca.gov/2008publications/CEC-300-2008-003/CEC-300-2008-003-D.PDF

KEMA. (2011). *European Renewable Distributed Generation Infrastructure Study – Lessons Learned from Electricity Markets in Germany and Spain.* Consultant Report. Prepared for the California Energy Commission. (CEC-400-2011-011). Retrieved from http://www.energy.ca.gov/2011publications/CEC-400-2011-011/CEC-400-2011-011.pdf

KEMA, Corfee, K., Karcher, M., Cleijne, H., Burgers, J., Faasen, C., . . . Gifford, J. (2010). *Feed-in Tariff Designs for California: implications for Project Finance, Competitive Renewable Energy Zones, and Data Requirements: Prepared for the California Energy Commission.* (CEC-300-2010-006).

Kenny, A. (2014a). *Nuclear energy is safe, clean, economic and sustainable.* Retrieved from http://www.nuclearafrica.co.za/pdf/information/NuclearAfrica%20A%20Kenny.pdf

Kenny, A. (21 January 2014b). Germany's green energy disaster. *The Citizen.* Retrieved from http://citizen.co.za/113777/germanys-green-energy-disaster/

Keohane, R. O., & Victor, D. G. (2010). *The Regime Complex for Climate Change.* (Discussion Paper, 2010-33). Cambridge, Mass.: Harvard Project on International Climate. Retrieved from http://belfercenter.ksg.harvard.edu/files/Keohane_Victor_Final_2.pdf

Kern, K., Jörgens, H., & Jänicke, M. (1999). *Die Diffusion umweltpolitischer Innovationen: Ein Beitrag zur Globalisierung von Umweltpolitik.* (FFU-Report, 99-11). Retrieved from http://www.polsoz.fu-berlin.de/polwiss/forschung/systeme/ffu/publikationen /1999/ kern_kristine_jaenicke_martin_joergens_helge_1999/rep-99-11.PDF

Kern, K., Jörgens, H., & Jänicke, M. (2001). *The diffusion of environmental policy innovations - a contribution to the globalisation of environmental policy.* Wissenschaftszentrum Berlin für Sozialforschung. (Discussion Paper, FSII 01-302). Berlin. Retrieved from http://papers.ssrn.com/sol3/papers.cfm?abstract_id=653583

KfW. (2014). *Current Topics: Renewable Energy.* Retrieved from https://www.kfw-entwicklungsbank.de/PDF/Entwicklungsfinanzierung/Issues-NEW/Energy-Renewable-Energy-EN-2015.pdf

KfW. (2015). KfW finanziert mit 300 Mio. EUR „grünes" Stromnetz in Südafrika. Retrieved from https://www.kfw.de/KfW-Konzern/Newsroom/Aktuelles/Pressemitteilun gen/Pressemitteilungen-Details_268992.html

KfW. (2016a). Förderprodukte Energie & Umwelt. Retrieved from https://www.kfw.de/inlandsfoerderung/Unternehmen/Energie-Umwelt/Förderprodukte/Förderprodukte-(S3).html

KfW. (2016b). *Solarkomplex Ouarzazate - Marokko: Strom aus der Wüste. Projektinformation.* Retrieved from https://www.kfw-entwicklungsbank.de/PDF/Entwick lungsfinanzierung/Länder-und-Programme/Nordafrika-Nahost/Projekt_Marokko_Solar_2016.pdf

KfW. (2016c). *Marokko: Vorreiter in Sachen Klimaschutz: Erstes Solarkraftwerk bei Ouarzazate in Betrieb genommen.* Retrieved from https://www.kfw-entwicklungs bank.de/Internationale-Finanzierung/KfW-Entwicklungsbank/News/News-Details_33 8561.html

King, K. (2005). Knowledge-based aid: A new way of networking or a new North-South divide? In D. Stone & S. Maxwell (Eds.), *Global knowledge networks and international development. Bridges across boundaries* (pp. 72–88). London, New York: Routledge.

King, G., Keohane, R. O., & Verba, S. (1994a). *Designing social inquiry: Scientific inference in qualitative research.* Princeton, N.J: Princeton University Press.

King, K., & McGrath, S. A. (2004). *Knowledge for development?: Comparing British, Japanese, Swedish and World Bank aid.* London, New York: HSRC Press; Cape Town, South Africa: Zed Books.

Kingdon, J. W. (1984). *Agendas, alternatives, and public policies.* Boston: Little, Brown.

Kingdon, J. W. (2003). *Agendas, alternatives, and public policies* (2nd ed.). New York: Longman.

Thurber, J. A. (Ed.). (2011). *Agendas, alternatives, and public policies* (Updated 2nd ed). Boston: Longman.

Kisby, B. (2007). Analysing policy networks: Towards an Ideational Approach. *Policy Studies, 28*(1): 71-90.

Klima-Allianz deutschland. (2013). *Die Energiewende klimafreundlich, zukunftsfähig, sozial gestalten.* Retrieved from https://germanwatch.org/en/download/7473.pdf

Klimaretter. (2007). Marokko importiert das EEG. Retrieved from http://www.klimaretter.info/klimacamp-blog-mainmenu-298?start=40

Klingler-Vidra, R., & Schleifer, P. (2014). Convergence More or Less: Why Do Practices Vary as They Diffuse? *International Studies Review, 16*(2): 264–274.

Klingler-Vidra, R. (2014). *All Politics is Local. Sources of variance in the diffusion of venture capital policies.* (Doctoral thesis). London: London School of Economics.

Knill, C., Heichel, S., & Arndt, D. (2012). Really a front-runner, really a Straggler? Of environmental leaders and laggards in the European Union and beyond — A quantitative policy perspective. *Energy Policy, 48*(0): 36–45. https://doi.org/10.1016/j.enpol.2012.04.043

Knill, C., & Lenschow, A. (1997). *Coping with Europe: The impact of British and German administrations on the implementation of EU environmental policy.* (EUI RSC 97:57). San Domenico (FI): European University Institute.

Knill, C., Schulze, K., & Tosun, J. (2010). Politikwandel und seine Messung in der vergleichenden Staatstätigkeitsforschung: Konzeptionelle Probleme und mögliche Alternativen. *Politische Vierteljahresschrift, 51*(3): 409-432. https://doi.org/10.1007/s11615-010-0022-z

Knill, C., & Tosun, J. (2009). Hierarchy, networks, or markets: how does the EU shape environmental policy adoptions within and beyond its borders? *Journal of European Public Policy, 16*(6): 873–894. https://doi.org/10.1080/13501760903088090

Kohler, D. & Kreith, F. (1985). Morocco Renewable Energy Development: Evaluation Report. Retrieved from http://pdf.usaid.gov/pdf_docs/pdaau034.pdf

Komoto, K., Mazakazu, I., van der Vleuten, Peter, Faiman, D., & Kurokawa, K. (Eds.). (2009). *Energy from the desert: Very large scale photovoltaic power-state of the art and into the Futire*. London, Sterling, VA: Earthscan.

Konrad Adenauer Stiftung. (n.d.). Büros weltweit. Retrieved from http://www.kas.de/wf/de/71.4782/

Konrad Adenauer Stiftung. (2013). *Wahrnehmung der deutschen Energiewende in Schwellenlaendern: Ergebnisse einer qualitativen Expertenbefragung in Brasilien, China und Suedafrika*. Berlin: Konrad Adenauer Stiftung. Retrieved from http://www.kas.de/wf/de/33.34940/

Konrad Adenauer Stiftung. (2014). *Wahrnehmungen der deutschen Energiewende in Schwellenländern: Teil 2 - Ergebnisse einer qualitativen Expertenbefragung in Russland und Indien*. Berlin: Konrad Adenauer Stiftung. Retrieved from http://www.kas.de/wf/doc/kas_38988-544-1-30.pdf?140930140631

Kopp, O., Engelhorn, T., Onischka, M., Ehrhart, K.-M., Pietrzyk, S., Klessmann, C...Grave, K. (2013). *Wege in ein wettbewerbliches Strommarktdesign für erneuerbare Energien*. Retrieved from https://www.mvv-energie.de/media/media/downloads/mvv_energie_gruppe_1/nachhaltigkeit_1/MVV_Studie_EE_Marktdesign_2013.pdf

Kowalzig, J. (2015). *Aufbruch in den Aufwuchs bis 2020?: Zwischenstand zur Klimafinanzierung aus Deutschland*. Oxfam Deutschland. (Kurz-Briefing, 20 November). Retrieved from http://www.deutscheklimafinanzierung.de/wp-content/uploads/2015/11/HH2016_Oxfam_KURZ_Überblick_Klimafinanzierung_ver14November15.pdf

KPCC Wire Services. (2009). LA Mayor Villaraigosa signs agreement with Berlin clean tech park. Retrieved from http://www.scpr.org/news/2009/12/14/9151/la-mayor-villaraigosa-signs-agreement-berlin-clean/

Krasner, S. D. (1982). Structural causes and regime consequences: regimes as intervening variables. *Int Org, 36*(2): 185–205. https://doi.org/10.1017/S0020818300018920

Krause, F., Bossel, H., & Müller-Reissmann, K.-F. (1980). *Energie-Wende: Wachstum und Wohlstand ohne Erdöl und Uran: Ein Alternativ-Bericht des Öko-Instituts, Freiburg* (3rd ed.). Frankfurt am Main: Fischer.

Krischer, O. (2010). *Hintergrundpapier: Kosten und Nutzen des EEG*. Retrieved from http://oliver-krischer.eu/fileadmin/user_upload/gruene_btf_krischer/2010/EEGUmlageHintergrundpapier.pdf

Kristov, L., & Keehn, S. (2013). From the Brink of Abyss to a Green, Clean, and Smart Future: The Evolution of California's Electricity Market. In F. P. Sioshansi (Ed.), *Evolution of global electricity markets. New paradigms, new challenges, new approaches* (pp. 297–329). Waltham, MA: Elsevier Science & Technology.

Kübler, K. (2014). Zur Vorreiterrolle Deutschlands. *Energiewirtschaftliche Tagesfragen: et, 64*(8): 30–32.

Kückmann, F. (10 December 2014). Gönner nennt EEG Exportschlager. *Neue Osnabrücker Zeitung*. Retrieved from http://www.noz.de/deutschland-welt/politik/artikel/529370/gonner-nennt-eeg-exportschlager

Kunzig, R. (15 October 2015). Germany Could Be a Model for How We'll Get Power in the Future. *National Geographic*. Retrieved from http://ngm.nationalgeographic.com/2015/11/climate-change/germany-renewable-energy-revolution-text

Ladi, S. (2004). Environmental Policy Transfer in Germany and Greece. In M. Evans (Ed.), *Policy transfer in global perspective* (pp. 79–92). Aldershot, Hants, England; Burlington, VT: Ashgate.

Ladi, S. (2005). *Globalisation, policy transfer and policy research institutes*. Cheltenham, UK, Northampton, MA: E. Elgar.

Ladi, S. (2011). Policy Change and Soft Europeanization: The Transfer of the Ombudsman Institution to Greece, Cyprus and Malta. *Public administration, 89*(4): 1643–1663. https://doi.org/10.1111/j.1467-9299.2011.01929.x

LADWP. (2008). *Post Workshop Comments of the Los Angeles Dpeartment of Water and Power*. Los Angeles Department of Water and Power. CEC Docket 03-RPS-1078. Retrieved from http://docketpublic.energy.ca.gov/PublicDocuments/Migration-12-22-2015/IEPR/2008%20IEPR/08-IEP-1/October/TN%2048564%2010-10-08%20LOS%20ANGELES%20DEPT%20OF%20WATER%20AND%20POWER%20RE%20POST%20WORKSHOP%20COMMENTS.pdf

LADWP. (2010). Solar Power in Los Angeles: German Delegation Visit. Retrieved from http://www.ladwpnews.com/external/content/document/1475/908175/1/Solar-

Lahn, G., Padgett, S., & Lavenex, S. (2009). *External European Union Governance in Energy and the Environment*. Seminar Report, University of Strathclyde, 16 September 2009. Retrieved from http://www.chathamhouse.org/sites/default/files/public/Research/Energy,%20Environment%20and%20Development/160909s_report.pdf

Lana, X., & Evans, M. (2004). Policy Transfer Between Developing Countries: The Transfer of the Bolsa-Escola Programme to Ecuador. In M. Evans (Ed.), *Policy transfer in global perspective* (pp. 190–210). Aldershot, Hants, England ; Burlington, Vt.: Ashgate.

Land, R. (2009). Energiewende international: Interview with Hermann Scheer. *Berliner Debatte Initial, 20*(2): 67–74.

Lascoumes, P., & Le Galès, P. (2007). Introduction: Understanding Public Policy through Its Instruments-From the Nature of Instruments to the Sociology of Public Policy Instrumentation. *Governance, 20*(1): 1-21.
https://doi.org/10.1111/j.1468-0491.2007.00342.x

Lauber, V., & Jacobsson, S. (2015). The politics and economics of constructing, contesting and restricting socio-political space for renewables – The German Renewable Energy Act. *Environmental Innovation and Societal Transitions, 18*: 147–163. https://doi.org/10.1016/j.eist.2015.06.005

Lauber, V., & Mez, L. (2004). Three decades of renewable electricity policies in Germany. *Energy & Environment, 15*(4): 599–623.

Lauber, V., & Pesendorfer, D. (2004). Success through continuity: renewable electricity policies in Germany. In I. de Lovinfosse & F. Varone (Eds.), *Renewable electricity policies in Europe. Tradable green certificates in competitive markets* (pp. 121–182). Louvain-la-Neuve: Presses universitaires de Louvain.

Lavenex, S., Lehmkuhl, D., & Wichmann, N. (2009). Modes of external governance: a cross-national and cross-sectoral comparison. *Journal of European Public Policy, 16*(6): 813–833. https://doi.org/10.1080/13501760903087779

Lavenex, S., & Schimmelfennig, F. (2009). EU rules beyond EU borders: theorizing external governance in European politics. *Journal of European Public Policy, 16*(6): 791–812. https://doi.org/10.1080/13501760903087696

Le Blond, J. (13 October 2015). Coal resurgence darkens Germans' green image. *Financial* Times Online. Retrieved from https://next.ft.com/content/719ea15e-68fa-11e5-a57f-21b88f7d973f

Lee, C. K., & Strang, D. (2006). The International Diffusion of Public-Sector Downsizing: Network Emulation and Theory-Driven Learning. *International Organization, 60*(4): 883–909.

Legrand, T. (2012). Overseas and over here: policy transfer and evidence-based policy-making. *Policy Studies, 33*(4): 329–348.

Lehr, U., Edler, D., O'Sullivan, M., Peter, F., & Bickel, P. (2015). *Beschäftigung durch erneuerbare Energien in Deutschland: Ausbau und Betrieb, heute und morgen.* Commissioned by BMWi. Retrieved from https://www.bmwi.de/BMWi/Redaktion/PDF/Publikationen/Studien/beschaeftigung-durch-erneuerbare-energien-in-deutschland,property=pdf,bereich=bmwi2012,sprache=de,rwb=true.pdf

Leidreiter, A. (2012). The German Energy Transition – a Blueprint for Other Countries?. Heinrich Böll Foundation. Retrieved from http://www.ps.boell.org/web/118-713.html

Le Matin. (03 January 2014). Energie nucléaire. Adoption de l'accord franco-marocain. *Le Matin.* Retrieved from http://lematin.ma/express/2014/energie-nucleaire-_adoption-de-l-accord-franco-marocain/194317.html

Lempert, L. B. (2007). Asking Questions of the Data: Memo Writing in the Grounded Theory Tradition. In A. Bryant & K. Charmaz (Eds.), *The SAGE Handbook of Grounded Theory* (pp. 245–264). Los Angeles, London, New Delhi, Singapore: SAGE Publications.

Lepenies, P. (2009). Lernen vom Besserwisser: Wissenstransfer in der "Entwicklungshilfe" aus historischer Perspektive. In H. Büschel & D. Speich (Eds.), *Entwicklungswelten Globalgeschichte der Entwicklungszusammenarbeit* (pp. 33–59). Frankfurt am Main: Campus.

Lepenies, P. (2014). "La rage de vouloir conclure": Wissensvermittlung als Entwicklungsengpass oder warum Experten so arbeiten, wie sie es tun. In A. Ziai (Ed.), *Stand und Perspektiven der Entwicklungstheorie. Im Westen nichts Neues? Stand und Perspektiven der Entwicklungstheorie* (1st ed., pp. 213–234). Nomos.

Levine, M. E., & Forrence, J. L. (1990). Regulatory Capture, Public Interest, and the Public Agenda: Toward a Synthesis. *Journal of Law, Economics, & Organization, 6:* 167–198.

Li, L. (2016). *Soft Power for Solar Power: Germany's New Climate Foreign Policy.* Germanwatch. Retrieved from http://germanwatch.org/en/download/14552.pdf

Liefferink, D., & Andersen, M. S. (1998). Strategies of the green' member states in EU environmental policy-making. *Journal of European Public Policy, 5*(2): 254–270.

Liefferink, D., Arts, B., Kamstra, J., & Ooijevaar, J. (2009). Leaders and laggards in environmental policy: a quantitative analysis of domestic policy outputs. *Journal of European Public Policy, 16*(5): 677–700. https://doi.org/10.1080/13501760902983283

Lindblom, C. E. (1959). The Science of "Muddling Through". *Public Administration Review, 19*(2): 79–88. https://doi.org/10.2307/973677

Lindebjerg, E. S. (2014). *Shedding light on policy diffusion: The ban on incandescent light bulbs.* (Master thesis). Geneva: Graduate Institute of International and Development Studies.

Linder, S. H., & Peters, B. G. (1989). Instruments of Government: Perceptions and Contexts. *Journal of Public Policy, 9*(1): 35-58. https://doi.org/10.1017/S0143814X00007960

Liptow, H., & Remler, S. (2012). *Legal Framework for Renewable Energy: Policy analysis for 15 developing and emerging countries.* Commissioned by GIZ. Retrieved from http://www.giz.de/Themen/de/dokumente/giz2012-en-legal-frameworks-for-renewable-energy.pdf

Litz, F. T. (2008). *Toward a constructive dialogue on federal and state roles in u.s. climate change policy.* Pew Center on Global Climate Change. (Solutions White Paper Series). Retrieved from http://www.c2es.org/docUploads/StateFedRoles.pdf

Livingstone, D. (2015). Germany, the United States, and Climate Leadership in a New Age of Oil. *AICGS Policy Report*, (63): 17–23.

Los Angeles Business Council. (2013). *Largest-in-the-nation Feed-In Tariff Solar Program Kicks Off.* Retrieved from http://www.labusinesscouncil.org/LargestintheNation-Feedin-Tariff-Solar-Program-Kicks-Off

Löschel, A., Erdmann, G., Staiß, F., & Ziesing, H.-J. (2014). *Stellungnahme zum ersten Fortschrittsbericht der Bundesregierung für das Berichtsjahr 2013.* Expertenkommission zum Monitoring-Prozess "Energie der Zukunft". Retrieved from http://www.bmwi.de/BMWi/Redaktion/PDF/M-O/monitoringbericht-energie-der-zukunft-stellungnahme-2013,property=pdf,bereich=bmwi2012,sprache=de,rwb=true.pdf

Löschel, A., Erdmann, G., Staiß, F., & Ziesing, H.-J. (2015). *Stellungnahme zum vierten Monitoring-Bericht der Bundesregierung für das Berichtsjahr 2014.* Expertenkommission zum Monitoring-Prozess "Energie der Zukunft". Retrieved from https://www.bmwi.de/BMWi/Redaktion/PDF/M-O/monitoringbericht-energie-der-zukunft-stellungnahme-2014,property=pdf,bereich=bmwi2012,sprache=de,rwb=true.pdf

Lovinfosse, I. de, Janeiro, L., Gephart, M., & Klessmann, C. (2013). *Lessons for the tendering system for renewable electricity in South Africa from international experience in Brazil, Morocco and Peru.* (Ecofys). Commissioned by GIZ. Retrieved from http://www.ecofys.com/files/files/ecofys-giz-2013-international-experience-res-tender ing.pdf

Lüthi, S. (2012). *Effective renewable energy policy: empirical insights from choice experiments with project developers.* (Doctoral thesis). St Gallen: Universität St Gallen.

Mahoney, J., & Goertz, G. (2004). The Possibility Principle: Choosing Negative Cases in Comparative Research. *American Political Science Review, 98*(4): 653–669.

Majone, G. (1991). Cross-National Sources of Regulatory Policymaking in Europe and the United States. *Journal of Public Policy, 11*(1): 79–106. https://doi.org/10.1017/S0143814X00004943

Makse, T., & Volden, C. (2011). The Role of Policy Attributes in the Diffusion of Innovations. *Journal of Politics, 73*(01): 108–124. https://doi.org/10.1017/S0022381610000903

Malgas, I., Nawaal Gratwick, K., & Eberhard, A. (2008). Moroccan independent power producers – African pioneers. *The Journal of North African Studies, 13*(1): 15–36. https://doi.org/10.1080/13629380701642662

Malnes, R. (1995). 'Leader' and 'Entrepreneur' in International Negotiations:: A Conceptual Analysis. *European Journal of International Relations, 1*(1): 87–112. https://doi.org/10.1177/1354066195001001005

March, J. G., & Olsen, J. P. (1989). *Rediscovering institutions: The organizational basis of politics*. New York: Free Press.

Marquardt, J. (2014). Energiewende Made in Germany? Konstruktion und Bedeutung eines energiepolitischen Nationenimage. *Zeitschrift für Umweltpolitik und Umweltrecht, 37*(1): 78-95.

Marquardt, J., Steinbacher, K., & Schreurs, M. (2015). Driving force or forced transition?: The role of development cooperation in promoting energy transitions in the Philippines and Morocco. *Journal of Cleaner Production.* https://doi.org/10.1016/j.jclepro.2015.06.080

Marsh, D., & Sharman, J. (2009). Policy diffusion and policy transfer. *Policy Studies, 30*(3): 269–288. https://doi.org/10.1080/01442870902863851

Marsh, D., & Sharman, J. (2010). Policy Diffusion and policy transfer. In M. Evans (Ed.), *New directions in the study of policy transfer* (pp. 32–51). London: Routledge.

Martin, B., & Fig, D. (2015). *Final Report - Findings of the African Nuclear Study*. Retrieved from https://za.boell.org/sites/default/files/african_nuclear_study.pdf

Matisoff, D. C., & Edwards, J. (2014). Kindred spirits or intergovernmental competition? The innovation and diffusion of energy policies in the American states (1990–2008). *Environmental Politics, 23*(5): 795–817. https://doi.org/10.1080/09644016.2014.923639

Mattauch, L., Creutzig, F., & Edenhofer, O. (2015). Avoiding carbon lock-in: Policy options for advancing structural change. *Economic Modelling, 50:* 49–63. https://doi.org/10.1016/j.econmod.2015.06.002

Mattes, A. (2012). *Ökostrom – starker Rückhalt in der Bevölkerung.* (DIW Wochenbericht, 7). Retrieved from http://www.diw.de/documents/publikationen/73/diw_01.c.392841.de/12-7.pdf

May, P. J. (1992). Policy learning and failure. *Journal of Public Policy, 12*(4): 331–354. Retrieved from http://www.jstor.org/stable/4007550

Mayntz, R., & Scharpf, F. W. (1995). Der Ansatz des akteurzentrierten Institutionalismus. In R. Mayntz & F. W. Scharpf (Eds.), *Gesellschaftliche Selbstregelung und politische Steuerung* (pp. 39–72). Frankfurt/Main, New York: Campus.

Mayring, P. (1983). *Qualitative Inhaltsanalyse: Grundlagen und Techniken.* Weinheim: Beltz.

Mayring, P. (2000). Qualitative Inhaltsanalyse. *Forum: Qualitative Social Research, 1*(2): Art. 20.

McCown, B. (30 December 2013). Germany's Energy Goes Kaput, Threatening Economic Stability. *Forbes Energy.* Retrieved from http://www.forbes.com/sites/brighammccown/2013/12/30/germanys-energy-goes-kaput-threatening-economic-stability/#504e9cee6859

McDonald, L. (2012). Educational Transfer to Developing Countries: Policy and Skill Facilitation. *International Conference on Education & Educational Psychology. Procedia – Social and Behavioral Sciences, 69:* 1817–1826. https://doi.org/10.1016/j.sbspro.2012.12.132

McKeown, T. J. (2004). Case Studies and the Limits of the Quantitative Worldview. In H. E. Brady & D. Collier (Eds.), *Rethinking social inquiry. Diverse tools, shared standards* (pp. 139–167). Lanham, MD: Rowman & Littlefield.

McVoy, E. (1940). Patterns of Diffusion in the United States. *American Sociological Review, 5*(2): 219–227.

MEMEE. (2008). *Secteur de l'Energie et des Mines. Principales réalisations (1999–2008). Défis et Perspectives.* Royaume du Maroc. Retrieved from http://www.mem.gov.ma/siteassets/pdfdocumentation/principalesrealisations.pdf

MEMEE. (2009). *Stratégie énergétique nationale: Horizon 2030.* Rabat: Ministère de l'Energie, des Mines, de l'Eau et de l'Environnement.

MEMEE. (2010). *Loi n°13-09 relative aux énergies renouvelables.* Rabat: Ministère de l'Energie, des Mines, de l'Eau et de l'Environnement.

MEMEE. (2013a). *Allocution du Dr. Abdelkader Amara: Conférence-débat sur la transition énergétique le mercredi 27 novembre 2013.* Retrieved from http://www.mem.gov.ma/SitePages/Discours/Disc271113.pdf

MEMEE. (2013b). Bilan des investissements dans le secteur de l'énergie et des mines. Retrieved from http://www.mem.gov.ma/SiteAssets/PdfStatistiques/BilanInvestisEMSept2013.pdf.

MEMEE. (2015). La loi n°58-15 amendement et complétant de la loi n° 13-09 Relative aux énergies renouvelables. Retrieved from http://www.mem.gov.ma/SitePages/TestesReglementaires/loi%20n13-09ver23dec15.pdf

MEMEE. (2016). Mr. Amara: 10.000 MW propres supplémentaites d'ici 2030. Retrieved from http://www.mem.gov.ma/SitePages/Activites2016/Ac21Janv16.aspx

Mena-Institut. (2007). Marokko beschließt Gesetz zur Förderung erneuerbarer Energien nach dem Vorbild des EEG. Retrieved from http://www.solarserver.de/news/news-7029.html

Merkel, A. (14 July 2011). Pressestatements von Bundeskanzlerin Angela Merkel nach dem Gespräch mit dem Präsidenten der Bundesrepublik Nigeria, Goodluck Ebele Jonathan in Abuja. Abuja. Retrieved from https://www.bundeskanzlerin.de/Content Archiv/DE/Archiv17/Mitschrift/Pressekonferenzen/2011/07/2011-07-14-statement-merkel-nigeria.html

Merkel, A. (2014). Regierungserklärung von Bundeskanzlerin Merkel: 29 January 2014. Retrieved from http://www.bundesregierung.de/Content/DE/Regierungserklaerung/2014/2014-01-29-bt-merkel.html

Meseguer, C. (2005). *Rational learning and bounded learning in the diffusion of policy innovations. Working Paper: Vol. 316.* Notre Dame: Helen Kellogg Institute for International Studies. Retrieved from https://kellogg.nd.edu/publications/workingpapers/WPS/316.pdf

Messner, D., & Morgan, J. (2013). *Germany needs an Energy Transformation foreign policy.* (The Current Column, 07 January 2013).

Messner, D., Schellnhuber, H. Joachim, & Morgan, J. (2014). *Globale Wende durch Energiewende-Club.* (Die aktuelle Kolumne, 28.04.2014). Retrieved from http://www.die-gdi.de/uploads/media/Deutsches_Institut_fuer_Entwicklungspolitik_Messner_Schellnhuber_Morgan_28.04.2014.pdf

Metayer, M., Breyer, C., & Fell, H.-J. (2015). *The projections for the future and quality in the past of the World Energy Outlook for solar PV and other renewable energy technologies.* Retrieved from http://energywatchgroup.org/wp-content/uploads/2015/09/EWG_WEO-Study_2015.pdf

Meyer, T.-P. (2015). *10 Jahre Erfahrung mit Förderansätzen: 4E/IKLU für globalen Klimaschutz.* (Materialien zur Klimafinanzierung, 2, February 2015). Retrieved from https://www.kfw-entwicklungsbank.de/PDF/Download-Center/Materialien/Nr.-2_10-Jahre-Erfahrung-mit-Förderansätzen_DE.pdf

Mihm, A., & Steltzner, H. (19 February 2013). "Energiewende könnte bis zu einer Billion Euro kosten": Umweltminister Altmaier. *Frankfurter Allgemeine Zeitung.* Retrieved from http://www.faz.net/aktuell/politik/energiepolitik/umweltminister-altmaier-energiewende-koennte-bis-zu-einer-billion-euro-kosten-12086525.html?printPaged.Article=true#pageIndex_2

Mintrom, M., & Vergari, S. (1998). Policy Networks and Innovation Diffusion: The Case of State Education Reforms. *The Journal of Politics, 60*(01): 126–148. https://doi.org/10.2307/2648004

Mishler, E. G. (1986). *Research interviewing: Context and narrative*. Cambridge, Mass: Harvard University Press.

Molenaers, N., Dellepiane, S., & Faust, J. (2015). Political Conditionality and Foreign Aid. *World Development, 75:* 2–12. https://doi.org/10.1016/j.worlddev.2015.04.001

Le Monde. (09 November 2015). L'Allemagne va-t-elle réussir son « tournant énergétique » ? *Le Monde.fr*. Retrieved from http://www.lemonde.fr/cop21/article/2015/11/09/l-allemagne-va-t-elle-reussir-son-tournant-energetique_4805782_4527432.html#oM9wyjS7XbDommqp.99

Moravcsik, A. (1997). Taking Preferences Seriously: A Liberal Theory of International Politics. *International Organization, 51*(04): 513–553. https://doi.org/10.1162/002081897550447

Morgan, J., Messner, D., & Schellnhuber, H. J. (2014). A Renewables Club to Change the World. Retrieved from http://www.wri.org/blog/2014/05/renewables-club-change-world

Morgan, J., & Weischer, L. (15 June 2013). The World Needs More Energiewende. *The European*. Retrieved from http://www.theeuropean-magazine.com/jennifer-morgan--2/6927-germanys-path-towards-renewable-energies

Morris, C. (2016). How Germany helped bring down the cost of PV. 20 January 2016. Retrieved from http://energytransition.de/how-germany-helped-bring-down-the-cost-of-pv/

Morse, J. M. (2007). Sampling in Grounded Theory. In A. Bryant & K. Charmaz (Eds.), *The SAGE Handbook of Grounded Theory* (pp. 229–243). Los Angeles, London, New Delhi, Singapore: SAGE Publications.

Mossberger, K. (2000). *The politics of ideas and the spread of enterprise zones*. Washington, DC: Georgetown University Press.

Mossberger, K., & Wolman, H. (2003). Policy Transfer as a Form of Prospective Policy Evaluation: Challenges and Recommendations. *Public Administration Review, 63*(4): 428–440. https://doi.org/10.1111/1540-6210.00306

Mouline, S. (2007). L'énergie éolienne au Maroc: Historique et nouvelles opportunités. Retrieved from http://base.d-p-h.info/es/fiches/dph/fiche-dph-7431.html

Mucciaroni, G. (1992). The Garbage Can Model & the Study of Policy Making: A Critique. *Polity, 24*(3): 459–482. https://doi.org/10.2307/3235165

Müller, F., & Kals, E. (2004). Q-Sort Technique and Q-Methodology—Innovative Methods for Examining Attitudes and Opinions. *Forum: Qualitative Social Research, 5*(2):

Müller, S. (05 May 2015). *Energiepartnerschaften der Bundesregierung: Informationsveranstaltung Exportinitiative am 5. Mai 2015 in Berlin.* Presentation. BMWi. Berlin. Retrieved from https://www.export-erneuerbare.de/EEE/Redaktion/DE/Downloads/Publikationen/Praesentationen/2015-05-05-iv-aegypten-07-bmwi.pdf?__blob=publicationFile&v=1

Münch, S. (2016). *Interpretative Policy-Analyse: Eine Einführung.* Wiesbaden: Springer VS.

Nedley, A. (2004). Policy Transfer and the Developing Country Experience Gap: Taking a Southern Perspective. In M. Evans (Ed.), *Policy transfer in global perspective* (pp. 165–189). Aldershot, Hants, England ; Burlington, Vt.: Ashgate.

NERSA. (2011). *Review of Renewable Energy Feed - In Tariffs: NERSA Consultation Paper.* Retrieved from http://www.nersa.org.za/Admin/Document/Editor/file/Electricity/Consultation/Documents/Review%20of%20Renewable%20Energy%20Feed-In%20Tariffs%20Consultation%20Paper.pdf

NERSA. (2014). *Discussion Document: Small-Scale Renewable Embedded Generation: Regulatory Framework for Distributors.* Retrieved from http://pqrs.co.za/wp-content/uploads/2015/01/Regulatory-framework-NERSA-Dec-2014.pdf

Nganga, J., Wohlert, M., Woods, M., Becker-Birck, C., Summer, J., & Rickerson, W. (2013). *Powering Africa through Feed-In Tariffs.* A Study for the World Future Council (WFC), the Heinrich Böll Stiftung (HBS) and Friends of the Earth England, Wales & Northern Ireland (FoE-EWNI). Nairobi. Retrieved from http://www.boell.de/worldwide/africa/africa-renewable-energy-feed-in-tarifs-survey-africa-16982.html

Nicholson-Crotty, S. (2009). The Politics of Diffusion: Public Policy in the American States. *The Journal of Politics, 71*(01): 192–205. https://doi.org/10.1017/S0022381608090129

Nimgaonkar, V. (2015). The Energiewende Paradox: Rising Emissions despite an Emphasis on Renewable Energy Policy. Retrieved from http://publicpolicy.wharton.upenn.edu/live/news/1056-the-energiewende-paradox-rising-german-emissions

Niskanen, W. A. (1971). *Bureaucracy and representative government.* Chicago: Aldine, Atherton.

Nitsch, J., & Wenzel, B. (2009). *Langfristszenarien und Strategien für den Ausbau erneuerbarer Energien in Deutschland: Leitszenario 2009.* Berlin: BMU. Retrieved from http://www.dlr.de/dlr/Portaldata/1/Resources/documents/BMU_Leitszenario2009_Langfassung.pdf

Nold, D. (2014). Deutsch-südafrikanische Energiepartnerschaft: Presentation, 28.11.2014. Retrieved from https://www.efficiency-from-germany.info/ENEFF/Redaktion/DE/Audioslideshows/2014/Suedafrika/Vortrag6/praesentation.pdf?__blob=publicationFile&v=2

Nye, J. S. (2004). *Soft power: The means to success in world politics* (12th ed.). New York: PublicAffairs.

Odendaal, N. (16 October 2015). First biowaste-produced renewable energy made available to BMW's Rosslyn plant. *Engineering News*. Retrieved from http://www.engineeringnews.co.za/article/first-biowaste-produced-renewable-energy-made-available-to-bmws-rosslyn-plant-2015-10-16/article_comments:1

OECD. (2015). *DAC-Prüfbericht über die Entwicklungszusammenarbeit: Deutschland 2015*. Paris: OECD.

Office of Governor Edmund G. Brown. (2011). The Governor's Conference on Local Renewable Energy Resource. Retrieved from https://www.gov.ca.gov/s_energyconference.php

Ogden, J., Walt, G., & Lush, L. (2003). The politics of 'branding' in policy transfer: the case of DOTS for tuberculosis control. *Social Science & Medicine, 57*(1): 179–188. https://doi.org/10.1016/S0277-9536(02)00373-8

Ohlhorst, D., Schreurs, M., & Gullberg, A. T. (2012). Norway - "Battery" for the German Energy Transition? Different National Interests in Energy Policies. *GAIA - Ecological Perspectives for Science and Society, 21*(4): 319-320.

Olson, M. (1971). *The logic of collective action: Public goods and the theory of groups*. Cambridge, Mass: Harvard University Press.

ONEE. (2014). Programme d'Electrification Rurale Global: L'électricité pour tous. Retrieved from http://www.one.org.ma/fr/pages/interne.asp?esp=2&id1=6&t1=1

ONEE. (2010). Projet Marocain Eolien Integré. Retrieved from http://www.one.org.ma/fr/pages/Program_mar_ener_eol.asp?esp=2&id1=8&id2=70&t2=1

ONEE. (2014a). *Chiffres Clés à fin 2014*. Retrieved from http://www.one.org.ma/FR/pages/interne.asp?esp=2&id1=4&id2=52&t2=1

ONEE. (2014b). Distribution de l'éléctricité. Retrieved from www.one.org.ma/FR/pages/interne.asp?esp=2&id1=4&id2=53&id3=41&t2=1&t3=1

Orenstein, M. A. (2008). *Privatizing pensions: The transnational campaign for social security reform*. Princeton: Princeton University Press.

Osianowski, R.-P. (1989). Sonderenergieprogramm Marokko. In GTZ (Ed.), *Sonderenergieprogramm 1982 - 1988. Zwischenbilanz und Perspektiven* (pp. 90–95). Eschborn: GTZ.

Osianowski, R.-P. (1997). Internationale Erfahrungen bei der Durchführung von SEP der GTZ in Marokko. In H. G. Brauch (Ed.), *Energiepolitik. Technische Entwicklung, politische Strategien, Handlungskonzepte zu erneuerbaren Energien und zur rationellen Energienutzung* (pp. 311–328). Berlin: Springer Verlag.

Ostrom, E. (2009). *A Polycentric Approach For Coping With Climate Change*. World Bank. (Policy Research Working Papers, October 2009). https://doi.org/ doi.org/10.1596/1813-9450-5095

Ostrom, E. (2012). Nested externalities and polycentric institutions: must we wait for global solutions to climate change before taking actions at other scales? *Economic Theory, 49*(2): 353–369. https://doi.org/10.1007/s00199-010-0558-6

Ostrowski, D. (2010). *Die Public Diplomacy der deutschen Auslandsvertretungen weltweit: Theorie und Praxis der deutschen Auslandsöffentlichkeitsarbeit.* Wiesbaden: VS Verlag für Sozialwissenschaften.

Pacific Environment. (2010). *Feed-in Tariffs: A Time for Real Action on Renewable Energy: Biographies.* Retrieved from http://pacificenvironment.org/downloads/Fit Conf_biographies4-3.pdf

Pahle, M., Knopf, B., Tietjen, O., & Schmid, E. (2012). *Kosten des Ausbaus erneuerbarer Energien: Eine Metaanalyse von Szenarien.* Umweltbundesamt. (Climate Change, N° 23). Retrieved from http://www.uba.de/uba-info-medien/4351.html

Pariente-David, S. & Zakou, A. (2012). *The MENA CSP Scale-up Project.* Joint Workshop of the African Development Bank and the World Bank Group, Tunis, 28th of June 2012. Retrieved from http://www.afdb.org/fileadmin/uploads/afdb/Documents/Generic-Documents/csp%20mena%20overview.pdf

Parkinson, G. (2016). New Low For Wind Energy Costs: Morocco Tender Averages $US30/MWh. Retrieved from http://reneweconomy.com.au/2016/new-low-for-wind-energy-costs-morocco-tender-averages-us30mwh-81108

Paton, C. (02 September 2015). Updated IRP may raise share of nuclear power. *Business Daily Live.* Retrieved from http://www.bdlive.co.za/business/energy/2015/09/02/updated-irp-may-raise-share-of-nuclear-power

Pegels, A. (2011). Pitfalls of policy implementation: The case of the South African feed-in tariff. In J. Haselip, I. Nygaard, U. Hansen, & E. Ackom (Eds.), *Technology Transfer Perspectives Series. Diffusion of renewable energy technologies: case studies of enabling frameworks in developing countries* (pp. 101–110). Roskilde: UNEP Risoe Centre.

Petersen, N. H. (1 April 2010). Einspeisevergütung: Ein Umstrittener Exportschlager. Energlobe. Retrieved from http://energlobe.de/archiv/politik/global/ein-umstrittener-exportschlager

Piccio, L. (2013). Amid DfID's pullout from South Africa, will other donors follow? Retrieved from https://www.devex.com/news/amid-dfid-s-pullout-from-south-africa-will-other-donors-follow-80871

Pöller, M. (2013). *White Paper P13008: Net-metering concept for Small Scale Embedded Generation in South Africa.* Prepared for GIZ. Retrieved from http://www.ameu.co.za/Portals/16/Conventions/Convention%202014%20Papers/Geeven%20Moodley,%20SALGA.PDF

Porter, M. E., & van der Linde, C. (1995). Green and Competitive: Ending the Stalemate. *Harvard Business Review, 73*(5): 120–134.

Prince, R. (2012). Policy transfer, consultants and the geographies of governance. *Progress in Human Geography, 36*(2): 188–203. https://doi.org/10.1177/0309132511417659

ProQuest. (n.d.). *ProQuest Social Science Database*. Retrieved from www.proquest.com

Putnam, R. D. (1988). Diplomacy and domestic politics: the logic of two-level games. *International Organization, 42*(03): 427–460. https://doi.org/10.1017/S0020818300027697

Quitzow, R., Röhrkasten, S., & Jänicke, M. (2016). *The German Energy Transition in International Perspective*. (IASS Study, March 2016). Retrieved from http://www.iass-potsdam.de/sites/default/files/files/iass_study_thegermanenergytransition_ininternationalperspective_en_0.pdf

Radaelli, C. M. (2009). Measuring policy learning: regulatory impact assessment in Europe. *J. Eur. Public Policy, 16*(8): 1145–1164. https://doi.org/10.1080/13501760903332647

Radaelli, C. M. (1995). The role of knowledge in the policy process. *Journal of European Public Policy, 2*(2): 159–183. https://doi.org/10.1080/13501769508406981

Radaelli, C. M. (2005). Diffusion without convergence: how political context shapes the adoption of regulatory impact assessment. *Journal of European Public Policy, 12*(5): 924–943. https://doi.org/10.1080/13501760500161621

Ragin, C. C. (1997). Introduction: Cases of "What is a case?". In C. C. Ragin & H. S. Becker (Eds.), *What is a case? Exploring the foundations of social inquiry* (pp. 1–18). Cambridge: Cambridge University Press.

Ralston, H. (2013). *Subnational partnerships for sustainable development: Transatlantic cooperation between the United States and Germany. New horizons in enviromental politics*. Cheltenham: Edward Elgar.

Ranis, G., Vreeland, J. R., & Kosack, S. (Eds.). (2006). *Globalization and the nation state: The impact of the IMF and the World Bank*. London, New York: Routledge.

Rathbun, B. C. (2008). Interviewing and Qualitative Field Methods: Pragmatism and Practicalities. In J. M. Box-Steffensmeier (Ed.), *The Oxford Handbook of Political Methodology* . Oxford: Oxford University Press.

Reichertz, J. (2007). Abduction: The Logic of Discovery in Grounded Theory. In A. Bryant & K. Charmaz (Eds.), *The SAGE Handbook of Grounded Theory* (pp. 214–228). Los Angeles, London, New Delhi, Singapore: SAGE Publications.

Reichertz, J. (2010). Abduction: The Logic of Discovery of Grounded Theory. *Forum: Qualitative Social Research, 11*(Art. 13):

REN21 Secretariat. (2014). *The First Decade: 2004-2014: 10 Years of Renewable Energy Progress*. Paris: REN21. Retrieved from http://www.ren21.net/Portals/0/documents/activities/Topical%20Reports/REN21_10yr.pdf

REN21 Secretariat. (2015). *Global Renewables Status Report*. Paris: REN21. Retrieved from http://www.ren21.net/wp-content/uploads/2015/07/REN12-GSR2015_Onlinebook_low1.pdf

Renewables 100 Institute. (2007). Dr. Hermann Scheer book release "Energy Autonomy" in Malibu. Retrieved from http://www.renewables100.org/index.php?id=129&tx_ttnews%5Byear%5D=2007&tx_ttnews%5Bmonth%5D=02&tx_ttnews%5Bday%5D=15&tx_ttnews%5Btt_news%5D=12&cHash=c2452f595ec62ccc406adf77bb18a2c1

Rennings, K. & Cleff, T. (2011). First and second mover strategy options for pioneering countries on environmental markets: From national lead market to combined lead market and lead supplier strategies. (Working Paper, N° 3). Retrieved from http://kooperationen.zew.de/fileadmin/user_upload/Redaktion/Lead_Markets/Werkstattberichte/WB3_Rennings_Cleff.pdf

Reuter, B. (12 March 2013). 1 Billion: Wo sich Altmaier bei der Energiewende verrechnet hat. *Green WiWo*. Retrieved from http://green.wiwo.de/1-billion-fur-die-energiewende-hat-sich-altmaier-verrechnet/

Reuters. (2010). Anti-Atombewegung feiert mit Groß-Demo ihr Comeback. Retrieved from http://de.reuters.com/article/deutschland-atom-demonstration-2zf-idDEBEE68I00R20100919

Rhys, J. (2013). *Current German Energy Policy - the "Energiewende": A UK and climate change perspective*. The Oxford Institute for Energy Studies. (Oxford Energy Comment, April 2013). Retrieved from https://www.oxfordenergy.org/wpcms/wp-content/uploads/2013/04/Current-German-Energy-Policy-A-UK-and-climate-concern-perspective.pdf

Richardson, J. (2000). Government, Interest Groups and Policy Change. *Political Studies, 48*(5): 1006–1025. https://doi.org/10.1111/1467-9248.00292

Rickerson, W., Baker, S. E., & Wheeler, M. (2008). Is California the next Germany? Renewable gas and California' s new feed- in tariff. *BioCycle, 49*(March): 56–61.

Rickerson, W., Bennhold, F., & Bradbury, J. (2008). *Feed-in Tariffs for Renewable Energy in the USA - a Policy Update*. Retrieved from http://www.wind-works.org/cms/uploads/media/Feed-in_Tariffs_and_Renewable_Energy_in_the_USA_-_a_Policy_Update.pdf

Riley, T. (2008). New Moroccan energy strategy hits familiar themes. Wikileaks. Retrieved from http://search.wikileaks.org/plusd/cables/08RABAT693_a.html

Risse-Kappen, T. (1994). Ideas do not float freely: transnational coalitions, domestic structures, and the end of the cold war. *International Organization, 48*(2): 185–214. https://doi.org/10.1017/S0020818300028162

Robertson, D. B. (1991). Political Conflict and Lesson-Drawing. *Journal of Public Policy, 11*(1): 55–78.

Roelf, W. (15 March 2016). South African nuclear regulator receives site applications from Eskom. *Reuters*. Retrieved from http://www.reuters.com/article/safrica-nuclear-idUSL5N16N1P2

Rogers, E. M. (2003). *Diffusion of innovations* (5th ed). New York: Free Press.

Röhrkasten, S. (2015). *Global Governance on Renewable Energy: Contrasting the Ideas of the German and the Brazilian Governments.* Wiesbaden: VS Verlag für Sozialwissenschaften.

Röhrkasten, S., & Westphal, K. (2012). *IRENA: Stay the Course!* German Institute for International and Security Affairs. (SWP Comments, N° 2012-C 37). Retrieved from http://www.swp-berlin.org/en/publications/swp-comments-en/swp-aktuelle-details/article/renewable_energy_irena.html

Röhrkasten, S., & Westphal, K. (2013). *IRENA and Germany's Foreign Renewable Energy Policy: Aiming at Multilevel Governance and an Internationalization of the Energiewende?.* Stiftung Wissenschaft und Politik. (SWP Woking Papers, N° FG8-2013/01). Berlin. Retrieved from http://www.swp-berlin.org/fileadmin/contents/products/arbeitspapiere/Rks_Wep_FG08WorkingPaper_2013.pdf

Roller, G., Lefèvre, M., Wirtz, J., Schmidt-Sercander, B., Eichhammer, W., Ragwitz, M…Mouline, S. (2007). Commissioned by GTZ. *Etude sur le cadre organisationnel, institutionnel et législatif pour la promotion des énergies renouvelables.* Retrieved from http://www.giz.de/en/worldwide/20191.html

Rosa Luxemburg Stiftung. (n.d.). Auslandsbüros. Retrieved from https://www.rosalux.de/weltweit/auslandsbueros.html

Rosatom. (2014). *Russia and South Africa signed the agreement on strategic partnership in nuclear energy.* Retrieved from http://www.rosatom.ru/en/presscentre/highlights/8ce8ca804590ef279dae9d9207a61cab

Rosenkranz, G., & WWF Germany. (2015). *Megatrends der globalen Energiewende.* WWF Germany and Lichtblick (ed.). Retrieved from http://www.energiewende beschleunigen.de/fileadmin/fm-wwf/lichtblick/Megatrends-der-globalen-Energiewende.pdf

Rose, R. (1991). What is Lesson-Drawing? *Journal of Public Policy, 11*(01): 3–30. https://doi.org/10.1017/S0143814X00004918

Rose, R. (2005). *Learning from comparative public policy: A practical guide.* London, New York: Routledge.

Rousselin, M. (2012). But Why Would They Do That? European External Governance and the Domestic Preferences of Rule Importers. *Journal of Contemporary European Research, 8*(4): 470–489.

Rzepka, G. (2013). Die Energiewende exportieren. Retrieved from https://www.giz.de/de/downloads/giz2013-de-akzente02-energiewende-exportieren.pdf

Sabatier, P. A. (1987). Knowledge, Policy-Oriented Learning, and Policy Change: An Advocacy Coalition Framework. *Science Communication, 8*(4): 649–692. https://doi.org/10.1177/0164025987008004005

Sabatier, P. A. (1988). An advocacy coalition framework of policy change and the role of policy-oriented learning therein. *Policy Sciences, 21*(2-3): 129-168. https://doi.org/10.1007/BF00136406

Sabatier, P. A., & Jenkins-Smith, H. C. (Eds.). (1993). *Theoretical lenses on public policy. Policy change and learning: An advocacy coalition approach*. Boulder, Colo: Westview Press.

Sabatier, P. A., & Weible, C. (2007). The Advocacy Coalition Framework: Innovations and Clarifications. In P. A. Sabatier (Ed.), *Theories of the policy process* (pp. 189–222). Boulder, Colo: Westview Press.

Sachs, J. (1988). *Conditionality, debt relief, and the developing country debt crisis*. (NBER working paper series: no. 2644). Cambridge, Mass.: National Bureau of Economic Research.

Saint-Martin, D. (2010). *Studying Policy Transfer in Time: Or, Why is the Politics of Lobbying Reform in the EU on "Fast Forward"?* Paper Prepared for the ECPR Fifth Pan-European Conference on EU Politics. Retrieved from http://www.jhubc.it/ecpr-porto/virtualpaperroom/097.pdf

SANEDI. (2014). *PRO Cogen & ESCO: Market Development for Co-/Trigeneration and ESCOs*. Retrieved from http://www.sanedi.org.za/wp-content/uploads/2013/12/Pro-Cogen_ESCO-Flyer-final.pdf

SARETEC. (13 June 2013). Nordex supports renewable energy training at SARETEC, CPUT. Retrieved from http://www.saretec.co.za/news/nordex-supports-renewable-energy-training-at-saretec-cput.html

Saul, U., & Seidel, C. (2011). Does leadership promote cooperation in climate change mitigation policy? *Climate Policy, 11*(2): 901–921. https://doi.org/10.3763/cpol.2009.0004

Schimmelfennig, F., & Sedelmeier. (2005). *The Europeanization of Central and Eastern Europe: Cornell studies in political economy*. Ithaca, NY: Cornell Univ. Press.

Schlager, E. (1999). A Comparison of Frameworks, Theories, and Models of Policy Processes. In P. A. Sabatier (Ed.), *Theoretical lenses on public policy. Theories of the policy process* (pp. 293–319). Boulder Colo: Westview Press.

Schmidt, V. A. (2008). Discursive institutionalism: The explanatory power of ideas and discourse. *Annual Review Of Political Science, 11:* 303–326. https://doi.org/10.1146/annurev.polisci.11.060606.135342

Schreurs, M. & Papadakis, E., (Eds.). (2007). *Historical dictionary of the green movement* (2nd ed). Lanham, Md: The Scarecrow Press.

Schreurs, M. (2013). Orchestrating a Low-Carbon Energy Revolution Without Nuclear: Germany's Response to the Fukushima Nuclear Crisis. *Theoretical Inquiries in Law, 14*(1): 83–108. https://doi.org/10.1515/til-2013-006

Schreurs, M. A. (2012). Breaking the impasse in the international climate negotiations: The potential of green technologies: Special Section: Frontiers of Sustainability. *Energy Policy, 48*(0): 5–12. https://doi.org/10.1016/j.enpol.2012.04.044

Schreurs, M. A., & Tiberghien, Y. (2007). Multi-Level Reinforcement: Explaining European Union Leadership in Climate Change Mitigation. *Global Environmental Politics, 7*(4): 19–46.

Schröder, G. (02 September 2002). Rede von Bundeskanzler Schröder auf dem Weltgipfel für nachhaltige Entwicklung in Johannesburg. Johannesburg. Retrieved from http://adrien.barbaresi.eu/corpora/speeches/BR/t/457.html

Schwarzenegger, A. (24 September 2007). *United Nations Speech by Governor Arnold Schwarzenegger*. Retrieved from http://www.un.org/webcast/climatechange/highlevel/2007/pdfs/schwarzenegger.pdf

Schwerhoff, G. (2013). *Leadership and International Climate Cooperation.* FEEM. (Nota di Lavoro, N° 97). Retrieved from http://www.feem.it/userfiles/attach/201311261311574NDL2013-097.pdf

Scott, J. M., & Steele, C. A. (2011). Sponsoring Democracy: The United States and Democracy Aid to the Developing World, 1988–20011. *International Studies Quarterly, 55*(1): 47–69. https://doi.org/10.1111/j.1468-2478.2010.00635.x

Seawright, J., & Gerring, J. (2008). Case Selection Techniques in Case Study Research: A Menu of Qualitative and Quantitative Options. *Political Research Quarterly, 61*(2): 294–308.

Sebitosi, A. B., & Pillay, P. (2008). Grappling with a half-hearted policy: The case of renewable energy and the environment in South Africa. *Energy Policy, 36*(7): 2513–2516. https://doi.org/10.1016/j.enpol.2008.03.011

Seel, J., Barbose, G. L., & Wiser, R. H. (2014). An analysis of residential PV system price differences between the United States and Germany. *Energy Policy, 69*(0): 216–226. https://doi.org/10.1016/j.enpol.2014.02.022

SEIA. (2014). *Informal Comments of the Solar Energy Industry Association on Guiding Principles and Program Elements for Successor Net Energy Metering Tariff.* Solar Energy Industry Association. Retrieved from ftp://ftp2.cpuc.ca.gov/PG&E20150130ResponseToA1312012Ruling/2014/05/SB_GT&S_0071689.pdf

Shahan, Z. (2013). Sunrun Supports Net Metering, Not Feed-in Tariffs or Value Of Solar Tariffs. Retrieved from http://cleantechnica.com/2013/08/26/sunrun-supports-net-metering-not-feed-in-tariffs-or-value-of-solar-tariffs/

Shipan, C. R., & Volden, C. (2008). The Mechanisms of Policy Diffusion. *American Journal of Political Science, 52*(4): 840–857. https://doi.org/10.1111/j.1540-5907.2008.00346.x

Sidki, W. (2011). *Le Programme de 1 Million de toits: Une solution énergétique possible pour l'introduction de la technologie photovoltaïque à grande échelle au Maroc.* Commissioned by GIZ. Retrieved from https://www.giz.de/de/downloads/giz2011-fr-programme-million-de-toits.pdf

SIE. (2014). The SIE and Dii Signed an Agreement of Cooperation. Retrieved from http://www.siem.ma/en/the-sie/news/item/the-sie-and-dii-signed-an-agreement-of-cooperation

Sierra Club. (2008). *Comments 2009 IEPR – Feed-In Tariffs to California Energy Commission (CEC) from Energy-Climate Committee, Sierra Club California.* (Docket No. 09-IEP-1G, No. 03-RPS-1078). Retrieved from http://www.energy.ca.gov/portfolio/documents/2008-12-01_workshop/comments/TN-49346_2008-12-10_Sierra_Club_Committee_Comments_Re_2009_IEPR_-_Feed-In_Tariffs.PDF

Sierra Club California, & Metropulos, J. (2009). *Comments on the Commission's Energy Division Feed-in Tariff Proposal of March 2009. Order Instituting Rulemaking to Continue Implementation and Administration of California Renewable Portfolio Standard Program.* (R. 08-08-009, Filed 21 August 2008). Sacramento: CPUC.

Sikkink, K. (1995). Nongovernmental Organizations and Transnational Issue Networks in International Politics. *Proceedings of the American Society of International Law, 89:* 413–415.

Simmons, B., & Elkins, Z. (2004). The Globalization of Liberalization: Policy Diffusion in the International Political Economy. *American Political Science Review, 98*(01): 171–189. https://doi.org/10.1017/S0003055404001078

Simmons, B., Dobbin, F., & Garrett, G. (2007). Introduction: The diffusion of liberalization. In B. Simmons, F. Dobbin, & G. Garrett (Eds.), *The global diffusion of markets and democracy* . Cambridge, New York: Cambridge University Press.

Sisney, J. & Garosi, J. (2015). 2014 GDP: California Ranks 7th or 8th in the World. California Legislative Analyst's Office. Retrieved from http://www.lao.ca.gov/LAOEconTax/Article/Detail/90

Skodvin, T., & Andresen, S. (2006). Leadership Revisited. *Global Environmental Politics, 6*(3): 13–27.

Smithwood, R. B. (2011). *Competition and Collaboration in Renewable Portfolio Standard Adoption and Policy Design: Lessons from New England.* The Fletcher School, Tufts University. (Energy, Climate, and Innovation Discussion Paper). Retrieved from http://fletcher.tufts.edu/CIERP/Publications/more/~/media/Fletcher/Microsites/CIERP/Publications/2011/Smithwood11SepUSenergyPolicy_Fsin.pdf

Solarify. (2014). Das EEG - einsamer Rekord: Mehr als 100 Erneuerbare-Energien-Gesetze weltweit. Retrieved from http://www.solarify.eu/2014/03/15/017-das-eeg-einsamer-rekord/

SolarPACES. (2009). *SolarPACES 2009: Program.* Retrieved from http://www.solarpaces2009.org/cms/upload/pdf/SolarPACES09-Program.pdf

Solingen, E. (2012). Of Dominoes and Firewalls: The Domestic, Regional, and Global Politics of International Diffusion 1. *International Studies Quarterly, 56*(4): 631–644. https://doi.org/10.1111/isqu.12034

Solorio, I., Öller, E., & Jörgens, H. (2014). The German Energy Transition in the Context of the EU Renewable Energy Policy. In A. Brunnengräber & M. R. Di Nucci (Eds.), *Im Hürdenlauf zur Energiewende* (pp. 189-200). Wiesbaden: Springer Fachmedien.

Sommerer, T. (2011). *Können Staaten voneinander lernen?: Eine vergleichende Analyse der Umweltpolitik in 24 Ländern* (1st ed.). Wiesbaden: VS Verlag für Sozialwissenschaften.

Sopher, P. (2014). *While Critics Debate Energiewende, Germany is Gaining a Global Advantage.* Environmental Defense Fund. (Energy Exchange). Retrieved from http://blogs.edf.org/energyexchange/2014/10/06/while-critics-debate-energiewende-germany-is-gaining-a-global-advantage/

South Africa-Germany Binational Commission. (2014). *Joint Report of the 8th South Africa Germany Bi-National Commission: Pretoria, 21 November 2014.* Retrieved from http://www.bmas.de/SharedDocs/Downloads/EN/PDF-Publikationen/joint-report-bi-national-commission-pretoria.pdf?__blob=publicationFile

SPD, & Bündnis 90/Die Grünen. (1998). *Aufbruch und Erneuerung – Deutschlands Weg ins 21. Jahrhundert: Koalitionsvereinbarung zwischen der Sozialdemokratischen Partei Deutschlands und BÜNDNIS 90/DIE GRÜNEN.* Bonn. Retrieved from https://www.gruene.de/fileadmin/user_upload/Bilder/Redaktion/30_Jahre_-_Serie/Teil_21_Joschka_Fischer/Rot-Gruener_Koalitionsvertrag1998.pdf

Stands, S. R. (2015). *Utility-Scale Renewable Energy Job Creation: An investigation of the South African Renewable Energy Independent Power Producer Procurement Programme (REIPPPP).* (Master thesis). Stellenbosch: Stellenbosch University.

Starke, P. (2013). Qualitative Methods for the Study of Policy Diffusion: Challenges and Available Solutions. *Policy Studies Journal, 41*(4): 561–582. https://doi.org/10.1111/psj.12032

Statistics South Africa. (01 March 2016). The economy: winners and losers of 2015. Retrieved from http://www.statssa.gov.za/?p=6233

Steinbacher, K. (2015). Drawing Lessons When Objectives Differ? Assessing Renewable Energy Policy Transfer from Germany to Morocco. *Politics and Governance, 3*(2): 34–50.

Steinbacher, K., & Pahle, M. (forthcoming). Leadership and the Energiewende: Current Actions and Available Options for German Leadership by Diffusion. *Global Environmental Politics,*

Steinbacher, K., & Pahle, M. (2015). *Leadership by diffusion and the German Energiewende.* (PIK Discussion Paper, February 2015). Retrieved from https://www.pik-potsdam.de/members/pahle/dp-ew-leadership-2015.pdf

Steinmeier, F.-W. (2016). Speech by Foreign Minister Steinmeier at the conference Berlin Energy Transition Dialogue: "From Negotiation to Action – Towards a global Energiewende". Retrieved from http://www.auswaertiges-amt.de/sid_E06CB5DB791D8FCC459187BCAAC1078B/EN/Infoservice/Presse/Reden/2016/160317_Energiewende.html?nn=481166

Steinmo, S. (2008). Historical institutionalism. In D. Della Porta & M. Keating (Eds.), *Approaches and methodologies in the social sciences. A pluralist perspective* (pp. 118–138). Cambridge New York: Cambridge University Press.

Stephan, B., Schurig, S., & Leidreiter, A. (2016). *What Place for Renewables in the INDCs?* World Future Council. Retrieved from http://www.worldfuturecouncil.org/file/2016/03/WFC_2016_What_Place_for_Renewables_in_the_INDCs.pdf

Steuwer, D. S. (2013). *Energy Efficiency Governance: The Case of White Certificate Instruments for Energy Efficiency in Europe*. Wiesbaden: Springer Fachmedien.

Stigler, G. J. (1971). The Theory of Economic Regulation. *The Bell Journal of Economics and Management Science, 2*(1): 3–21. https://doi.org/10.2307/3003160

Stigson, B., Babu, S. P., Bordewijk, J., Haavisto, P., Morgan, J., Moosa, V., . . . Yun, S.-J. (2013). *Sustainability - Made in Germany: The Second Review by a Group of International Peers, commissioned by the German Federal Chancellery*. German Council for Sustainable Development. Berlin. Retrieved from http://www.nachhaltigkeitsrat.de/uploads/media/20130925_Peer_Review_Sustainability_Germany_2013_01.pdf

Stone, D. A. (1988). *Policy paradox and political reason*. Glenview, Ill: Scott, Foresman.

Stone, D. (1999). Learning Lessons and Transferring Policy across Time, Space and Disciplines. *Politics, 19*(1): 51–59. https://doi.org/10.1111/1467-9256.00086

Stone, D. (2000). Non-Governmental Policy Transfer: The Strategies of Independent Policy Institutes. *Governance, 13*(1): 45–70. https://doi.org/10.1111/0952-1895.00123

Stone, D. (2001). *Learning lessons, policy transfer and the international diffusion of policy ideas*. Centre For The Study Of Globalisation And Regionalisation. (CSGR Working Paper, 69/1). Retrieved from http://wrap.warwick.ac.uk/2056/1/WRAP_Stone_wp6901.pdf

Stone, D. (2004). Transfer agents and global networks in the 'transnationalization' of policy. *Journal of European Public Policy, 11*(3): 545–566. https://doi.org/10.1080/13501760410001694291

Stone, D., & Maxwell, S. (Eds.). (2005). *Global knowledge networks and international development: Bridges across boundaries*. London, New York: Routledge.

Stopper, W. (2014). *Exportförderung von Technologien der Energiewende in Deutschland: Die Exportinitiativen Erneuerbare Energien & Energieeffizienz. Presentation, 27 February 2014*. Berlin. Retrieved from http://www.dena.de/fileadmin/user_upload/Veranstaltungen/2014/Dialogforum_27.02.14/DF_Technologien_weltweit_07_Stopper_BMWi.pdf

Stoutenborough, J. W., & Beverlin, M. (2008). Encouraging Pollution-Free Energy: The Diffusion of State Net Metering Policies. *Social Science Quarterly, 89*(5): 1230–1251. Retrieved from 10.1111/j.1540-6237.2008.00571.x

Strang, D., & Macy, M. W. (2001). In Search of Excellence: Fads, Success Stories, and Adaptive Emulation. *American Journal of Sociology, 107*(1): 147–182. https://doi.org/10.1086/323039

Strauss, A., & Corbin, J. (1994). Grounded Theory Methodology: An Overview. In N. K. Denzin & Y. S. Lincoln (Eds.), *Qualitative research. Handbook of qualitative research* (pp. 273–285). Thousand Oaks, CA: SAGE Publications.

Strauss, A. L., & Corbin, J. M. (1998). *Basics of qualitative research: Techniques and procedures for developing grounded theory* (2nd ed). Thousand Oaks, CA: SAGE Publications.

Strebel, F. (2011). Inter-governmental institutions as promoters of energy policy diffusion in a federal setting. *Energy Policy, 39*(1): 467–476. https://doi.org/10.1016/j.enpol.2010.10.028

Subtil Lacerda, J., & van Den Bergh, Jeroen C. J. M. (2014). International Diffusion of Renewable Energy Innovations: Lessons from the Lead Markets for Wind Power in China, Germany and USA. *Energies, 7*(12): 8236–8263. https://doi.org/10.3390/en7128236

Sustainable Energy Africa. (2013). *The Impact of Energy Efficiency and Renewable Energy on City Revenue: Resulting impact on service delivery to the poor.* Westlake. Retrieved from http://www.sustainable.org.za/uploads/resources/resource_25.pdf

Sustainable Energy Africa. (2015). *State of Energy in South African Cities 2015.* Westlake. Retrieved from http://www.sustainable.org.za/uploads/resources/resources_25.pdf

Svensson, J. (2003). Why conditional aid does not work and what can be done about it? *Journal of Development Economics, 70*(2): 381–402. https://doi.org/10.1016/S0304-3878(02)00102-5

Tambulasi, R. (2013). Why can't you lead a horse to water and make it drink?: The learning oriented transfer of health sector reforms and bureaucratic interest in Malawi. In P. Carroll & R. Common (Eds.), *Policy transfer and learning in public policy and management. International contexts, content and development* (pp. 80–104). Milton Park, Abingdon, Oxon: Routledge.

Tänzler, D., & Wolters, S. (2014). Energiewende und Außenpolitik: Gestaltungsmacht auf dem Prüfstand. *Zeitschrift für Außen- und Sicherheitspolitik, 7*(2): 133–143.

Task Team on South-South Cooperation. (2011). *Scaling Up Knowledge sharing for Development: A Working Paper for the G-20 Development Working Group, Pillar 9.* Retrieved from http://www.oecd.org/g20/summits/cannes/Scaling-Up-Knowledge-sharing-for-Development.pdf

Tews, K. (2002). *Der Diffusionsansatz für die vergleichende Policy-Analyse. Wurzeln und Potenziale eines Konzepts: Eine Literaturstudie.* (FFU-Report, 02-2002). Retrieved from http://userpage.fu-berlin.de/ffu/download/rep_2002-02.pdf

Tews, K. (2005). The diffusion of environmental policy innovations: cornerstones of an analytical framework. *European Environment, 15*(2): 63–79. https://doi.org/10.1002/eet.378

Tews, K., Busch, P.-O., & Jörgens, H. (2003). The diffusion of new environmental policy instruments. *European Journal of Political Research, 42*(4): 569–600. https://doi.org/10.1111/1475-6765.00096

Tews, K., & Jänicke, M. (2005). *Die Diffusion umweltpolitischer Innovationen im internationalen System* (1st ed.). *Forschung Politik*. Wiesbaden: VS Verlag für Sozialwissenschaften.

Tews, K. (2015). Europeanization of Energy and Climate Policy. *The Journal of Environment & Development, 24*(3): 267-291.

The Economist Intelligence Unit. (2016). *Democracy Index 2015: Democracy in an age of anxiety*. Retrieved from http://www.eiu.com/public/topical_report.aspx?campaignid=DemocracyIndex2015

The Economist. (28 July 2012). Energiewende: Germany's energy transformation. *The Economist (US)*, (404.8795): 46.

The Economist. (14 December 2014). What has gone wrong with Germany's energy policy. The Economist Online. Retrieved from http://www.economist.com/blogs/economist-explains/2014/12/economist-explains-10

Thelen, K. (1999). Historical Institutionalism in Comparative Politics. *Annu. Rev. Polit. Sci., 2*(1): 369–404. https://doi.org/10.1146/annurev.polisci.2.1.369

Thelen, K., & Steinmo, S. (1992). Historical institutionalism in comparative politics. In S. Steinmo, K. Thelen, & F. Longstreth (Eds.), *Structuring politics. Historical institutionalism in comparative analysis* (pp. 1–32). Cambridge: Cambridge University Press.

Tietjen, O., Amazo Blanco, A.L., Pfefferle, T. (2015) *Renewable energy auctions: Goal-oriented policy design*. Eschborn: GIZ. Retrieved from http://www.record.org.za/resources/downloads?task=callelement&format=raw&item_id=648&element=f85c494b-2b32-4109-b8c1-083cca2b7db6&method=download&args[0]=904d99b702e2cecc7d7e2cccfb0feb92

Torney, D. (2015). *European climate leadership in question: Policies toward China and India. Earth system governance*. Cambridge, Mass.: MIT Press.

TREC Development Group. (2003). *Trans-Mediterranean Renewable Energy Cooperation "TREC" for development, climate stabilisation and good neighbourhood*. Retrieved from http://www.dlr.de/tt/Portaldata/41/Resources/dokumente/institut/system/publications/Ammanpaper-small.pdf

Trieb, F. (2004). *A Renewable Energy and Development Partnership EU-MENA for large-scale solar thermal power & desalination in the Middle East and in North Africa: Thematic background paper for the MENAREC conference in Sana'a 21/22 of April 2004 in preparation of the international conference renewables2004, 1-4 June 2004*. Retrieved from http://www.menarec.org/resources/Sanaa+paper+and+annex_15-04-2004.pdf

Trieb, F., & Müller-Steinhagen, H. (2007). *Sustainable Electricity and Water for Europe, Middle East and North Africa*. Retrieved from http://www.dlr.de/tt/Portaldata/41/Resources/dokumente/institut/system/publications/Trieb_EUMENA_Power_and_Water-2007-10-10-FT2.pdf

Tylor, E. B. (1889). On a Method of Investigating the Development of Institutions; Applied to Laws of Marriage and Descent. *The Journal of the Anthropological Institute of Great Britain and Ireland, 18:* 245–272. https://doi.org/10.2307/2842423

Uekötter, F. (2014). *The greenest nation?: A new history of German environmentalism.* Cambridge, Mass.: MIT Press.

Under 2 MOU. (2015). Background. Retrieved from http://under2mou.org/?page_id=228

Underdal, A. (1992). *Leadership in international environmental negotiations: Designing feasible solutions.* CICERO. (Working Papers, N° 8). Retrieved from http://brage.bibsys.no/xmlui/bitstream/handle/11250/192260/-1/CICERO_Working_Paper_1992-08.pdf

Underdal, A. (1994). Leadership theory: rediscovering the arts of management. In W. I. Zartman (Ed.), *International Multilateral Negotiation: Approaches to the Management of Complexity* (pp. 178–197). San Francisco: Jossey-Bass Publishers.

UNDP. (2015a). Briefing note for countries on the 2015 Human Development Report: South Africa. Retrieved from http://hdr.undp.org/sites/all/themes/hdr_theme/country-notes/ZAF.pdf

UNDP. (2015b). Human Development Report: Work for Human Development: Country Profile Morocco. Retrieved from http://hdr.undp.org/en/countries/profiles/MAR

UNFCCC. (2013). Statement by UNFCCC Executive Secretary on crossing of 400 ppm CO2 threshold. Retrieved from http://unfccc.int/press/press_releases_advisories/items/7365.php

UNFCCC. (2015). *Paris Agreement As contained in the report of the Conference of the Parties on its twenty-first session.* FCCC/CP/2015/10/Add.1. Retrieved from http://unfccc.int/files/meetings/paris_nov_2015/application/pdf/paris_agreement_english_.pdf

US Energy Information Administration. (2015). *Average monthly residential electricity consumption, prices, and bills by state*. eia. Retrieved from https://www.eia.gov/tools/faqs/faq.cfm?id=97&t=3

Unruh, G. C. (2000). Understanding carbon lock-in. *Energy Policy, 28*(12): 817–830. https://doi.org/10.1016/S0301-4215(00)00070-7

Ürge-Vorsatz, D., Rezessy, S., & Antypas, A. (2004). Renewable Electricity Support Schemes in Central Europe: A Case of Incomplete Policy Transfer. *Energy & Environment, 15*(4): 699–721. https://doi.org/10.1260/0958305042259783

USAID. (1981). *Morocco Project 608-0159. Renewable Energy Development: Evaluation Report Phase I.* Retrieved from http://pdf.usaid.gov/pdf_docs/Xdaaj661a.pdf

Vagliasindi, M., & Besant-Jones, J. (2013). *Power market structure: Revisiting policy options. Directions in development. Energy and mining.* Washington, D.C: World Bank.

Van den Berg, Johan. (15 October 2013). *The true cost of electricity options and the choices for South Africa: Presentation to the Fossil Fuel Foundation. South Africa's Electricity Supply Conference.* Retrieved from http://www.fossilfuel.co.za/conferences/2013/electricityConference/Session-B/Johan-van-den-Berg.pdf

Vasseur, M. (2014). Convergence and Divergence in Renewable Energy Policy among US States from 1998 to 2011. *Social Forces, 92*(4): 1637–1657. https://doi.org/10.1093/sf/sou011

Vennesson, P. (2008). Case studies and process tracing: theories and practices. In D. Della Porta & M. Keating (Eds.), *Approaches and methodologies in the social sciences. A pluralist perspective* (pp. 223–239). Cambridge New York: Cambridge University Press.

Vogel, D. (1997a). Trading up and governing across: transnational governance and environmental protection. *Journal of European Public Policy, 4*(4): 556–571. https://doi.org/10.1080/135017697344064

Vogel, D. (1997b). *Trading up: Consumer and environmental regulation in a global economy* (2nd ed.). Cambridge, Mass.: Harvard University Press.

Volden, C., Ting, M., & Carpenter, D. (2008). A Formal Model of Learning and Policy Diffusion. *American Political Science Review, 102*(03): 319–332. https://doi.org/10.1017/S0003055408080271

Volkery, A., & Jacob, K. (2003). *Pioneers in environmental policy-making: Colloquium report.* (FFU-Report, 05-2003). Retrieved from http://www.polsoz.fu-berlin.de/polwiss/forschung/systeme/ffu/publikationen/2003/volkery_axel_jacob_klaus_2003/rep_2003-05.pdf

Walker, J. (1969). The Diffusion of Innovations among the American States. *The American Political Science Review, 63*(3): 880–899.

Walsh, W. V., Meiers-De Pastino, R., & Southern California Edison. (2014). *Southern California Edison Company's (U 338-E) Opening Post-Workshop Comments.* (Rulemaking 14-07-002). Retrieved from http://www3.sce.com/sscc/law/dis/dbattach5e.nsf/0/2B1246EDA9BD5C0388257D64007D5790/$FILE/R1407002_NEM%20Tariffs%20Successor_SCE%20Opening%20Post%20Workshop%20Comments.pdf

Wambach, A. (2014). Wirksame Klimapolitik. *Energiewirtschaftliche Tagesfragen: et, 64*(9): 37–39.

Wang, S. (2013). What Can the United States Learn from Germany's Energiewende?. Clean Coalition. Retrieved from http://www.clean-coalition.org/what-can-the-united-states-learn-from-germanys-energiewende/

WBGU. (2012). *Globale Energiewende Finanzieren.* Wissenschaftlicher Beirat Globale Umweltfragen der Bundesregierung. (Politikpapier, No 7). Retrieved from http://www.wbgu.de/fileadmin/templates/dateien/veroeffentlichungen/politikpapiere/pp2012-pp7/wbgu_pp7_dt.pdf

Weale, G. (2014). *The German Energiewende: Californian PUC Thought Leaders Meeting. RWE AG.* Retrieved from http://www.cpuc.ca.gov/uploadedFiles/CPUC_Public_Website/Content/About_Us/Organization/Divisions/Policy_and_Planning/Thought_Leaders_Events/140605_RWETheEnergiewendeFINALWeale.pdf

Weidner, H., & Mez, L. (2008). German Climate Change Policy: A Success Story With Some Flaws. *The Journal of Environment & Development, 17*(4): 356–378. https://doi.org/10.1177/1070496508325910

Weidner, H., Jänicke, M., & Jörgens, H. (Eds.). (2002). *Capacity building in national environmental policy: A comparative study of 17 countries*. Berlin, New York: Springer-Verlag.

Weimann, J. (2012). Atomausstieg und Energiewende: Wie sinnvoll ist der deutsche Alleingang? *Energiewirtschaftliche Tagesfragen: et ; Zeitschrift für Energiewirtschaft, Recht, Technik und Umwelt, 62*(2):

Weischer, L. & Morgan, J. (2013). *The Solar Economy Club: leadership club approach to international climate policy: A short study commissioned by the Green Party Parliamentary Group in the German Bundestag.* Retrieved from http://www.gruene-bundestag.de/fileadmin/media/gruenebundestag_de/themen_az/klimaschutz/politik_der_unterschiedlichen_geschwindi/WRI_Study_Climate_Clubs_Greens.pdf

Weiss, C. H. (1979). The Many Meanings of Research Utilization. *Public Administration Review, 39*(5): 426–431. https://doi.org/10.2307/3109916

Weiss, J. (2014). *Solar Energy Support in Germany: A Coser Look.* Prepared for Solar Energy Industries Association. Retrieved from http://www.seia.org/sites/default/files/resources/1053germany-closer-look.pdf

Wejnert, B. (2014). *Diffusion of democracy: The past and future of global democracy*. Cambridge: Cambridge University Press.

Wentzel, M., & Pouris, A. (2007). The development impact of solar cookers: A review of solar cooking impact research in South Africa. *Energy Policy, 35*(3): 1909–1919. https://doi.org/10.1016/j.enpol.2006.06.002

Wesoff, E. (21 February 2012). Feed-In Tariff for PV in Palo Alto, Calif. Imminent. *greentech media.* Retrieved from http://www.greentechmedia.com/articles/read/Feed-in-Tariff-for-PV-in-Palo-Alto-Ca-Imminent

Westphal, K. (2012). *Die Energiewende global denken.* (SWP Aktuell, N° A37). Berlin. Retrieved from http://www.swp-berlin.org/de/publikationen/swp-aktuell-de/swp-aktuell-detail/article/die_energiewende_global_denken.html

Wettmann, R. W. (2016). *Le tournant énergétique en Allemagne - état des lieux en 2015/2016: La double sortie du nucléaire et du fossile.* Paris: Friedrich Ebert Stiftung Bureau de Paris. Retrieved from http://www.fesparis.org/tl_files/fesparis/pdf/publication/Energiewende%20Wettmann%20final.pdf

Weyland, K. G. (2004a). Conclusion. In K. G. Weyland (Ed.), *Learning from foreign models in Latin American policy reform* (pp. 241–280). Washington, DC: Woodrow Wilson Center Press.

Weyland, K. G. (2004b). Learning from Foreign Models in Latin American Policy Reform: An Introduction. In K. G. Weyland (Ed.), *Learning from foreign models in Latin American policy reform* (pp. 1–34). Washington, DC: Woodrow Wilson Center Press.

Weyland, K. G. (Ed.). (2004c). *Learning from foreign models in Latin American policy reform*. Washington, DC: Woodrow Wilson Center Press.

Weyland, K. G. (2005). Theories of Policy Diffusion Lessons from Latin American Pension Reform. *World Politics, 57*(02): 262–295. https://doi.org/10.1353/wp.2005.0019

Weyland, K. G. (2006). *Bounded rationality and policy diffusion: Social sector reform in Latin America*. Princeton, N.J: Princeton University Press.

Wieczorek-Zeul, H. (2009 (2001)). Economic Co-operation for Renewable Energy: Speeches at the International Impusle Conference for the Creation of an International Renewable Energy Agency powered by EUROSOLAR, Berlin 8 June 2001. In EUROSOLAR & WCRE (Eds.), *The Long Road to IRENA. From the Idea to the Foundation of the International Renewable Energy Agency* (pp. 34–42). Bochum: Ponte Press Verlags-GmbH.

Wienges, S. (2009). *Governance in global policy networks: Individual strategies and collective action in five sustainable energy-related type II partnerships*. Frankfurt am Main, New York: Peter Lang.

Wiesegart, K., DuBois, A., Sommer, D., Weisheng, W., & Yang, W. (2011). *Options for the Establishment of a South African Wind Energy Centre (SAWEC): with Lessons Learnt from China and Germany. Final Report June 2011*. Commissioned by GIZ. Retrieved from http://crses.sun.ac.za/files/research/publications/technical-reports/SAWEC-Report-final%20draft-06-2011.pdf

Windhoff-Héritier, A., Kerwer, D., Knill, C., Lemkuhl, D., Teutsch, M., & Douillet, A.-C. (Eds.). (2001). *Differential Europe: The European Union impact on national policymaking*. Lanham, Md: Rowman & Littlefield.

Wolman, H. (1992). Understanding Cross National Policy Transfers: The Case of Britain and the US. *Governance, 5*(1): 27–45.

Wolman, H. (2009). *What We Know About What Transfers, How It Happens, and How to Do It*. Presentation to Gastein European Forum for Health Policy. Retrieved from https://gwipp.gwu.edu/files/downloads/Working_Paper_038_PolicyTransfer.pdf

Wolman, H., & Page, E. (2002). Policy Transfer among Local Governments: An Information–Theory Approach. *Governance, 15*(4): 577. https://doi.org/10.1111/1468-0491.00198

World Bank. (1998). *Knowledge for development: World Development Report*. Washington, D.C.: World Bank

World Bank. (2006). *Assessment of the World Bank Group/GEF Strategy for the Market Development of Concentrating Solar Thermal Power*. Retrieved from http://siteresources.worldbank.org/GLOBALENVIRONMENTFACILITYGEFOPERATIONS/Resources/Publications-Presentations/SolarThermal.pdf

World Bank. (2008a). *Illustrative Investment Programs for the Clean Technology Fund.* Retrieved from http://siteresources.worldbank.org/INTCC/Resources/Illustrative_Investment_program_May_15_2008.pdf

World Bank. (2008b). *Maroc. Mission de suivi du Projet de Politique de Développement pour le Secteur de l'Energie (PPD) Energie.* Retrieved from https://www-cif.climateinvestmentfunds.org/sites/default/files/meeting-documents/morocco_july_16_27_2008_aide_memoire_french.pdf

World Bank. (2009a). *Implementation Completion and Results Report on a loan in the amount of US$100 Million equivalent to the Kingdom of Morocco for an Energy Sector Development Policy Loan.* Retrieved from http://www-wds.worldbank.org/external/default/WDSContentServer/WDSP/IB/2010/04/19/000334955_20100419025244/Rendered/PDF/ICR12230P099611C0disclosed041151101.pdf

World Bank. (2009b). *MENA Regional Concentrated Solar Power Scale-up Program: Joint Workshop of the World Bank Group and the African Development Bank.* Retrieved from https://www.esmap.org/sites/esmap.org/files/8312009114352_MENACSPScaleupWorkshop.pdf

World Bank. (2011). *Project Appraisal Document on a Proposed Loan n the Amount of US$200 Million and a proposed Clean Technology Fund Loan in the Amount of US$97 Million to the Moroccan Agenca for Solar Energy (MASEN).* Retrieved from http://www-wds.worldbank.org/external/default/WDSContentServer/WDSP/IB/2011/11/18/000406484_20111118174227/Rendered/INDEX/09725281.txt

World Bank. (2016a). World Bank Indicators: Annual per capita emissions of C02 in metric tons. Data: Carbon Dioxide Information Analysis Center, Environmental Sciences Division, Oak Ridge National Laboratory, Tennessee, United States. Retrieved from http://data.worldbank.org/indicator/EN.ATM.CO2E.PC

World Bank. (2016b). World Development Indicators. World DataBank. Retrieved from data.worldbank.org

World Bank. (2016c). World Development Indicators: Access to electricity (% of population): World Bank, Sustainable Energy for All (SE4ALL). Database from World Bank, Global Electrification database. Retrieved from http://data.worldbank.org/indicator/EG.ELC.ACCS.ZS

Wurzel, R., & Connelly, J. (Eds.). (2011a). *The European Union as a leader in international climate change politics*. London, New York: Routledge.

Wurzel, R., & Connelly, J. (2011b). Introduction: European Union political leadership in international climate politics. In R. Wurzel & J. Connelly (Eds.),*The European Union as a leader in international climate change politics* (pp. 3–20). London, New York: Routledge.

WWF South Africa. (2015). *A review of the local community development requirements in South Africa's renewable energy procurement programme.* WWF-South Africa. Retrieved from http://awsassets.wwf.org.za/downloads/local_community_development_report_20150618.pdf

Yin, R. K. (1981). The Case Study as a Serious Research Strategy. *Science Communication, 3*(1): 97–114. https://doi.org/10.1177/107554708100300106

Yin, R. K. (2009). *Case Study Research: Design and Methods* (4th Edition). Los Angeles, London, New Delhi, Singapore, Washington DC: SAGE Publications.

Young, O. R. (1991). Political Leadership and Regime Formation: On the Development of Institutions in International Society. *International Organization, 45*(3): 281–308. https://doi.org/10.2307/2706733

Zahariadis, N. (2007). The Multiple Streams Framework: Structure, Limitations, Prospects. In P. A. Sabatier (Ed.), *Theories of the policy process* (pp. 93–128). Boulder, Colo: Westview Press.

ZEIT ONLINE. (27 March 2011). Grün-roter Triumph in Baden-Württemberg. *ZEIT ONLINE.* Retrieved from http://www.zeit.de/politik/deutschland/2011-03/landtagswahl-baden-wuerttemberg-ergebnisse

Zimmermann, J.-R. (15 April 2015). Energiewende: Mehr als 230.000 neue Jobs!. *neue energie – Das Magazin für Erneuerbare Energien.* Retrieved from https://www.neueenergie.net/wirtschaft/markt/energiewende-mehr-als-230000-neue-jobs

Zisenwine, D. (2011). Mohammed VI and Moroccan foreign policy. In B. Maddy-Weitzman & D. Zisenwine (Eds.). *Contemporary Morocco. State, politics and society under Mohammmed VI* (pp. 70–81). Milton Park, Abingdon, New York: Routledge.

Zöllner, O. (2009). German Public Diplomacy. In N. Snow & P. M. Taylor,(Eds.). *Routledge handbook of public diplomacy* (pp. 262–269). New York, Milton Park, Abingdon, Oxon: Routledge